T0200068

TIMELINES OF SCIENCE

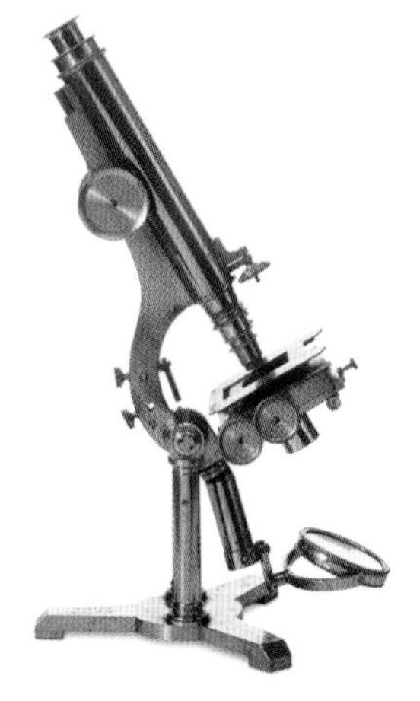

TIMELINES OF SCIENCE

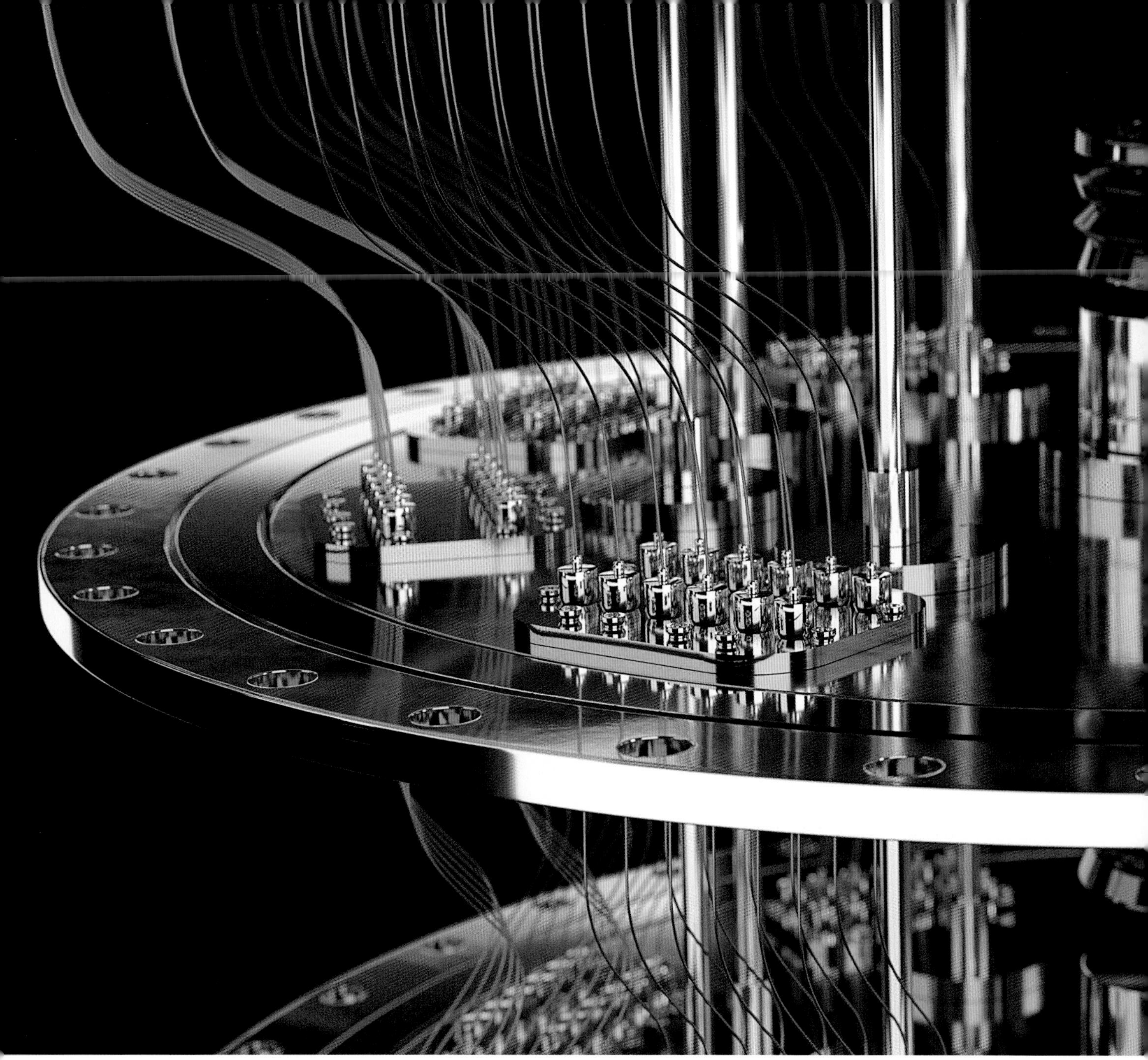

Produced for DK by
cobalt id
www.cobaltid.co.uk

Editors Marek Walisiewicz, Kay Celtel
Designers Paul Tilby, Darren Bland, Paul Reid

Senior Editor Peter Frances
Senior Art Editor Duncan Turner
Editors David and Sylvia Tombesi-Walton
Illustrator Priyal Mote
Managing Editor Angeles Gavira Guerrero
Managing Art Editor Michael Duffy
Production Editor Andy Hilliard
Production Controller Laura Andrews
Senior Jackets Designer Akiko Kato
Jackets Design Development Manager Sophia MTT
Art Director Karen Self
Design Director Phil Ormerod
Associate Publishing Director Liz Wheeler
Publishing Director Jonathan Metcalf

First published in Great Britain in 2023 by
Dorling Kindersley Limited
DK, One Embassy Gardens, 8 Viaduct Gardens,
London, SW11 7BW

The authorized representative in the EEA is
Dorling Kindersley Verlag GmbH. Arnulfstr. 124,
80636 Munich, Germany

10 9 8 7 6 5 4 3 2 1
001-334000-June/2023

A CIP catalogue record for this book
is available from the British Library.
ISBN: 978-0-2416-0097-9

Printed and bound in China

For the curious
www.dk.com

This book was made with Forest Stewardship Council™ certified paper – one small step in DK's commitment to a sustainable future.
For more information go to www.dk.com/our-green-pledge

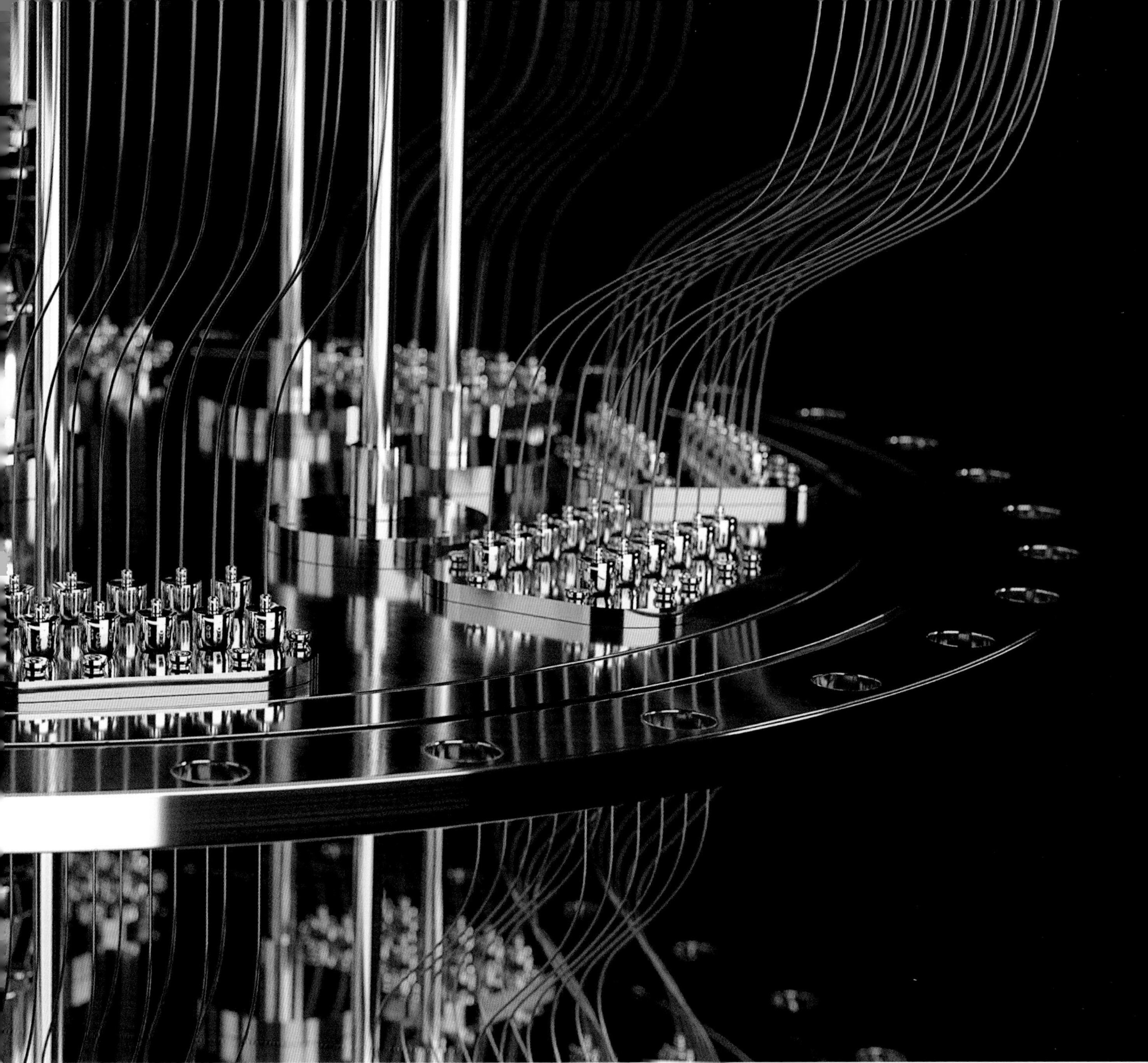

CONTRIBUTORS

Tony Allen has written many works on history for the general reader and was Series Editor for the 24-volume *Time-Life History of the World.*

Jack Challoner gained a physics degree at Imperial College, London, and has been a science writer since 1991, with more than 40 books to his name.

Julian Emsley is a chemist, mathematics teacher, and author who specializes in writing about the impact of chemistry and chemicals in the world.

Hilary Lamb is an award-winning science journalist, editor, and author. She has worked on previous DK titles including *The Physics Book, Simply Quantum Physics*, and *Simply Artificial Intelligence*.

Dr Douglas Palmer is a Cambridge-based writer on palaeontology and earth science. He has written many popular science books and works part-time for the Sedgwick Museum, Cambridge.

Philip Parker is a historian specializing in the medieval world. He has written numerous historical books and atlases.

Bea Perks has a degree in zoology, a PhD in clinical pharmacology, and over 20 years' experience in biomedical writing and publishing.

Giles Sparrow is a Fellow of the Royal Astronomical Society with degrees in Astronomy and Science Communication. He is the author of more than two dozen books about space and astronomy.

Martin Walters is an author and naturalist with a special interest in birds, botany, and conservation. He has written and contributed to many books, both for adults and children.

Marcus Weeks is a musician and author who has written and contributed to many books on philosophy, the arts, and the history of the ancient world.

CONSULTANT

Robin McKie has been science editor of *The Observer* for the past 40 years and is the author of several books on genetics and the origins of modern humans.

Half-title page Achromatic compound microscope used by Charles Darwin, 1847

Title page Polish-French physicist and chemist Marie Curie in her Paris laboratory, c. 1905

Above Components of a quantum computer, 2020

CONTENTS

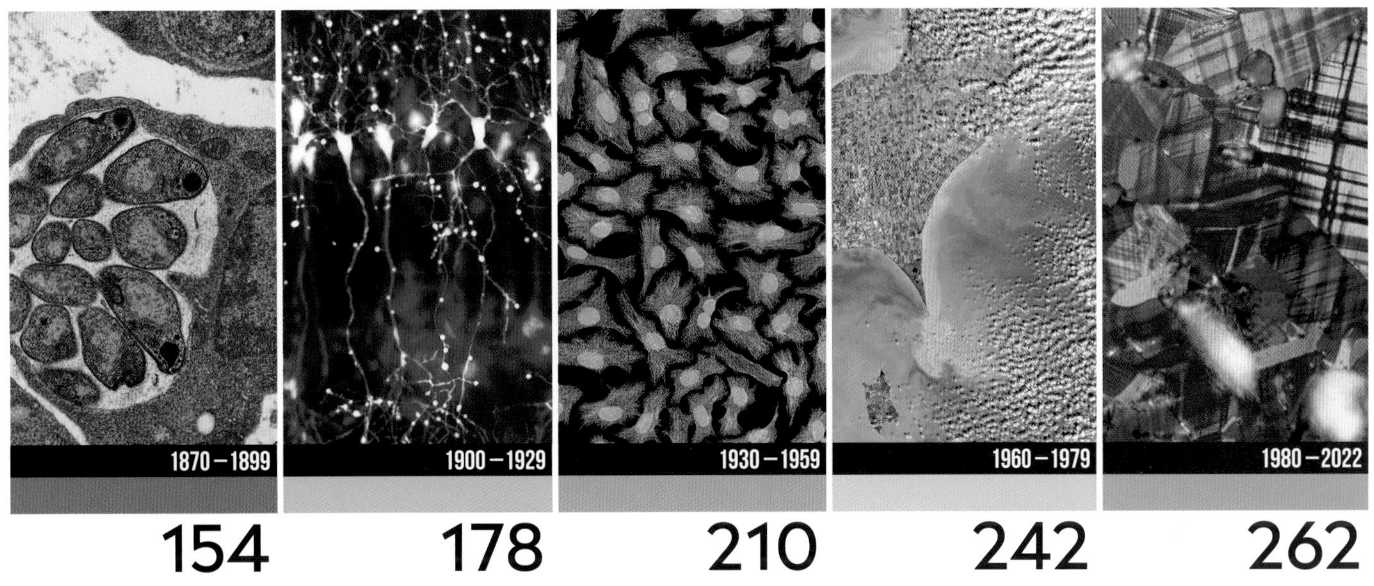
1870 – 1899
154
1900 – 1929
178
1930 – 1959
210
1960 – 1979
242
1980 – 2022
262

These apparent hills and valleys in the Carina Nebula, some 7,600 light-years away, are the margins of a star-forming region. The image, captured in 2022 by NASA's James Webb telescope, shows the exceptional ability of the instrument to gather infrared light.

Stone hand axes were in use for a period of 1.5 million years

c. 1.7 MYA

HAND AXES

Made by chipping flakes from a rock with a hard hammerstone and then shaping them with softer bone or antler hammers, hand axes were first produced in East Africa. Known as Acheulean tools, the axes had two faces and, usually, a wider base that served as a grip, making them very versatile.

△ **Acheulean** hand axe, 700,000–200,000 BCE

c. 2.6 MYA

OLDOWAN STONE TECHNOLOGY

Oldowan tools were made by striking stones to shear off flakes, so creating sharp edges for cutting and scraping. Probably first made by *Homo habilis*, a hominin species, they originated in East Africa and were later spread beyond the continent by *Homo erectus*.

△ **Oldowan** stone tools

3.3 MYA

c. 1.8 MYA The oldest evidence of house construction is a stone and grass hut found at Olduvai Gorge, Tanzania

c. 3.3 MYA

FIRST TOOL USERS

The oldest known tools were discovered in 2011 at the Lomekwi 3 site on a dried-up river bed in Kenya. Around 150 stones had been modified by striking them against another rock to create primitive hammers or cutting tools, possibly for scraping or shattering animal bones. The tools may have been created by an Australopithecine species such as *Australopithecus afarensis* which predates the evolution of our genus, *Homo*.

◁ ***Australopithecus afarensis* "Lucy"** (reconstruction)

c. 460,000 BCE

EARLY SPEARS

The earliest known stone-tipped spears were made at Kathu Pan in southern Africa. The stones had sharp edges that made them effective hunting weapons. They were hafted by thinning them at the base and attaching them to a wooden shaft. They exemplify our human ancestors' increasingly sophisticated use of technology.

◁ **Stone spear** tips

c. 325,000 BCE The Levallois stone-working technique emerges: shaped flakes are struck from a rock core to create tools

c. 24,000 BCE

c. 420,000 BCE The Clacton Spear, the earliest known worked wooden implement, is made

c. 62,000 BCE Stone arrowheads found in Sibudu Cave, South Africa, are the earliest evidence of bows and arrows

▽ **Jebel Irhoud** skull (computer reconstruction)

c. 350,000 BCE

HOMO SAPIENS EMERGES

Remains found at Jebel Irhoud in Morocco in the 1960s were identified in 2017 as those of the very earliest members of our own species, *Homo sapiens*. The find reversed theories that the species had evolved in East Africa. The first *Homo sapiens* had longer brain cases and more pronounced brow ridges than modern humans; their brain size and resulting versatility allowed them to spread and eventually replace all other hominins.

△ **Deliberately set fire**

c. 790,000 BCE

CONTROLLED FIRE

Although early hominins may have opportunistically used natural fires caused by lightning strikes for warmth or to drive off wild animals, the earliest known controlled use of a deliberately set fire was at Gesher Benot Ya'aqov, Israel. A hearth found there contained charred flint, as well as remains of olive wood, wild grape, and wild barley, indicating that it may have been used for cooking. Cooked food allowed hominins to digest more efficiently, giving them access to greater levels of nutrition.

c. 24,000 BCE
FIRED EARTH

Discovered in a cave in the Czech Republic, the small figurine of a woman known as the "Venus of Dolna Vestonice" is among the oldest examples of ceramic technology in the world. The figurine was shaped from clay and powdered bone and carries the fingerprint of a young child who probably picked it up before it was fired in a kiln.

▷ **Venus** figurine

c. 9500 BCE
CROP PLANTS

Villagers in the region of Abu Hureyra (in modern Syria) domesticated the first plant crops, marking the beginnings of agriculture. They selectively bred rye and einkorn (a type of wheat) – grasses that they had previously gathered in the wild. The seeds were eaten, but a portion was retained for planting the next crops close to human settlements.

▽ **Einkorn** wheat

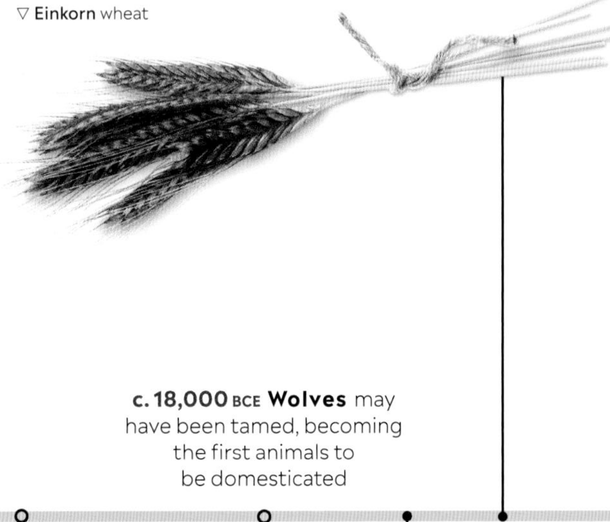

24,000 BCE

c. 18,000 BCE **Wolves** may have been tamed, becoming the first animals to be domesticated

c. 18,000 BCE **The oldest known** clay pots are made – vessels from Xianrendong Cave, Jiangxi, China

c. 14,000 BCE
LAND BRIDGE

From around 70 million years ago, a series of land bridges that appeared during periods of glaciation linked Asia and North America, enabling both mammals and dinosaurs to cross between the continents. Around 20,000 years ago, Beringia – the frozen landscape of the land bridge – reached its largest extent, and by 14,000 BCE humans had established themselves in North America. As glaciers melted and the sea levels rose over the next several thousand years, the land bridge was lost.

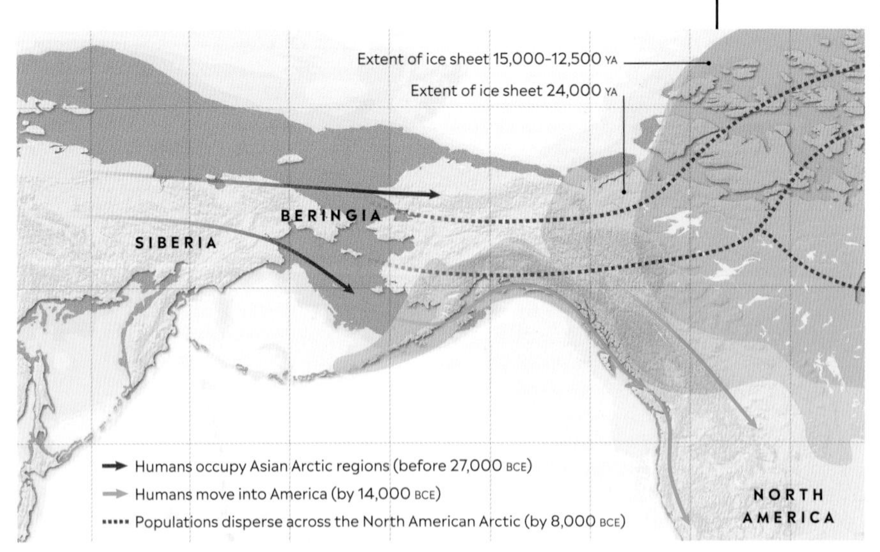
Extent of ice sheet 15,000–12,500 YA
Extent of ice sheet 24,000 YA
BERINGIA
SIBERIA
NORTH AMERICA
Humans occupy Asian Arctic regions (before 27,000 BCE)
Humans move into America (by 14,000 BCE)
Populations disperse across the North American Arctic (by 8,000 BCE)

△ **Beringia** land bridge

The last mammoth population lived on Wrangel, off the Siberian coast

c. 9000 BCE

MAMMOTH EXTINCTION

Woolly mammoth, which had been hunted by humans throughout northern Eurasia for millennia, became extinct on the mainland. The mammoths' range shrank as temperatures rose from the end of the Ice Age, leaving only a few populations on isolated islands. These died out around 2000 BCE.

▷ **Mammoth,** artist's impression

8000 BCE The walls and watchtower built to protect Jericho in the Middle East are the earliest example of building technology being used for military purposes

7401 BCE

c. 8000 BCE The oldest surviving boat, the Pesse Canoe from the Netherlands, is made by hollowing out a platform from a large log

△ **Gobekli Tepe** ruins

△ **Asiatic mouflon** sheep

c. 8500 BCE

DOMESTICATION OF ANIMALS

Villagers in the Near East began taming animals and then selectively breeding them for desired characteristics, such as docility. They domesticated Asiatic mouflon (the ancestors of sheep), bezoar goats, aurochs (ancestral cattle), and wild boar (pigs), providing ready sources of milk, meat, and skins that helped to spread the agricultural way of life.

c. 9500 BCE

BUILDING TECHNOLOGY

Construction began on the world's first monumental structure, at Gobekli Tepe in central Anatolia (Turkey). Probably a temple, the structure features 20 circular stone enclosures and the earliest known megaliths – massive stone blocks worked and carved with stone tools.

c. 4500 BCE
MEGALITHS

People in western and north-western Europe began to erect megaliths, huge stone monuments that required considerable engineering expertise to transport and raise. The stones at Carnac in France are believed to be among the very earliest. Megaliths may have served as tombs, temples, or astronomical observatories.

▷ **Carnac's Neolithic** standing stones

7400 BCE

c. 6500 BCE **Hump-back zebu cattle** are domesticated in the Indus Valley (Pakistan)

c. 5500 BCE **The earliest irrigation canals** are built at Choga Mami (Iraq) to carry the waters of the Tigris to fields

c. 5000 BCE **Evidence of copper smelting** from slag found in Belovode Serbia marks the true beginnings of metallurgy

△ **Çatal Höyük** archaeological site, near Konya, Turkey

c. 7400 BCE
FIRST TOWNS AND CITIES

In Central Anatolia (Turkey), farming peoples built Çatal Höyük, the first settlement to reach the size of a town. The settlement marked the transition in human development from hunting and gathering subsistence to increasing agriculture and animal domestication. Its 8,000 inhabitants grew wheat and peas, kept sheep and cattle, and lived in tightly clustered, rectangular, mud-brick houses.

c. 6000 BCE
THE ARD PLOUGH

The ard plough was developed and spread rapidly through West and South Asia. Created by adding a long pole and crossbeam to a hoe that scratched out a furrow for seed, it was often drawn by cattle. The ard broke up the soil rather than turning it over and land still had to be cleared by hand. Even so, it greatly increased the efficiency of agriculture.

△ **Ard plough**

c. 3200 BCE
INVENTION OF THE WHEEL

Wheels in the form of circular disks connected to a rod-like axle allowed far greater loads to be carried than was possible manually. The oldest known wheel came from the Ljubljana Marshes in Slovenia and was made of ash with an oak axle, but the depiction of what may be a wheeled cart appeared on a pot created in Poland around 400 years earlier. Before long, wheels also appeared in Mesopotamia.

△ **Bronze Age wheel** unearthed in a dig

c. 4000 BCE Wetfield cultivation of rice is established in China

3001 BCE

▷ **Early Bronze Age** axe head

3500 BCE The development of the potter's wheel in Mesopotamia leads to a rise in quantity and quality of ceramics

3100 BCE The first boats powered by sails rather than oars appear on Egypt's River Nile

c. 4500 BCE
BRONZE OBJECTS

The first known bronze objects were made in Serbia, where smiths at a settlement at Pločnik smelted tin with copper. The strong bronze metal was produced in ventilated furnaces capable of reaching very high temperatures. The Pločnik smiths fashioned ornaments, but smiths in the Near East (where the techniques to make bronze were probably developed independently) made durable weapons.

△ **Clay tablet** with early cuneiform script

"What kind of a scribe is a scribe who does not know Sumerian?"

SUMERIAN PROVERB

c. 3200 BCE
WRITING SYSTEMS

Full writing systems were developed in Egypt (hieroglyphic) and Sumeria (cuneiform). At first used to keep administrative records of taxation and commercial transactions, they were soon employed to record rulers' achievements, laws, religious texts, epic poems, and histories on materials ranging from clay and stone to papyrus, animal skin, and eventually paper.

c. 2700 BCE
CHINESE MEDICINE

Legends tell that the ancient ruler Shennong established Chinese medicine by personally testing hundreds of herbs for their medicinal properties. His successor Huangdi (the "Yellow Emperor") was said to have written the *Huangdi Neijing*, the oldest Chinese medical textbook. This rejected explanations that diseases were caused by demonic influences and suggested that dietary, lifestyle, and other environmental influences caused imbalances in the body that could be cured by acupuncture and herbal remedies. Elements of this Chinese traditional medicine are still in use in the 21st century.

▷ **Emperor Shennong**

3000 BCE

c. 3000 BCE **Faience** (a paste made of crushed silica and lime) is invented in Egypt as a decorative device in jewellery

△ **Step Pyramid** of Djoser

c. 2680 BC
THE FIRST PYRAMID

In Egypt, Pharaoh Djoser's chief architect Imhotep designed the Step Pyramid at Saqqara as a tomb for his master. Previous Egyptian rulers had been buried in mastabas, single-storey rectangular brick structures, but by placing six successively smaller mastabas one on top of the other, Imhotep created a stepped structure in the form of a pyramid. At around 62 m (203 ft) in height and surrounded by a large enclosing wall, it provided the model for the later, larger pyramids at Giza.

27th century BCE
IMHOTEP

The chief minister to Egyptian Pharaoh Djoser, Imhotep was a scribe, architect of the Step Pyramid, and by tradition a physician who was later worshipped as a medical deity. His many accomplishments led him to be labelled as the first scientist.

> "Harvest, Gazelle Feast, Piglet Feast, Ubi-bird Feast, Weaving-place of Ninazu, Ninazu Festival, Akitu, Festival of Sulgi, Lofty Festival"
>
> *MONTH NAMES FROM THE CALENDAR OF SHULGI, c. 2025 BCE*

c. 2025 BCE
CALENDAR DEVISED

The first known standardized calendar, the Umma Calendar of Shulgi, was devised in the Mesopotamian city of Ur. A lunisolar calendar with 12 months of 29 or 30 days and an intercalary month added every few years to prevent it becoming out of step with the seasons, it helped to regulate farming and religious activities.

◁ **Copper** figure of Shulgi

c. 2550 BCE **The Great Pyramid** is built at Giza in Egypt using 2.3 million limestone blocks

c. 2500 BCE **The technique of gold granulation** is developed in Egypt

2001 BCE

c. 2500 BCE **The first known map** of a specific area is created; it shows field plots in an area between two hills at Nuzi in Mesopotamia

c. 2500 BCE **Steering oars** for boats are developed in Egypt

△ **The Great Bath** at Mohenjo-Daro

2600 BCE
WATER SUPPLIES

The Indus Valley civilization cities of Mohenjo-Daro and Harappa (in modern Pakistan) were provided with the world's first public water and sewage systems. Houses on streets laid out on a grid were supplied with water from wells; most residences had baths and latrines. Waste water was drained into brick channels and then into a main sewer that ran through the city. At Mohenjo-Daro the huge "Great Bath" was filled with water and probably played a role in religious rituals.

c. 2300 BCE
WEIGHTS AND MEASURES

Sargon of Akkad, who conquered the city-states of Mesopotamia to create the world's earliest empire, imposed a standard system of weights and measures for the first time. It was based on the *gur*, the length of one side of a standard cube, and ensured merchants could trade more securely throughout the growing empire.

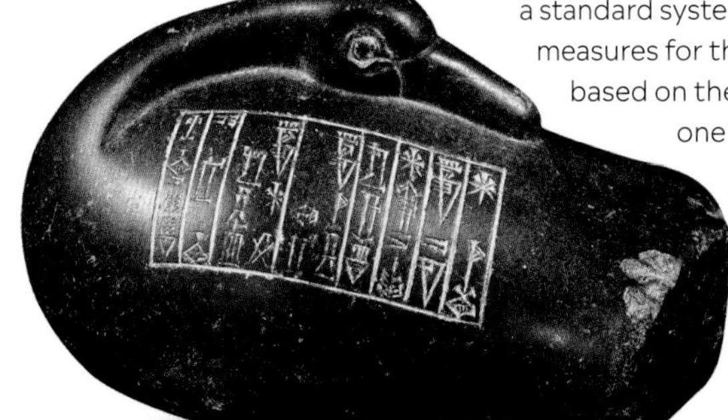

△ **Weight** in the form of a duck

▽ **Shaduf in a garden,** Tomb of Ipuy

c. 2000 BCE

EARLY IRRIGATION TECHNOLOGY

The *shaduf*, a device to assist in irrigation, was invented at about the same time in Mesopotamia and Egypt. This simple but effective piece of engineering consisted of a long pole mounted on a pivot; at one end of the pole was a bucket, and at the other a weight. When pushed down, the counterweight lifted the bucket filled with water to a higher level. It allowed water to be easily raised from riverbanks, ditches, and wells and moved to irrigate fields without having to haul it by hand.

2000 BCE

c. 1830 BCE Babylonian astronomers begin to record their observations of the heavens

c. 1825 BCE The Kahoun papyrus, from Egypt, is the world's oldest work on gynaecology

c. 1800 BCE The first version of what has become known as Pythogoras's theorem is produced in Babylon

c. 1800 BCE

ALPHABETICAL SCRIPT

Stoneworkers of Semitic origin in Egypt's Sinai Desert invented the world's first alphabetical script with each sign representing a letter rather than a syllable or word. Consisting of about 20 letters, whose shapes were adapted from Egyptian hieroglyphs, versions of the script may have influenced the development of the Phoenician alphabet.

△ **Sandstone sphinx** inscribed in a Semitic language

c. 1800 BCE

IRON AND STEEL

The first small objects of smelted iron were created in Anatolia (Turkey), possibly by the Hittite culture. Iron's high melting point (around 1500 °C, or 2730 °F) made the hard material very difficult to produce, and it did not come into common use until c. 1200 BCE. Later on, the addition of carbon during smelting produced an even harder alloy, steel, which the Romans used in their legionaries' swords.

△ **Roman infantry** sword (spatha)

The Rhind mathematical papyrus is more than 5 m (16 ft) long

c. 1800–1650 BCE

BABYLONIAN MATHEMATICS

The Babylonians developed a sophisticated understanding of mathematics, revealed in over 400 cuneiform tablets unearthed by archaeologists. Using a sexagesimal system (based on multiples of 6 and 60), their mathematicians compiled multiplication tables, developed rules for the calculation of the areas and volumes of regular shapes and solids, and derived an estimate of the square root of 2.

△ **A cuneiform** mathematical text

c. 1650–1550 BCE

EGYPTIAN EQUATIONS

The Egyptian Rhind papyrus is the world's earliest mathematical treatise. It covers linear equations and the volumes of pyramids and cylinders. Written by the scribe Ahmes – the first mathematician whose name we know – the papyrus contains the solutions to 84 mathematical problems, especially involving the manipulation of fractions, rather than statements of axioms or general principles.

△ **Rhind** mathematical papyrus

1651 BCE

c. 1800 BCE

FERMENTATION AND BEER

The Egyptians were the first to control fermentation to produce beer on a large scale. Yeast – a single-celled organism – added to beer mash, converted sugars in the mash to alcohol, carbon dioxide, and other products. Although beer was known earlier (the Sumerians had a beer goddess, Ninkasi, around 3000 BCE), the Egyptians produced it on a far greater scale. Workers on the pyramids at Giza were given a ration of about 6 litres (10 pints) a day, which they drank through straws from communal ceramic vessels.

◁ **Ancient brewing scene** from a funerary chapel

▽ **Cuneiform Venus** tablet from 17th-century BCE Babylon

c. 1650 BCE

PHASES OF VENUS

Babylonian astronomers began to compile detailed tables of the phases of Venus during the reign of King Ammisaduqa in the 17th century BCE. Their observations, which showed times for the rising and setting of the planet over a 21-year period, survived on cuneiform tablets dating from the 8th century BCE.

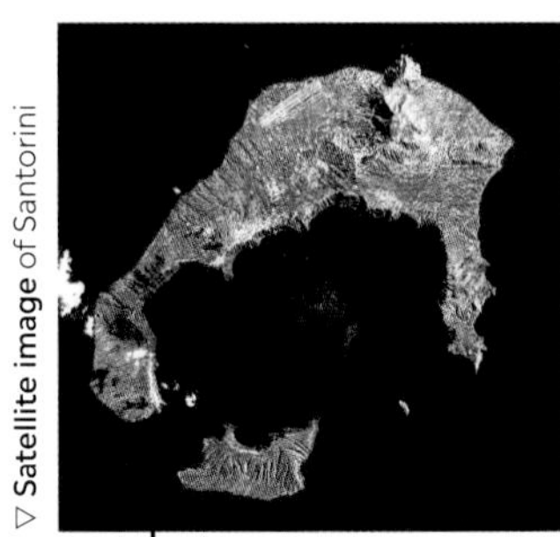

▽ **Satellite image** of Santorini

c. 1600 BCE

ERUPTION OF THERA

A massive volcanic eruption destroyed the Minoan settlement of Akrotiri on the Aegean island of Thera (now Santorini). A tsunami hit parts of the eastern Mediterranean and volcanic ash fell across the region. The eruption may have caused a temporary fall in temperatures and contributed to the collapse of Minoan civilization.

1650 BCE

c. 1500 BCE Pewter, an alloy of tin with copper, antimony, and lead is produced in the Near East

1560 BCE The Ebers Papyrus from Egypt contains a large number of spells and herbal remedies for a wide range of ailments, including depression, toothache, and kidney disease

c. 1600 BCE

SURGICAL MANUAL

The world's oldest surviving surgical treatise was written in Egypt. Now known as the Edwin Smith papyrus for its 19th-century discoverer, it describes 48 cases of trauma, beginning with head injuries (and containing the first specific reference to the brain), and working down the body to the toes. It contains detailed descriptions of symptoms and prescribed treatments including sutures, poultices, and immobilization of the patient.

◁ **The Edwin Smith** papyrus

c. 1500 BCE

GLASS VESSELS

The earliest glass vessels were produced in Egypt during the reign of Tuthmosis I. Previously, small glass objects had been formed by fusing vitreous materials such as silica and quartz at high temperatures. Later, glass workers learned to remelt large glass chunks and shape them in moulds to form vessels such as cups and bowls.

◁ **Ancient Egyptian** glass bottle

It took over 50 kg (120 lb) of murex snails to make one gram (3/100 oz) of Tyrian purple

△ **Egyptian clay** clepsydra inscribed with the name of Amenophis III

c. 1375 BCE

WATER CLOCK

The clepsydra, or water clock, was invented by the Egyptians. It comprised a vessel filled with a set quantity of water that drained at a constant rate throughout the day. The time was told by measuring the level of the remaining water level against a series of scales on the inside of the vessel.

c. 1200 BCE Babylonian perfume maker Tapputi-Belatekallim is recorded to have used a still and various solvents to produce her perfumes

c. 1200 BCE

TEXTILE DYES

The Phoenicians perfected the art of dyeing textiles, creating a special shade of purple made from the dehydrated mucus glands of murex snails. The dye, named Tyrian purple, was so expensive that its use was confined to elites, such as Roman emperors.

▽ **Dye-makers** in ancient Phoenicia

c. 1050 BCE In the region of Byblos, the Phoenicians devise an alphabetic script, an ancestor of modern European scripts

c. 1010 BCE The first iron-rimmed wheels are devised in Celtic Europe

1001 BCE

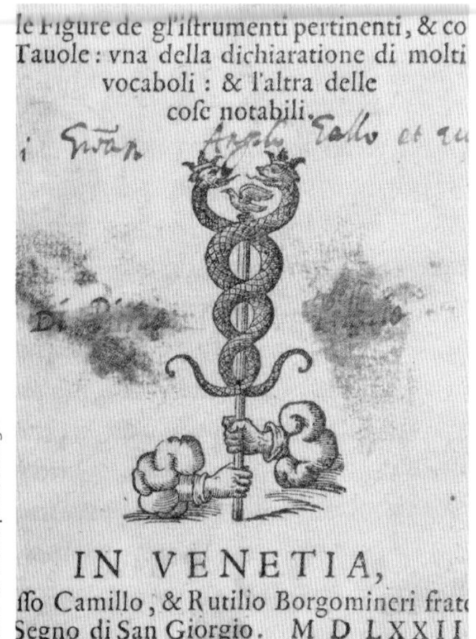

le Figure de gl'iſtrumenti pertinenti, & co
Tauole: vna della dichiaratione di molti
vocaboli: & l'altra delle
coſe notabili.

IN VENETIA,
ſſo Camillo, & Rutilio Borgomineri frat
Segno di San Giorgio. M D LXXII

▷ **Rod of Asclepius** design

c. 900 BCE

ANCIENT HEALING

The Greek healer Aesculapius (thought to have been a historical figure) began his elevation to divine status. As a god of healing, he inspired a clan - the Asclepiads - who claimed a knowledge of healing and built temples in his name. Physicians adopted his symbol of rod entwined by serpents, which remains an emblem of the medical profession to this day.

c. 650 BCE

CLEAR GLASS

The first manual of glassmaking techniques was produced in Assyria. At about the same time, the Phoenicians discovered how to make clear glass, extending the aesthetic appeal of glass objects and increasing demand for glass vessels and containers.

◁ **Glass** alabastron

1000 BCE

c. 800 BCE **In India, the *Sulba Sutras*** contain solutions to practical problems such as the square root of two

c. 900 BCE **The process for producing cast iron** is discovered in China, but is only adopted on a large scale around 550 BCE

Water is released at the top

Central shaft rotates

Screw action pulls water upwards

▷ **Archimedes** screw

△ **Babylonian** clay tablet map of the world

c. 700 BCE

THE ARCHIMEDES SCREW

An inscription dating from the 7th century BCE indicates that the Assyrians had developed a form of screw pump by that time. A hollow cylinder enclosed a spiral-shaped gear that, when rotated, caused water to be pushed from the bottom to the top of the cylinder, where it was released. The device was later described by the Greek mathematician Archimedes, who saw one in Egypt around 234 BCE.

c. 600 BCE

EARLY MAPPING

The earliest surviving map of the world was created on a clay tablet in Babylon. It made no attempt to be geographically precise but showed the city at the centre of the world and marked neighbouring cities as well as a world-bounding river.

c. 580 BCE

NATURAL PHILOSOPHY

Thales, from the Greek settlement at Miletus (now in western Turkey), proposed that water is the basic material of the Universe. He was the first philosopher to speculate that the operations of the world have natural rather than divine causes and was said to have predicted an eclipse. He also devised several geometric theorems.

△ **Thales** of Miletus

STATIC ELECTRICITY

Neutral materials are those that contain an equal number of positive and negative electric charges (respectively known as nuclei and electrons). Static electricity is created when these charges are made unequal. This can be done, for example, by rubbing one material against another, leaving an excess of positive charge on one and an excess of negative charge on the other. The "unbalanced" charges remain in place until they can be released, such as through a spark of electricity.

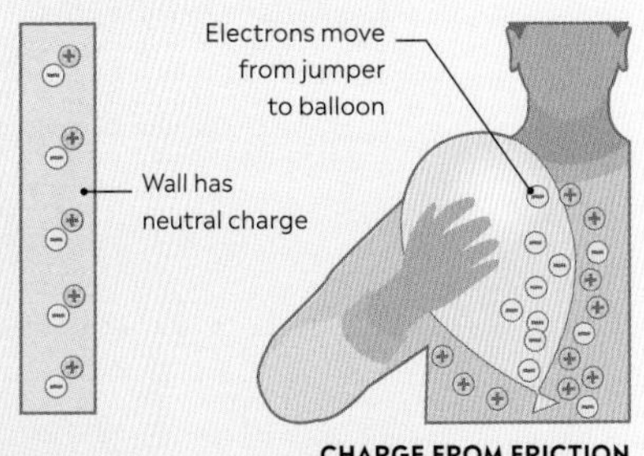

Electrons move from jumper to balloon
Wall has neutral charge
CHARGE FROM FRICTION

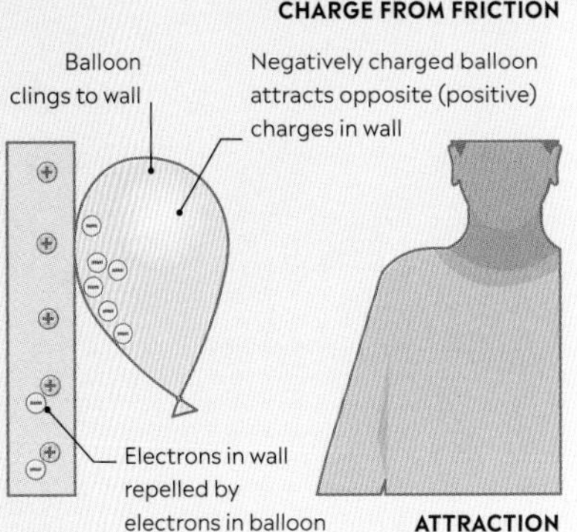

Balloon clings to wall
Negatively charged balloon attracts opposite (positive) charges in wall
Electrons in wall repelled by electrons in balloon
ATTRACTION

Balloon-and-wall trick

Rubbing a balloon against a jumper transfers electrons to the balloon's surface. Now with a negative electric charge, the balloon attracts positive charges, causing it to stick to a wall.

530 BCE Eupalinos of Samos excavates a water tunnel through a hillside in Samos

530 BCE Greek philosopher Pythagoras sets out his theory of the relationship between the length of sides in a right-angled triangle

c. 500 BCE Greek philosopher Heraclitus of Ephesus suggests that the Universe is in a constant state of flux

c. 500 BCE The *Zhou Bi Suan-Jing*, the first major Chinese mathematical treatise, is published

501 BCE

c. 550 BCE

ORIGIN OF MATTER

The Greek philosopher Anaximander of Miletus suggested that the basic element that makes up the Universe is *apeiron*, or "the infinite", a substance that appeared before anything else. He also formulated a theory of evolution, speculating that humanity had developed from sea creatures.

◁ **Anaximander** shown in a Roman mosaic

"The infinite is the universal cause of the generation and destruction of the Universe."

ANAXIMANDER OF MILETUS, c. 550 BCE

c. 450 BCE

FOUR ROOTS

The Greek philosopher Empedocles of Acragas first put forward the theory that everything in nature was made up of four roots (elements), Earth, Air, Fire, and Water, and two basic forces or principles - Love and Strife. He rejected Parmenides' idea that change was impossible and suggested that the interplay between roots and forces was what caused change in the Universe.

△ **The four roots** of Empedocles

c. 420 BCE

THE CONCEPT OF ATOMS

Greek philosopher Democritus of Abdera developed the idea that the Universe is composed of an infinite number of tiny objects that cannot be divided or modified. He called these particles atoms ("uncuttable") and believed that their different shapes determined the type of matter they made up.

△ **Democritus**

500 BCE

c. 480 BCE **Parmenides of Elea,** a Greek philosopher, teaches that change is a logical impossibility

c. 400 BCE **Greek scientist Philolaus of Croton** suggests that Earth is not at the centre of the cosmos, but travels (with the planets and the Sun) around a "central fire"

c. 400 BCE

HUMOURS OF THE BODY

The Greek physician Hippocrates of Cos believed the body had four basic substances, or humours - blood, phlegm, yellow bile, and black bile - and that imbalances between these were the cause of disease. Hippocrates and his followers prescribed diets, exercises, and medicine to restore the balance of the humours. For example, cold baths were used to increase phlegm to combat fevers caused by yellow bile. The four humours were also later linked to four temperaments.

△ **Depiction of melancholy,** linked to black bile.

> "When the veins are excluded from the air by the phlegm and do not receive it, the man loses his speech and intellect, and the hands become powerless."
>
> *HIPPOCRATES OF COS,* ON THE SACRED DISEASE, *c. 400 BCE*

▷ **Celestial** spheres

c. 375 BCE

CELESTIAL SPHERES

The Greek astronomer Eudoxus of Cnidus developed a theory of celestial spheres to account for the observed irregularity of the motions of some planets. In his system, Earth sat at the centre of the Solar System, with the Sun, Moon, planets, and stars all rotating around it. The Sun and Moon moved within three spheres; the other five planets (the Greeks only knew of Venus, Mercury, Mars, Saturn, and Jupiter) had four to which they were attached at the Poles and within which their movement was guided. The stars were said to exist along the final, 27th celestial sphere.

Eudoxus of Cnidus believed there were 27 celestial spheres

c. 360 BCE Greek astronomer Heraclides of Pontus teaches that the world rotates on its axis each day

351 BCE

c. 385 BCE Archytas of Tarentum's (now in Italy) theory of harmonics establishes the relationship between the pitch of a note and the length of the string or pipe producing it

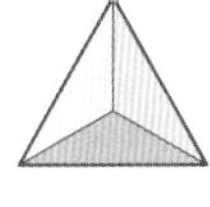

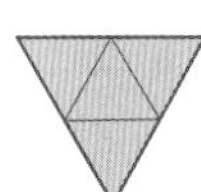

△ **A terahedron** has four triangular faces.

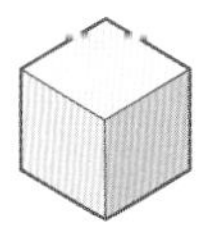

△ **A cube** has six square faces.

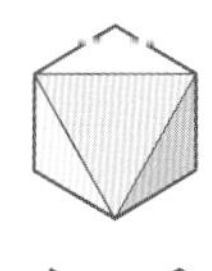

△ **An octahedron** has eight triangular faces.

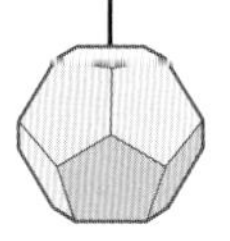

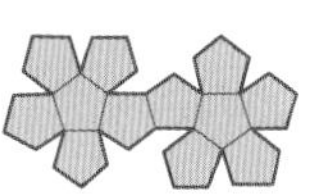

△ **A dodecahedron** has 12 pentagonal faces.

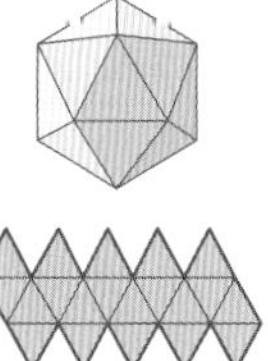

△ **An icosahedron** has 20 triangular faces.

c. 360 BCE

PLATONIC SOLIDS

The Greek philosopher Plato developed the idea that all matter was composed of five regular polyhedra. He linked these to the traditional four-element theory, with the tetrahedron linked to fire, the octahedron to air, the icosahedron to water, and the hexahedron (or cube) to earth, and the dodecahedron associated with the Universe (its 12 faces linked to the 12 constellations). Plato believed different combinations of the polyhedra created different elements.

Plato's solids
The faces of the five Platonic solids are made up of regular shapes (triangles, squares, or pentagons). Plato saw in this symmetry a building block of the Universe.

428–347 BCE

PLATO

A student of Socrates, Plato established a school of philosophy in Athens called the Academy. In his philosophy, ideal forms were thought to be reflected in inferior earthly equivalents.

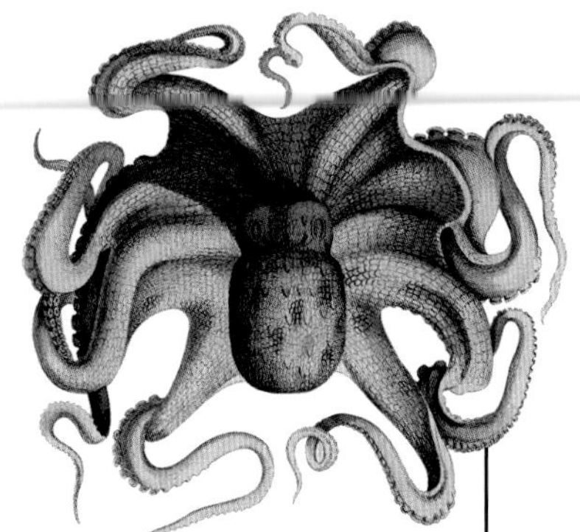

▷ *Octopus vulgaris*

c. 350–322 BCE

ZOOLOGICAL SYSTEM

The Greek philosopher Aristotle pioneered the science of zoology in his *History of Animals*, in which he described the structure and behaviour of over 500 animals, attempting to identify the common attributes of species and dividing them into "blooded" and "bloodless". Many of his descriptions came from direct observation, for example, of octopi at Lesbos.

c. 350–322 BCE

ARISTOTLE'S ACHIEVEMENTS

Aristotle was one of the most significant thinkers of all time, and played a key role in the development of the scientific method (*see below*). In *Metaphysics*, he examined the difference between the substance and essence of a thing, while in *Mechanics*, he founded the science of motion, believing that an object would fall infinitely far in a vacuum. He proposed that geological changes, such as the formation of mountains, must occur over vast time periods and is recognized as the first systematic biologist.

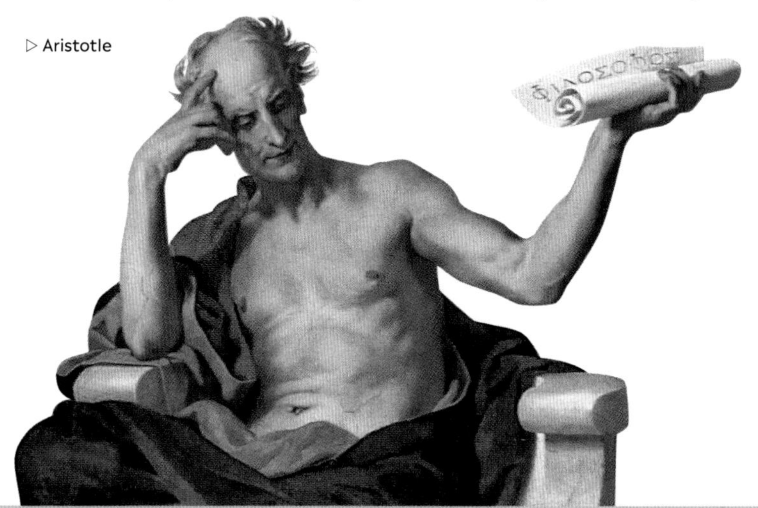

▷ Aristotle

350 BCE

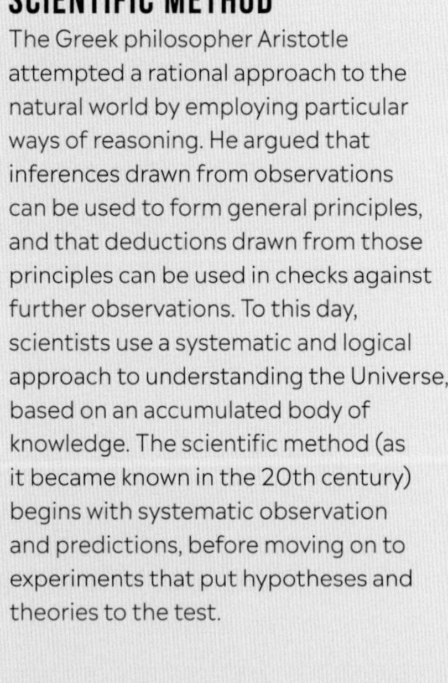

SCIENTIFIC METHOD

The Greek philosopher Aristotle attempted a rational approach to the natural world by employing particular ways of reasoning. He argued that inferences drawn from observations can be used to form general principles, and that deductions drawn from those principles can be used in checks against further observations. To this day, scientists use a systematic and logical approach to understanding the Universe, based on an accumulated body of knowledge. The scientific method (as it became known in the 20th century) begins with systematic observation and predictions, before moving on to experiments that put hypotheses and theories to the test.

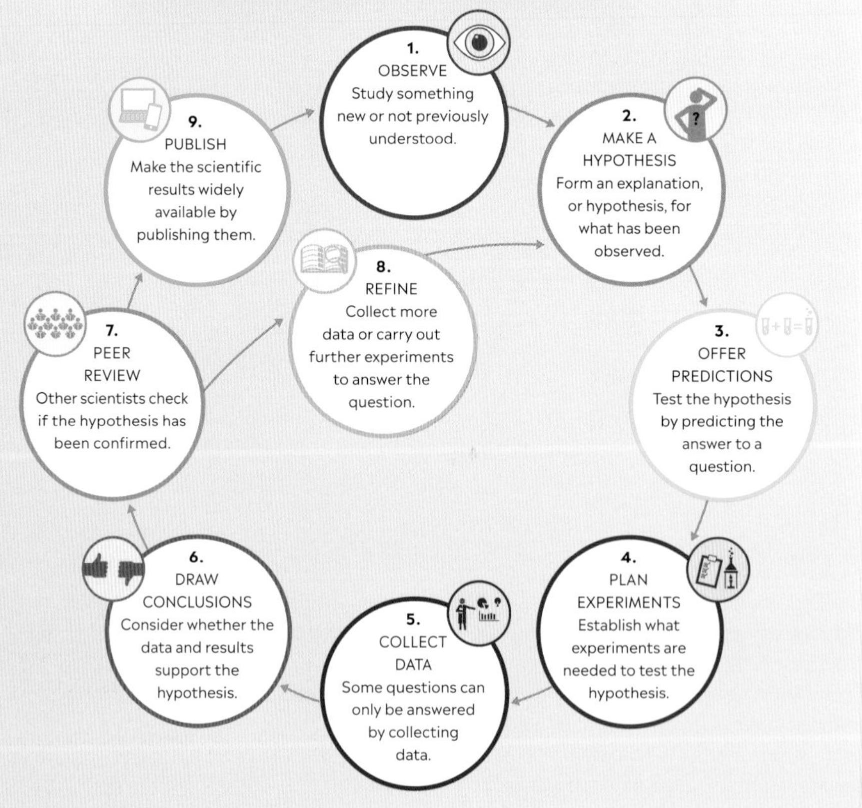

1. OBSERVE Study something new or not previously understood.
2. MAKE A HYPOTHESIS Form an explanation, or hypothesis, for what has been observed.
3. OFFER PREDICTIONS Test the hypothesis by predicting the answer to a question.
4. PLAN EXPERIMENTS Establish what experiments are needed to test the hypothesis.
5. COLLECT DATA Some questions can only be answered by collecting data.
6. DRAW CONCLUSIONS Consider whether the data and results support the hypothesis.
7. PEER REVIEW Other scientists check if the hypothesis has been confirmed.
8. REFINE Collect more data or carry out further experiments to answer the question.
9. PUBLISH Make the scientific results widely available by publishing them.

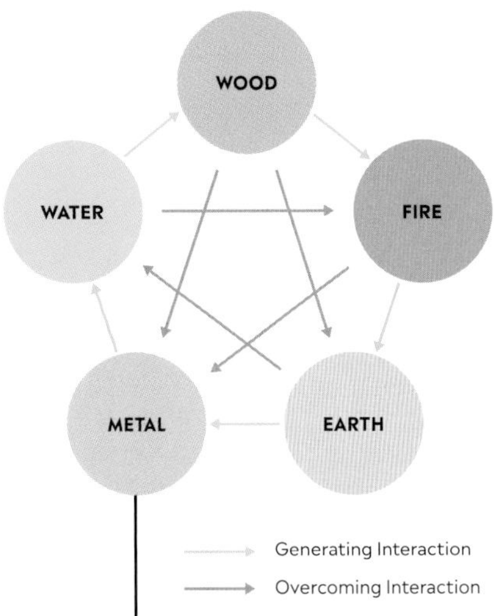

c. 330–270 BCE

CHINESE SCHOOL OF NATURALISTS

Zou Yan, a scholar from the Chinese state of Qi, combined two pre-existing theories to establish the Naturalist School, which formed the basis of much later Chinese thought. According to the *wuxing* or Five Element Theory, there were five basic elements in nature: metal, wood, fire, earth, and water). Zou Yan mixed this with the idea that there are two essential cosmic principles, the *yin* (feminine and earth-like) and the *yang* (masculine and heavenly), suggesting that *yinyang* and *wuxing* interact in cycles to create change.

c. 325–265 BCE

EUCLID OF ALEXANDRIA

The Greek mathematician and scientist Euclid taught in the city of Alexandria. He is best known for *Elements*, his treatise on geometry, and major contributions to optics and astronomy.

c. 310 BCE **Greek physician Praxagoras** describes the difference between arteries and veins

c. 305 BCE **Strato of Lampsacus,** a Greek philosopher, develops a theory of vacuum

301 BCE

c. 310 BCE **The physician Herophilus of Chalcedon** (in Turkey) identifies the brain as the seat of the nervous system

c. 305 BCE **Euclid** develops his theory of optics

△ Page from *Enquiry into Plants*

c. 325–287 BCE

EARLY CLASSIFICATION

Head of the Lyceum, the school founded by Aristotle in Athens, Theophrastus wrote the *Enquiry into Plants*, containing the first consistent classification system for plants and minerals. He categorized plants according to their size, practical uses, and means of reproduction and divided them into six broad groups: trees, two types of shrubs, herbs, fruit-bearing plants, and those producing gums and resins.

▽ The Library of Alexandria

c. 301 BCE

LIBRARY OF ALEXANDRIA

Pharaoh Ptolemy I of Egypt and his successor Ptolemy II founded the Great Library of Alexandria. It contained hundreds of thousands of scrolls of precious manuscripts. The library's heads, such as the geographer and astronomer Eratosthenes of Cyrene, were leading scholars, and the Mouseion, the research centre attached to it, became one of the ancient world's principal seats of learning.

c. 300 BCE

TIDES

The Greek navigator Pytheas made a voyage into the Atlantic, stopping at Land's End in Cornwall, England, and reaching as far as Thule (which may have been Iceland). He was the first to consider that the Moon had an effect on the tides, and he observed how summer days grew longer as he travelled north.

△ **The trireme** of Pytheas

300 BCE

c. 250 BCE Ctesibius of Alexandria (in Egypt) develops the Ctesibian pump, a two-chamber force pump with pistons

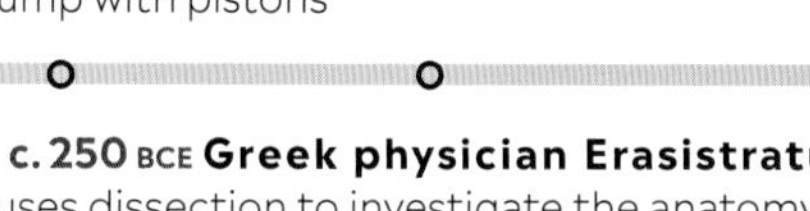

c. 250 BCE Greek physician Erasistratus uses dissection to investigate the anatomy of the nervous system, distinguishing between sensory and motor nerves

c. 240 BCE The vertical-wheeled water mill is invented in Alexandria, Egypt

c. 250 BCE

STATICS AND HYDROSTATICS

The Greek mathematician Archimedes performed pioneering work in statics (the science of bodies at rest) and hydrostatics (the study of liquids). He realized that objects displace a quantity of water equal to their own weight (allegedly when dropping a golden crown into his own bath) and invented a complex block-and-tackle pulley system that could move a ship.

c. 250 BCE

CIRCLES AND SPHERES

In his work *On the Measurement of a Circle,* Archimedes presented methods for calculating the area and circumference of a circle, using the constant pi (π). He also studied conic sections, levers, and the principle of a centre of gravity. He was killed around 212 BCE when, distracted by work, he ignored the command of a Roman soldier.

△ **Archimedes** at work

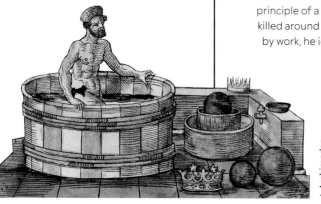

△ **Archimedes**

> "Do not disturb my circles."
>
> *ALLEGED LAST WORDS OF ARCHIMEDES TO THE ROMAN SOLDIER ABOUT TO KILL HIM, c. 212 BCE*

c. 250 BCE

CIRCULATION OF THE BLOOD

The Greek physician Erasistratus of Kos was among the first to formulate a theory of blood circulation. He realized that the heart acted as a kind of pump to which both veins and arteries were connected. He proposed that blood passed through the body in the veins and (erroneously) that the *pneuma* (air or life-force) travelled via the arteries. He achieved fame early in his career by identifying the chronic ailment suffered by Antiochus, the son of the Seleucid ruler Nicator I, as psychosomatic.

◁ **Antiochus** ill in bed, attended by Erasistratus; painting by Jacques-Louis David

c. 210 BCE **Apollonius of Perga** (now in Turkey) describes the properties of conic sections

200 BCE **Chinese texts** describe a magnetic lodestone that can be used to devise a type of compass

193 BCE **The first large concrete** building, the Porticus Aemilia, is built in Rome

160 BCE **Greek astronomer Hipparchus of Nicaea** describes the precession of the equinoxes

161 BCE

△ ***Eratosthenes Teaching in Alexandria*** by Bernardo Strozzi

c. 240 BCE

EARTH'S SIZE

The Greek astronomer Eratosthenes of Cyrene made the first accurate calculation of the circumference of the Earth. He used shadows to observe the difference in the angle at which sunlight hit Earth in Alexandria and Syene (*see panel, right*).

ERATOSTHENES' METHOD

Eratosthenes understood that Earth is round rather than flat. He was able to estimate its circumference by taking measurements at the towns of Syene (modern-day Aswan) and Alexandria. He knew that on the summer solstice, the Sun shone directly down a well in Syene. On the same day, he calculated the angle of the Sun's rays as they cast a shadow on one side of a column in Alexandria. By measuring the distance between Syene and Alexandria (achieved by carefully counting steps), he was able to calculate Earth's circumference as 250,000 stadia or 39,250 km (24,389 miles) – remarkably close to the actual figure of 40,075 km (24,901 miles).

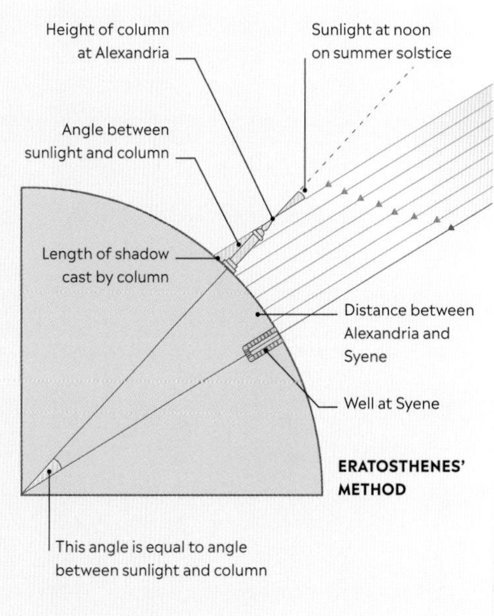

ERATOSTHENES' METHOD

c. 150 BCE

DISTANCE TO THE MOON

Using an ingenious method, the Greek astronomer Hipparchus of Nicaea made the first accurate estimates of the distance from Earth to the Moon. He had two people observe a solar eclipse from different latitudes, near the Hellespont (in modern Turkey) and at Alexandria (in Egypt). When the Sun was fully eclipsed at the Hellespont, a sliver of Sun could be seen at Alexandria. The amount (angle) of sun visible allowed Hipparchus to use trigonometry to calculate the Moon's distance from Earth as 77 times Earth's radius.

◁ **Hipparchus** depicted in an engraving

c. 90 BCE **Posidonius of Apamea** (now in Syria) uses the relative position of the star Canopus at Alexandria and Rhodes to attempt a calculation of the size of Earth

134 BCE

STAR MAP

Hipparchus is said to have been inspired by the appearance of a new star (probably a supernova or comet) in the constellation Scorpio to compile the first comprehensive star catalogue. It contained the locations of 850 stars, referenced by a latitude and longitude system and arranged into 46 constellations, together with their magnitudes (brightness as seen from Earth). Compiled over decades of observations, the original version was lost, but the text was transmitted and improved on by the 1st-century CE astronomer Ptolemy of Alexandria.

◁ **Hipparchus** observing the stars

◁ **Antikythera** mechanism

c. 100 BCE

THE ANTIKYTHERA MECHANISM

Recovered in 1901 from a shipwreck off a Greek island, the Antikythera mechanism was an ancient calculating device. Although much had eroded away, around one third of the device remained. Modern computerized reconstructions suggest that its 30 surviving bronze gear wheels were used to calculate eclipses and the movements of the Moon and other heavenly bodies. A unique survival, it shows the sophistication of Greek astronomy and engineering.

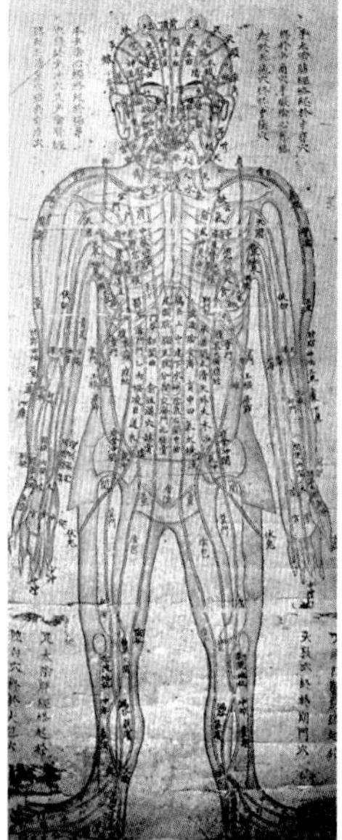

c. 90 BCE

ACUPUNCTURE

The first written reference to the use of needles for acupuncture was made in the *Qiji* (*Records of the Grand Historian*) of Sima Qian. The insertion of needles at key pressure points to control the *qi* or life force of the patient became an essential component of Chinese medical practice.

◁ **Chinese** acupuncture chart

Traditional Chinese medicine recognizes 361 acupuncture points

c. 60 BCE

A NEW THEORY OF HEALTH

The Greek doctor Asclepiades of Bithynia rejected the theory of humours and taught instead that invisible particles (similar to the atoms of Democritus) flowed through the body and that disruptions to these caused disease. He prescribed diet, massage, and exercise to restore their proper circulation.

c. 50 BCE **Glassblowing** using a long tube is invented in Syria

△ **Asclepiades** of Bithynia

45 BCE **The Julian calendar** is introduced

c. 40 BCE **Roman author Marcus Terentius Varro** identifies the link between stagnant water and malaria

c. 15 BCE **Roman architect Vitruvius** writes *De Architectura* (*On Architecture*), the most important ancient work on the subject

1 BCE

44 BCE

DARK SKIES

A huge eruption of Etna, a volcano in Sicily, threw up debris into the atmosphere, partially blocking the Sun and causing a widespread decline in temperatures. Trees were blighted by frost in Italy; the Nile failed to rise in Egypt, provoking a terrible famine in 43–41 BCE; and crop failures were recorded as far away as China. Phenomena such as a red comet and three suns seen in the sky were blamed on the volcano's effects. The climatic change, affecting food production and increasing disease, may have contributed to the political instability that sparked civil war in Italy.

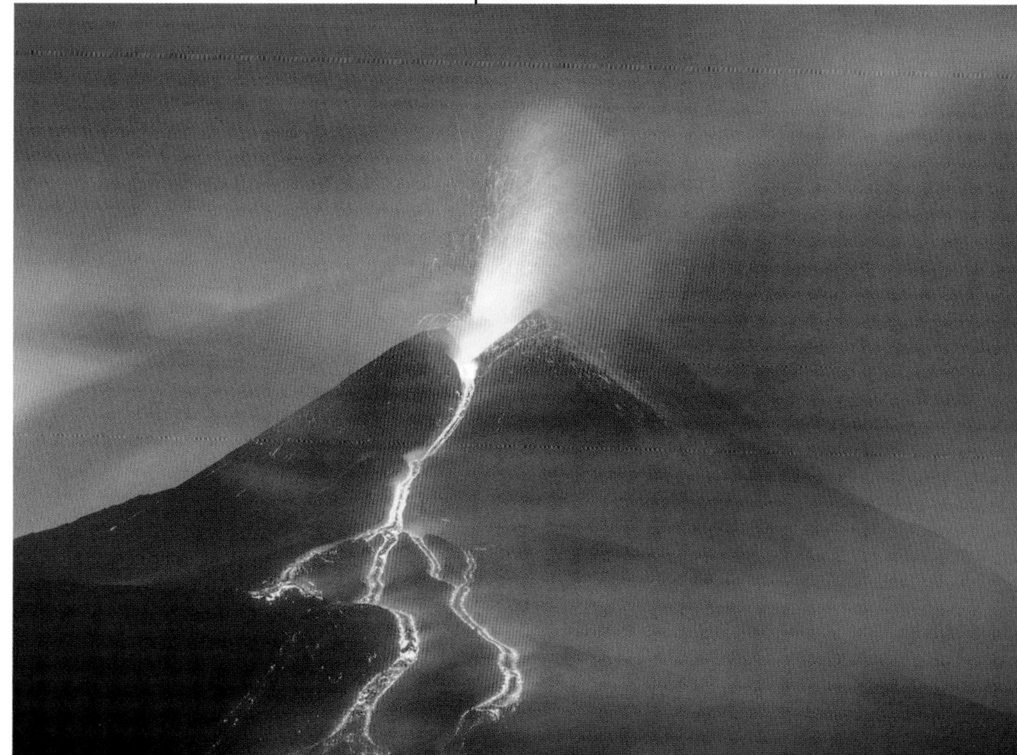

▷ **Lava flows** on Mount Etna

c. 50–70 CE

DIOSCORIDES' PHARMACOPOEIA

Pedanius Dioscorides, a Greek physician in the Roman army, spent around 20 years compiling his *De Materia Medica*, a comprehensive pharmacopoeia describing the medicinal uses of hundreds of herbs and plants. It remained a key text on the identification and properties of plants until the 19th century.

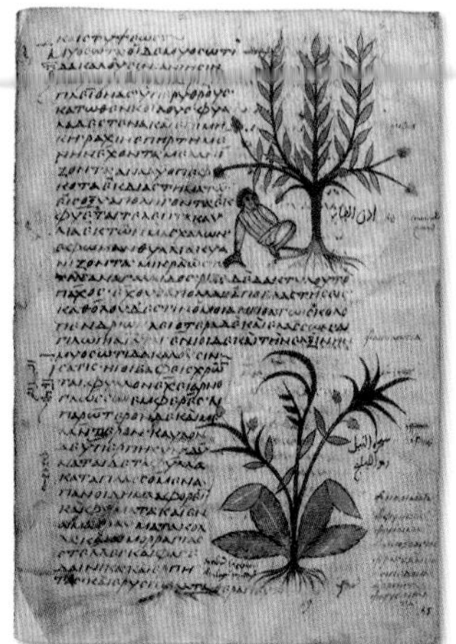

▷ ***De Materia Medica***, page from a 9th-century edition

78–139 CE

ZHANG HENG

Chinese polymath Zhang Heng served at the Imperial Court of the Han dynasty in Nanyang. As well as working as Chief Astronomer, he became known as an engineer and inventor and was a respected poet.

1 CE

1st century CE Indian physician Charaka compiles the *Caraka Samhita*, a compendium of Ayurvedic medicine

77 CE Roman historian Pliny the Elder publishes the first books of his 37-volume *Naturalis Historia* (*Natural History*)

c. 25–50 CE Roman scholar Aulus Cornelius Celsus writes *De Medicina* (*On Medicine*), an encyclopaedic treatise of medical knowledge

c. 40–70 CE

STEAM POWER

A pioneer of experimental science, Greek engineer Hero of Alexandria wrote several treatises on mechanics, physics, and mathematics based on observation and the results of his experiments. In the course of his work, he invented numerous devices, including the aeolipile or "Hero engine", in which a brass sphere is caused to rotate by steam power, the earliest known example of a steam engine.

◁ **Hero's aeolipile** (reconstruction)

△ **Cai Lun,** inventor of the papermaking process

105 CE

INVENTION OF PAPER

According to tradition, Cai Lun, a Chinese eunuch at the imperial court, invented the papermaking process in 105 CE. Although primitive forms of paper probably existed, Cai Lun introduced the use of bamboo, hemp, rags, fishnets, and tree bark macerated and beaten to a pulp to form a smooth, even writing surface.

169 CE

GALEN'S ANATOMY

In 169 CE, Emperor Marcus Aurelius appointed Claudius Galen, Rome's foremost surgeon, as his personal physician. Extensive surgical experience had given Galen a thorough knowledge of human anatomy and medical science. His work, informed by the traditional theory of the four humours, was influential well into the 1400s.

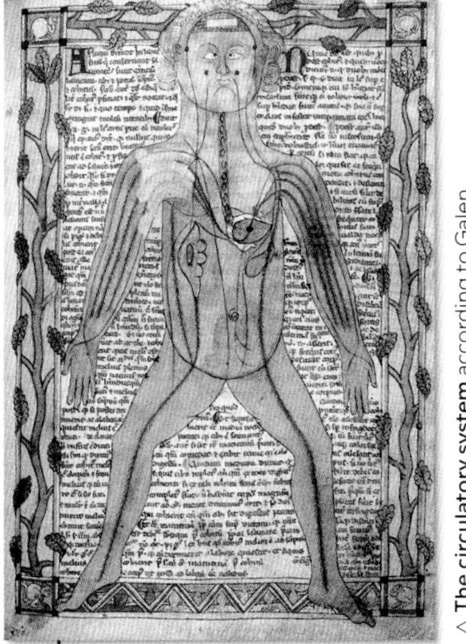

△ **The circulatory system** according to Galen

132 CE

THE FIRST SEISMOMETER

Zhang Heng invented his seismometer, a device capable of detecting even distant earthquakes and indicating the direction of their origin. The urn-shaped device contained a sensitive pendulum that, when disturbed, deposited a metal ball into one of eight frog-shaped receptacles.

△ **Zhang Heng's seismometer** (reconstruction)

c. 105–135 CE Greek physicians Aretaeus of Cappadocia, and Rufus and Soranus of Ephesus write landmark treatises on anatomy and disease

199 CE

c. 120 CE Chinese scientist Zhang Heng concludes from his observation of eclipses that the Moon reflects the light from the Sun

185 CE Chinese astronomers record the observation of a "guest star", later identified as a supernova, for the first time

Ptolemy's *Almagest* catalogues over 1,000 stars and 48 constellations

c. 150 CE

A GEOCENTRIC UNIVERSE

In his *Almagest*, Ptolemy of Alexandria provided a detailed account of his astronomical observations, from which he derived his theory of the motions of the heavenly bodies. This was based largely on the idea of celestial spheres, but using mathematics, Ptolemy created a model of a geocentric Solar System, with an immovable Earth at its centre.

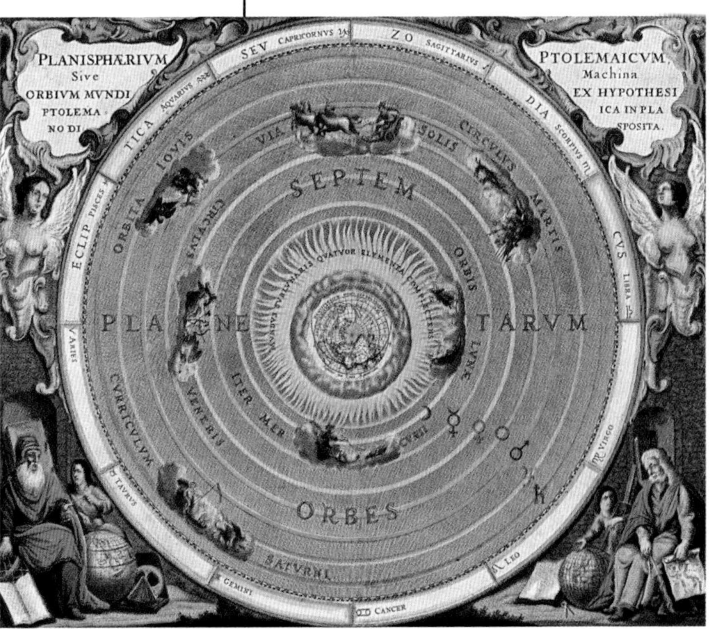

△ **Ptolemy's geocentric Universe**, 17th-century engraving

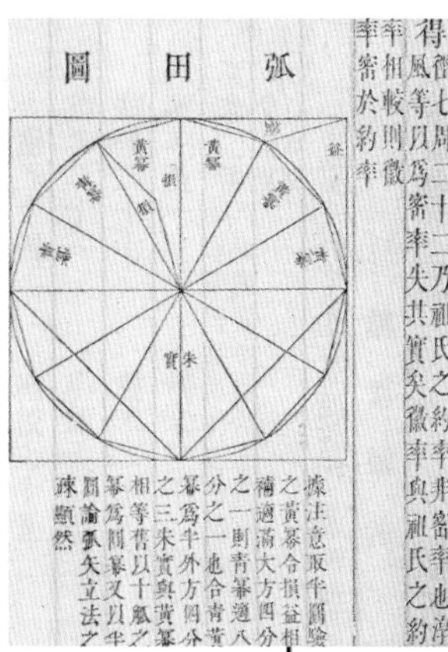

▽ **Diagram** from *The Nine Chapters on the Mathematical Art*

c. 200 CE

A CHINESE APPROACH TO MATHEMATICS

The Nine Chapters on the Mathematical Art, a compilation of the work of generations of Chinese scholars from the 10th century BCE onwards, was completed around 200 CE. It presented mathematical problems together with solutions and methodologies, and so was a more practical approach to numbers than theoretical Greek treatments.

c. 300 CE

NATURAL PEST CONTROL

In the citrus groves of China, farmers discovered a new way to protect their crops from herbivorous insects around the beginning of the 4th century CE. Yellow citrus ants, also known as weaver ants, are natural predators of these pests and were encouraged to thrive in the citrus trees, providing a natural form of pest control.

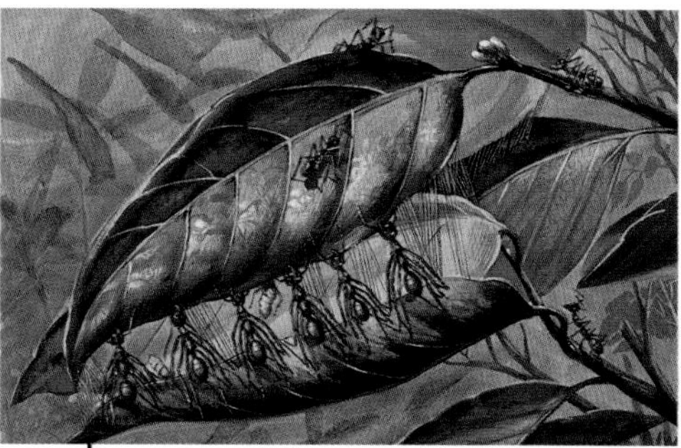

△ **Weaver ants** (*Oecophylla smaragdina*)

200 CE

263 CE In China, Liu Hui writes *Haidao Suanjing* (*The Sea Island Mathematical Manual*), a commentary on *The Nine Chapters on the Mathematical Art*

c. 300 CE Chinese mathematician Sun Zi gives instructions for the use of counting rods in the *Sun Zi Suan Ching*

c. 200 CE In the Bakhshali manuscript, Indian mathematicians introduce the use of a simple dot as a placeholder, anticipating a symbol for zero

c. 300 CE Greek alchemist Zosimos describes arsenic for the first time in his digest of the writings of ancient Egyptian alchemists

DIOPHANTI ALEXANDRINI ARITHMETICORVM LIBRI SEX.
ET DE NVMERIS MVLTANGVLIS LIBER VNVS.
LVTETIAE PARISIORVM,
Sumptibus SEBASTIANI CRAMOISY, via Iacobæa, sub Ciconiis.
M. DC. XXI.
CVM PRIVILEGIO REGIS.

▽ **Book Six of Diophantus'** *Arithmetica*, 17th-century edition

c. 250 CE

THE EMERGENCE OF ALGEBRA

Greek mathematician Diophantus of Alexandria wrote a series of books with the collective title *Arithmetica*, of which only six have survived. The text consists of a collection of some 130 problems solved using equations. In these, Diophantus introduced an innovative system of notating unknown quantities, thus laying the foundations of algebra.

Fermat noted his "Last Theorem" in the margin of a copy of Diophantus's *Arithmetica*

△ *The Seven Liberal Arts*, by Giovanni di Ser Giovanni c. 1460

c. 410–420 CE

THE SEVEN DISCIPLINES

In his allegorical work *On the Marriage of Philology and Mercury*, Martianus Capella distinguished the separate disciplines of grammar, dialectic, rhetoric, geometry, arithmetic, astronomy, and music, the seven "liberal arts" of the medieval education system. In the part on astronomy, he describes a geocentric model of the Universe in which the Sun and three planets orbit Earth, but Mercury and Venus orbit the Sun.

c. 309 CE Chen Zhuo collates the work of earlier Chinese astronomers into a unified system with his catalogue of almost 1,500 stars

346 CE Indian mathematicians are the earliest known users of a decimal system for arithmetic

c. 475 CE Chinese mathematician Zu Chongzi writes *Zhui Shu* (*Method of Interpolation*) in which he calculates π to seven decimal places (3.1415926)

499 CE

before 499 CE

THE MAYAN CALENDAR

By the 5th century CE, Mayan astronomers had developed a calendrical system based on cycles, or "counts", of several different lengths running concurrently. The Tzolkin, or count of days, consisted of a cycle of 13 numbered days, and a cycle of 20 named days, covering a 260-day period; alongside this was the Haab, a cycle of 18 20-day months, plus 5 nameless days, making a 365-day year. The combination of these counts enabled the pinpointing of any particular day within a 52-year period. The Aztecs adopted a similar system.

▷ **Aztec calendar stone**

c. 510 CE

THE LIGHT OF THE MOON

The Indian mathematician and astronomer Aryabhata presented several groundbreaking ideas in his magnum opus, the *Aryabhatiya*. Among the more controversial suggestions, generally rejected by contemporary astronomers, was the theory that Earth rotates on its axis, causing the cycle of night and day. He also proposed that the Moon, and other planets that shine in the night sky, are not a source of light themselves but in fact reflect light from the Sun.

◁ **The Moon** reflecting sunlight

The *Brahmasphutasiddhanta* was written in Sanskrit verse with no mathematical notation

500 CE

c. 520 CE Byzantine Greek philosopher John Philoponus' theory of impetus, similar to the concept of inertia, breaks with Aristotelian thinking

c. 560 CE Alexander of Tralles (now Aydin, Turkey) writes *Twelve Books on Medicine*, including a description of psychiatric disorders such as melancholia

6th century CE In the *Brihat-Samhita*, Indian astronomer Varāhamihira describes the periodic appearance of comets

610 CE Chinese physician Chao Yuanfang compiles a comprehensive medical treatise on many diseases, including smallpox

542 CE

BUBONIC PLAGUE

Bubonic plague spreads across the Byzantine Empire in the early 6th century, reaching the capital, Constantinople, in 542. Roman historian Procopius recorded the outbreak, detailing the symptoms, such as buboes (swollen lymph nodes) in the groin, armpit, or neck.

△ **Depiction** of bubonic plague from the 14th century

c. 570 CE

ARITHMETICAL CALCULATIONS

One of the many commentaries on the 2nd-century CE Chinese mathematical treatise *The Nine Chapters on the Mathematical Art* identified 14 different methods of arithmetical calculation. Among these was the first known description of calculation using an abacus, something only mentioned in passing in earlier texts.

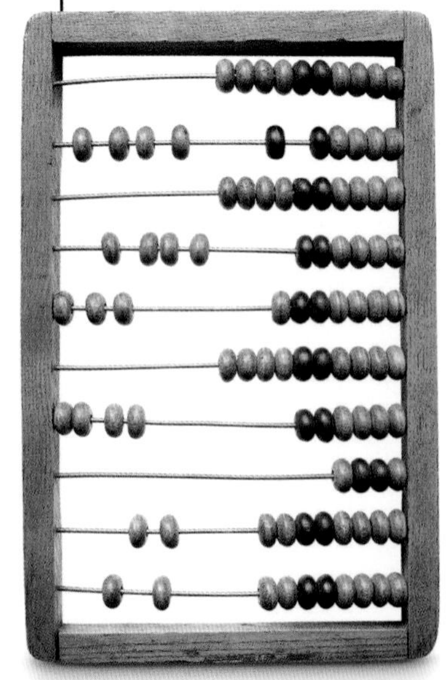

△ **Chinese** abacus

△ **Page** from the *Brahmasphutasiddhanta*

628 CE

WORKING WITH ZERO AND NEGATIVE NUMBERS

In the *Brahmasphutasiddhanta,* his revision of a standard Indian text on astronomy and mathematics, Indian mathematician Brahmagupta contributed his own ideas on algebra and geometry, and most importantly outlined for the first time rules for calculations using the figure zero and those involving negative numbers. He is also credited as the first to describe gravity as a force of attraction.

c. 660 CE Greek physician Paul of Aegina compiles *The Epitome of Medicine* from earlier sources, including Galen, but also describes innovative techniques such as cauterization

c. 725 CE A comprehensive astronomical survey led by Yi Xing leads to reform of the Chinese calendar

749 CE

c. 725 CE English monk Bede writes the influential treatise *De temporum ratione (On the Reckoning of Time)*

672/3-735 CE

BEDE

The scholar Bede, a monk in the Kingdom of Northumbria, England, is best known for his *Ecclesiastical History of the English People*, but he also used his knowledge of astronomy to refine the calendar.

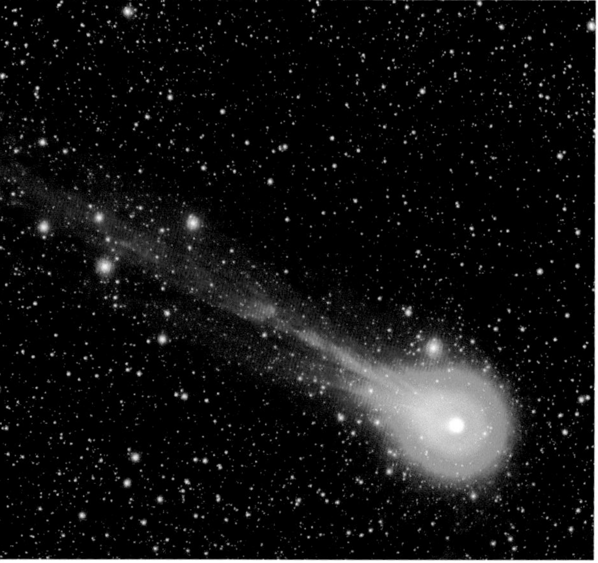

△ **Comet** and its tail

635 CE

COMET TAILS

From their observations, Chinese astronomers reached the conclusion that the tails of comets are not determined by the direction of travel of the comet, but instead always point away from the Sun. Scientists have subsequently found that these comet tails are streams of gases and dust released from a comet's nucleus by the effects of solar radiation.

▷ **Astrolabe** from 11th-century Spain

c. 750 CE

THE ASTROLABE

Invented and developed in Hellenistic Greece, the astrolabe was an invaluable aid to astronomical calculations. It was enthusiastically adopted by Islamic astronomers, who further refined the design. The first Islamic treatise on the astrolabe, written by Ibrahim al-Fazari, appeared in Baghdad in the middle of the 8th century.

▷ **Jabir ibn Hayyan** (Geber)

c. 770–99 CE

CLASSIFICATION OF SUBSTANCES

Known as the father of Arabic chemistry, Jabir ibn Hayyan (also known by his Latin name Geber) was the purported author of numerous treatises on alchemy. Texts attributed to him include theories about the classification of substances into, for example, metals and non-metals, and the distinction between acids and alkalis and their respective properties (*see below*).

750 CE

c. 770 CE Al-Asmai, a scholar of the Basra school in Iraq, pioneers the Islamic study of zoology and animal anatomy

762 CE Baghdad, the first planned Islamic city, is founded; it quickly becomes a centre of scholarship and scientific research

805 CE Jabril ibn Bukhtishu succeeds his grandfather Jurjis as physician at the court of al-Mansur and founds the first hospital in Baghdad

c. 810 CE The Bayt al-Hikmah or "House of Wisdom", a public library and academy, is founded in Baghdad

ACIDS AND BASES

Substances capable of giving up hydrogen ions (H^+) are called acids, while those that can accept them are called bases. Acids and bases react to neutralize each other, producing a salt and water. Common acids are found, for example, in lemon juice and vinegar, while bases are in baking powder and bleach. Acids and bases are detected using indicators (such as litmus paper), and their strength is defined by the pH scale.

The pH scale

The relative strength of an acid or base is shown on the pH scale, which indicates the concentration of hydrogen ions in a solution. The range runs from 0 (the most acidic) to 14 (a strong base).

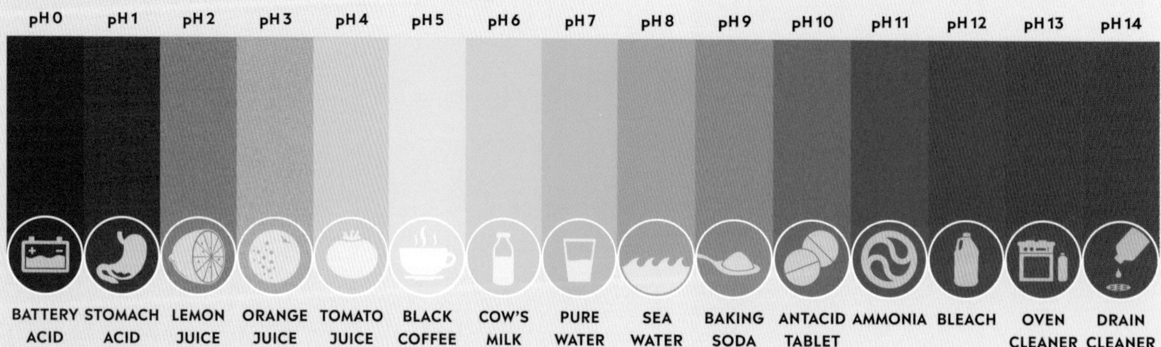

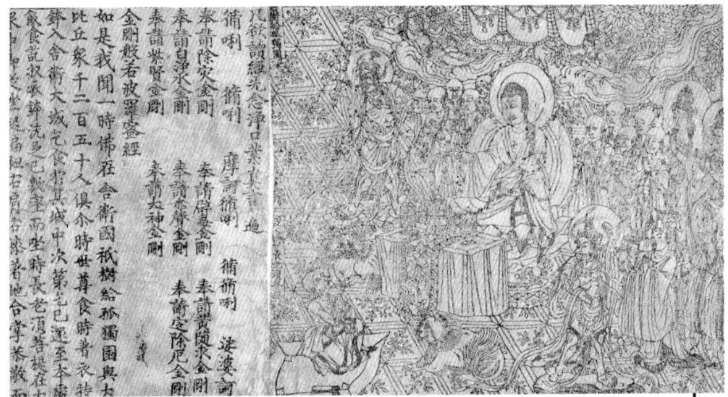

△ The *Diamond Sutra* from China

868 CE

WOODBLOCK PRINTED BOOKS

The earliest known printed book, a Chinese edition of the *Diamond Sutra*, a Buddhist text translated from Sanskrit, was discovered in China in 1907. Dated 11 May 868, the woodblock printed book is probably not the first of its kind, as the quality of printing suggests the techniques were well practised, but is the oldest to have survived.

c. 890 CE

DISTILLATION OF ALCOHOL

The Persian polymath and physician al-Razi, also known by his Latin name Rhazes, was a staunch advocate of experimental science. In the course of his experiments, he perfected a technique for distilling alcohol from wine using a vessel known as an alembic, in which the wine was heated. The vapours from the wine then condensed in the spout of the alembic, and the resulting distillation, the alcohol (from the Arabic *al kuhl*, meaning "essence"), was collected.

◁ Early Arabic alembic

899 CE

c. 820–845 CE At the House of Wisdom, al-Kindi translates and comments on Greek and Indian texts and writes on various scientific subjects

c. 855 CE Alchemists in China describe the discovery of gunpowder, made from a mixture of sulphur, carbon, and saltpetre

876 CE A specific symbol for zero, rather than a simple space or dot, is recorded for the first time on an inscription by Indian mathematicians

▽ Al-Khwarizmi's algebra

"We should not be ashamed to acknowledge truth from whatever source it comes."

AL-KINDI, ON FIRST PHILOSOPHY, *c. 840 CE*

c. 830 CE

CALCULATING WITH ALGEBRA

Among the scholars at the House of Wisdom in Baghdad was the Persian polymath al-Khwarizmi, whose *Compendious Book on Calculation by Completion and Balancing* became a foundational text in the emerging discipline of algebra. In it, he explains the method for solving linear and quadratic equations by the technique of balancing the sides of the equation. The modern word for this discipline, algebra, comes from the Arabic *al-jabr* (completion) in the title of his treatise.

c. 900–925 CE

DOUBTS ABOUT GALEN

Respected in many scientific disciplines, Persian scholar al-Razi was best known as the foremost physician in the Islamic world and the author of many important medical treatises that offered new insights into fields such as pediatrics, obstetrics, and mental disorders. More controversially, in *Doubts about Galen* he questioned the Greek's theory of the four humours.

◁ **Al-Razi medical treatise,** 13th-century translation

c. 998 CE

DISAPPEARING SEAS

Persian scholar al-Biruni suggested that at some time all land was once entirely covered by the sea. He reached this conclusion from fossil evidence: in the Arabian desert, he found shells and bones of obviously marine creatures, presumably stranded there when the waters receded.

△ **Cowrie shell** fossil

900

c. 910–c. 932 CE In Kairouan, Tunisia, Jewish physician Isaac Israeli ben Solomon writes influential studies of ailments and their remedies

976 CE The Hindu-Arabic number system appears for the first time in Europe, in the Spanish *Codex Vigilanus*

c. 990 CE Arab Andalusian physician al-Zahrawi publishes his *Method of Medicine*, which becomes a standard medical textbook in medieval Europe

984 CE

REFRACTION OF LIGHT

A Persian mathematician working in Baghdad, Ibn Sahl made a particular study of optics, and in his treatise on the subject, *On Burning Mirrors and Lenses*, he described the properties of curved mirrors and lenses. He is also credited with being the first to propose a law of refraction, which he derived from his mathematical analysis of his experimental findings.

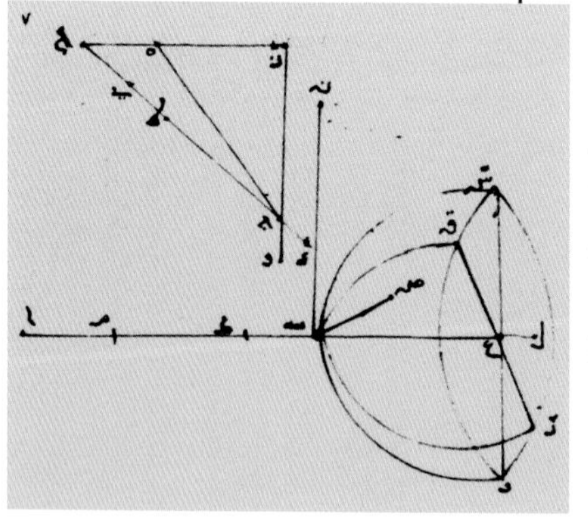

△ **Diagram** from Ibn Sahl's treatise

Al-Razi was the first physician to describe smallpox and measles as distinct diseases

1011–21
A THEORY OF VISION

Written over a period of ten years, the seven-volume *Book of Optics* by the Arab scholar Ibn al-Haytham (sometimes Latinized as Alhazen) presented a novel theory of vision. Supported by evidence from his experiments, al Haytham rejected the notion proposed by Ptolemy that sight resulted from rays of light emitted from the eye. Instead, he proposed a theory that light entering the eye from outside causes the sensation of vision.

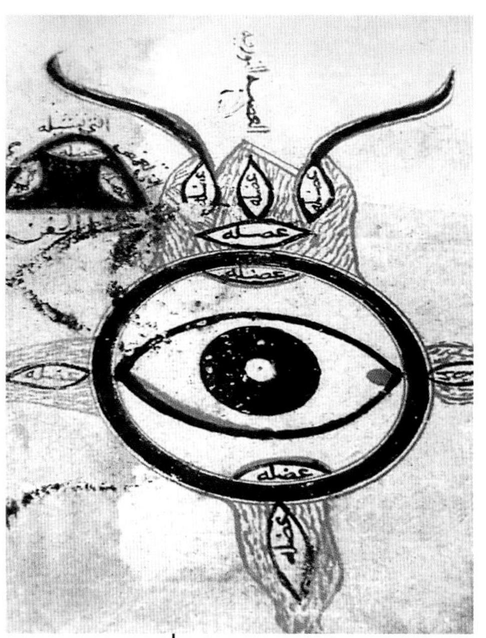

△ Diagram from the *Book of Optics*

980 CE–1037
IBN SINA (AVICENNA)

Persian philosopher, physician, and scientist Ibn Sina, better known in the West as Avicenna, was a prolific and influential writer whose works include the *Canon of Medicine* and an encyclopedia of science and philosophy, *The Book of Healing*.

c. 998 CE Gerbert of Aurillac (in France) is one of the first European scholars to study Classical and Islamic scientific texts; he introduces the abacus to Europe

c. 1030 Persian astronomer al-Biruni discusses the rotation of Earth, and suggests it may orbit the Sun

c. 1040 Pi Sheng begins printing with clay blocks, each bearing a single Chinese character

1049

△ The Crab Nebula

▽ *The Canon of Medicine*

1025
THE CANON OF MEDICINE

The magnum opus of Ibn Sina, *The Canon of Medicine* was a comprehensive encyclopedia of contemporary Islamic medical knowledge collated over a period of 20 years. Rigorously covering anatomy, physiology, medicines, and the diagnosis and treatment of specific diseases, it was a standard medical textbook for hundreds of years.

1006
EXPLODING STARS OBSERVED

The first detailed description of a supernova, the explosion of a star which we now know to have been the massive SN 1006, was given by Ali ibn Ridwan from Cairo. A further stellar explosion, the supernova that formed the Crab Nebula, SN 1054, was observed by Chinese and Arab astronomers in 1054.

1088
FIRST UNIVERSITIES FOUNDED

The great medieval universities developed from ecclesiastical centres of learning at the end of the 11th century. The first to be founded was the University of Bologna, in 1088, followed by similar institutions in Oxford (1096) and Paris (1150).

▽ **Medieval students** at Bologna

1088
DREAM POOL ESSAYS

In his *Dream Pool Essays*, written in Song China, Shen Kuo described the magnetic needle compass for the first time, and suggested it could be used for navigation. In connection with this, he also explained his discovery of the phenomenon of magnetic declination, the angle of difference between true (or geographic) and magnetic north.

◁ **Song-era compass**

The first recorded sighting of Halley's Comet was in 240 BCE

1050

1121 Persian astronomer al-Khazini proposes a theory that gravity varies according to the distance from Earth's centre

1073 Omar Khayyam establishes an observatory at Isfahan, planning to revise the Persian calendar

1094 Su Song publishes a detailed description of the sophisticated water-powered astronomical clock he had constructed in Kaifeng, China

1066
HALLEY'S COMET RECORDED

The comet now known as Halley's comet (named after the English astronomer Edmond Halley) is visible with the naked eye every 75–79 years, when its orbit is closest to Earth. Astronomers from ancient times noted its appearance, which was popularly seen as a portent of ill fortune in many cultures. It was particularly visible in 1066 and is featured as a significant event in the Bayeux Tapestry, which depicts the Norman invasion of England and Battle of Hastings.

◁ **Bayeux Tapestry** panel with Halley's comet visible at the top

TIMEKEEPING

Today, accurate timekeeping is achieved using atomic clocks, which have a margin of error as little as one second every 100 million years. Instead of using a pendulum or other mechanism, atomic clocks keep time by monitoring the tiny natural oscillations of atoms. A caesium-133 atom at absolute zero (0 K) oscillates more than 9 billion times every second. These oscillations are extremely consistent, making them ideal for accurate timekeeping.

How an atomic clock works

Subjecting atoms to microwave radiation at a specific frequency causes them to switch between energy states X and Y. This process is used to calculate a second.

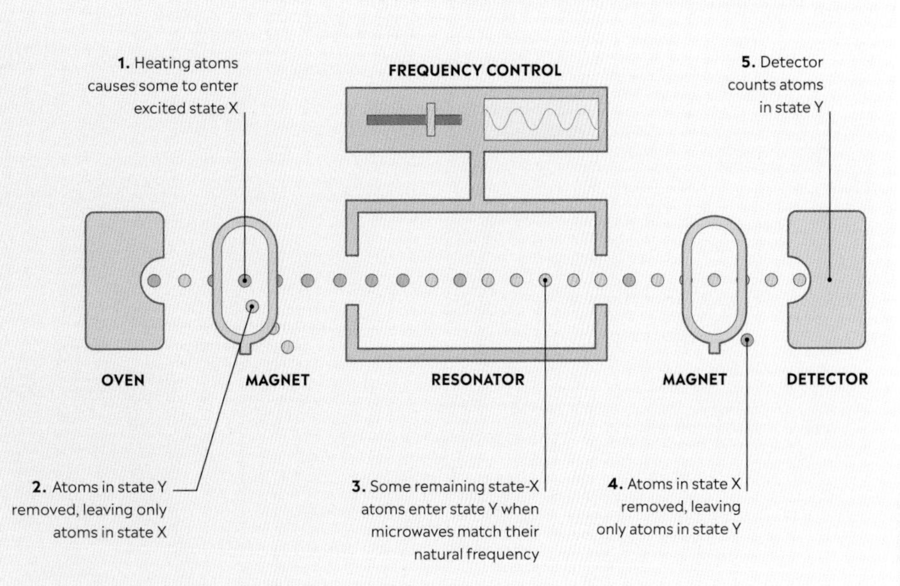
1. Heating atoms causes some to enter excited state X
FREQUENCY CONTROL
5. Detector counts atoms in state Y
OVEN
MAGNET
RESONATOR
MAGNET
DETECTOR
2. Atoms in state Y removed, leaving only atoms in state X
3. Some remaining state-X atoms enter state Y when microwaves match their natural frequency
4. Atoms in state X removed, leaving only atoms in state Y

1126 Archbishop Raymond of Toledo (in Spain), initiates a programme to translate Arabic, Greek, and Hebrew scientific texts into Latin

1150 Indian mathematician Bhaskara II shows that any number has two square roots, one positive and one negative, and that dividing a number by zero produces infinity

c. 1150 The three Trotula texts on women's medicine (one attributed to Trota of Salerno, a woman physician practising in Salerno, Italy) are published

c. 1155 Chinese cartographers produce the world's first printed map, of western China

c. 1170–1200 Moses Maimonides, a Sephardic Jewish polymath, is exiled in Egypt, where he writes a number of influential medical treatises

1199

c. 1150

ARISTOTELIANISM REVIVED

Ibn Rushd, also known in Europe as Averroës, was largely responsible for bringing the ideas of Aristotle to an Islamic readership, and subsequently to European scholars. He embarked on a series of commentaries on Aristotle's works, clarifying and explaining, and reconciling them with Islamic theology. He also added some ideas of his own, including an elaboration of Aristotle's theory of motion, which distinguishes between the weight of a body and its mass.

▷ Ibn Rushd's commentaries

1031–1095

SHEN KUO

A Chinese statesman and polymath, Shen Kuo is best known for his encyclopedic *Dream Pool Essays*, a wide-ranging collection covering all aspects of contemporary Chinese technology and science, including several original contributions.

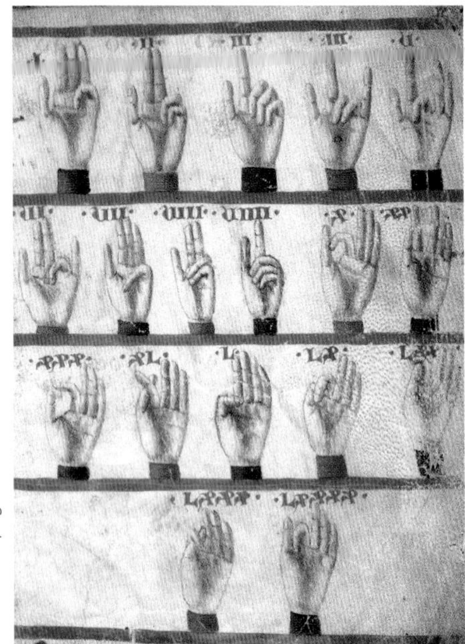
▷ *Liber Abaci* page

1202
ADOPTION OF INDO-ARABIC NUMERALS

In his ground-breaking mathematical treatise *Liber Abaci* (*Book of Calculation*), Leonardo of Pisa (in Italy), later known as Fibonacci, advocated the use of the simpler Indo-Arabic number system, rather than the cumbersome Roman numerals then in use in Europe. He is also credited with introducing to Europe the number series now known as the Fibonacci sequence.

▷ Song Ci

1247
FORENSIC SCIENCE

Forensic science was in its infancy when Chinese physician Song Ci wrote *Xi Yuan Ji Lu* (*Collected Cases of Injustice Rectified*), which includes the earliest known description of forensic entomology. He also reportedly solved a murder case by observing that flies were attracted to minute traces of blood on the perpetrator's sickle.

1200

13th-century Chinese medical texts include references to circadian cycles in humans

c. 1248 Islamic physician Ibn al-Baytar compiles his pharmacopoeia

1220–35 English bishop Robert Grosseteste describes the true nature of colour in *De Luce* (*On Light*)

1242 Arab physician Ibn al-Nafis describes the circulation of blood between the heart and lungs in *Sharh Tashrih al-Qanun*

c. 1260 Italian surgeons Ugo and Teodorico Borgognoni use wine to disinfect wounds and narcotic-soaked sponges to anaesthetize patients

1267
AN EMPIRICAL APPROACH

Translations of Classical and Islamic scholarly texts became more widely available in Europe in the 13th century, prompting Roger Bacon to write his *Opus Majus* (*Greater Work*). An overview of current knowledge of all the sciences, his work placed an emphasis on an empirical, experimental approach to their study.

▷ **Bacon conducting an experiment,** from Michael Maier's *Symbola Aureae*

1220–92
ROGER BACON

Roger Bacon was a Franciscan friar and philosopher, born in Somerset, England. He studied at the University of Oxford, and later taught there and at the University of Paris. He wrote on subjects as diverse as philosophy, linguistics, and the sciences.

"Experimental science is the queen of sciences, and the goal of all speculation"

ROGER BACON, OPUS TERTIUM, *C. 1267*

c. 1272

TABLES OF PLANETARY MOTION

The first set of astronomical tables compiled in Christian Europe were commissioned in Spain by Alphonso X of Castile and prepared by Jewish astronomers in Toledo from calculations by al-Zarqali. These Alphonsine Tables enabled the user to calculate accurately the positions of the planets, and their eclipses, at any given time. The tables were later translated into Latin, becoming the definitive reference until the 16th century.

▷ Alphonsine Tables

1276

CALENDAR REFORM

Guo Shoujing was already renowned as a civil engineer and astronomer when he was commissioned by Emperor Kublai Khan to reform the Chinese calendar. To gather the necessary information, Guo set about building 27 new observatories and equipping them with sophisticated astronomical devices, many of his own invention.

▷ **Gaocheng observatory,** China

1299

1269 French physicist Pierre de Maricourt describes magnetic poles and the laws of magnetic attraction and repulsion

1290

THE OBLIQUITY OF THE ECLIPTIC

French astronomer William of St Cloud, writing in 1290, described his observations of the Sun's position throughout the year, enabling him to plot the apparent course of the Sun, the ecliptic. Because Earth's axis is tilted in relation to a line perpendicular to its orbital plane, its equator is also titled in relation to the ecliptic, a phenomenon known as the obliquity of the ecliptic. St Cloud was able to correctly calculate the angle of tilt at about 23.4°.

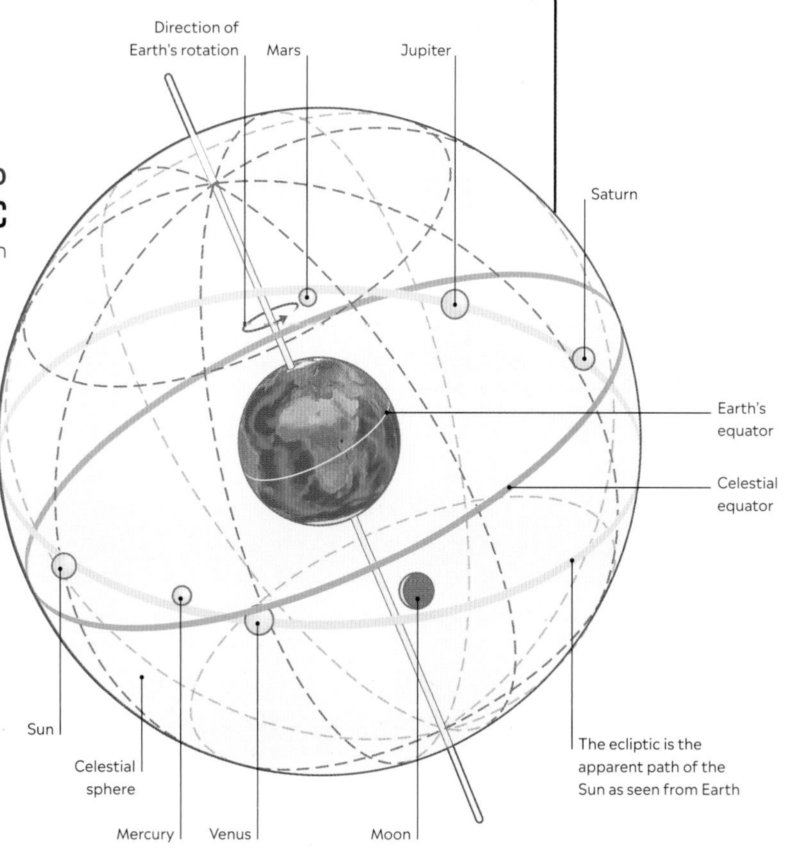

Celestial sphere and the ecliptic
To the observer on Earth, the Sun appears to travel around the sky over the course of the year, tracing a path known as the ecliptic. The position of the ecliptic can be plotted on an imaginary sphere centred on Earth known as the celestial sphere.

1310

COLOURS OF THE RAINBOW

Dominican friar Theodoric of Freiberg, Germany, provided the first satisfactory explanation of the phenomenon of rainbows. He experimented with shining light into lenses, crystal spheres, and mirrors to simulate sunlight passing through water droplets and thus successfully identified the processes of refraction and reflection that form the rainbow.

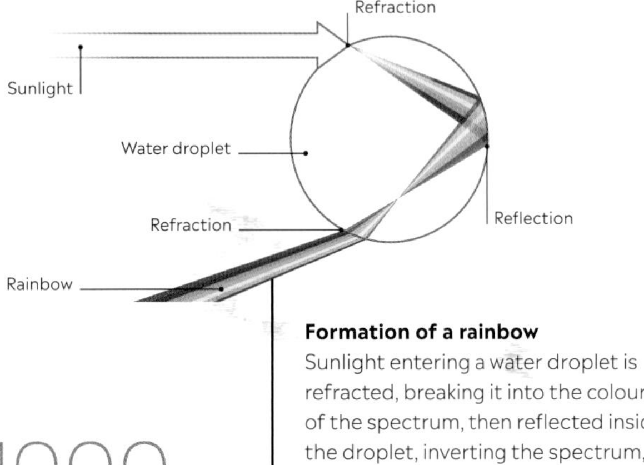

Formation of a rainbow
Sunlight entering a water droplet is refracted, breaking it into the colours of the spectrum, then reflected inside the droplet, inverting the spectrum, and refracted again when leaving.

△ Eye examination from *Cyrurgia*

1306-20

A HANDBOOK OF SURGERY

An experienced surgeon in Montpelier and Bologna, with an extensive knowledge of anatomy, Henri de Mondeville began work on *Cyrurgia* (*Surgery*), one of the earliest surgical textbooks, which remained unfinished at the time of his death.

1323 William of Ockham publishes *Summa Logicae* (*Handbook of Logic*), advocating an empirical approach to the acquisition of knowledge

1300

c. 1300 The first weight-driven clocks appear in western Europe following the invention of an efficient escapement

1315 Mondino de Luzzi performs his first public dissections in Bologna, Italy. The following year he publishes *Anatomia*, the first specifically anatomical textbook

c. 1285-1349

WILLIAM OF OCKHAM

William of Ockham (or Occam) was an English philosopher, scientist and theologian associated with "Ockham's razor", the principle of seeking an explanation in terms of the fewest assumptions.

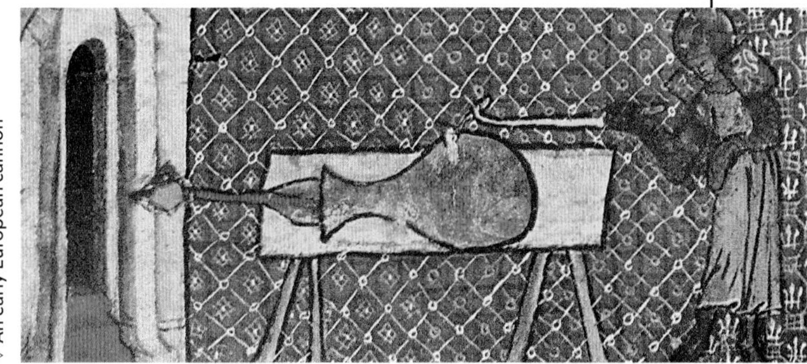

▽ An early European cannon

c. 1340

WIDESPREAD USE OF CANNONS

Although rudimentary weapons making use of gunpowder had been developed in China as early as the 12th century, true cannons, firing a projectile from a cylinder, emerged only gradually and were not widely used in battle until well into the 14th century. Similar weapons also appeared in the Middle East and Europe at around the same time, and saw their first significant use in the Hundred Years' War (1337-1453).

△ **Burying victims** of the Black Death in Tournai, France; illustration from Gilles de Muisit's annals

1346

THE BLACK DEATH HITS EUROPE

The bubonic plague pandemic spread from Asia into Europe, causing the deaths of around half of the population. It triggered an inevitable decline in scientific and technological progress that lasted for more than a century, but prompted a surge in the study of medical sciences.

1349

1348 French physician Guy de Chauliac notes for the first time the distinction between bubonic and pneumonic plagues when the Black Death breaks out in Avignon, France

c. 1349 English philosopher John of Dumbleton suggests that during contraction or expansion (such as condensation and rarefaction) a substance retains the same number of parts

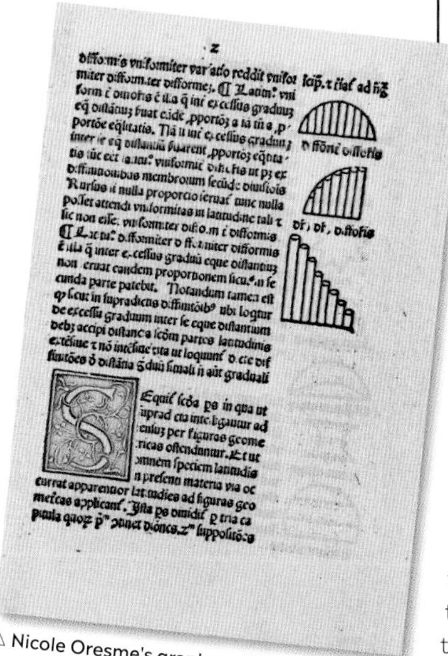

△ Nicole Oresme's graphs

> "It is pointless to do with more what can be done with fewer."
>
> WILLIAM OF OCKHAM, *SUMMA LOGICAE, 1323*

c. 1349

GRAPHICAL REPRESENTATION

French mathematician Nicole Oresme's most important contribution was the development of a system to graphically represent the change of a function. Resembling modern bar charts, his graphs used rectangular coordinates, which he called latitude and longitude, to plot the distribution of quantities, such as the change of velocity in relation to time. Anticipating René Descartes' coordinate system by 300 years, this work helped to lay the foundations of analytic geometry.

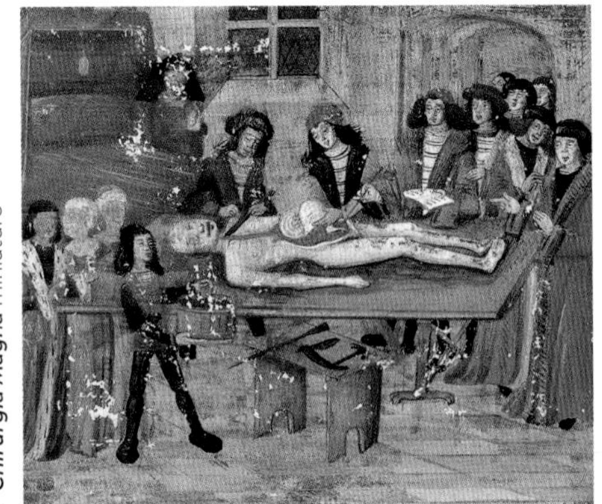
▽ *Chirurgia Magna* miniature

1363

A COMPREHENSIVE ENCYCLOPEDIA OF MEDICINE

One of the standard medieval textbooks, the seven-volume *Chirurgia Magna* (*Great Surgery*), was completed by French physician Guy de Chauliac just five years before his death in 1368. Although regarded in its day as a comprehensive and authoritative treatise, it was essentially an academic rather than practical guide and did not include the many new ideas pioneered by Italian physicians.

By 1351, the Black Death had claimed an estimated 25 million lives in Europe

1350

c. 1375 Indian mathematician Madhava founds the Kerala school of astronomy and mathematics

1376 Virdimura is the first woman to qualify as a surgeon at the school of medicine in Salerno, Sicily

1364

DE DONDI'S ASTRARIUM

Built over a period of about 16 years, Italian engineer Giovanni de Dondi's highly complex astronomical clock, known as the astrarium, was hailed as a marvel of the age. The clockwork mechanism was housed in a frame with seven faces, with dials and indicators for the Sun, Moon, and planets that provided information on their positions, as well as showing important dates in the religious and legal calendars. De Dondi also published the *Planetarium*, a description of its construction detailed enough to enable a modern reconstruction.

◁ **Astrarium** (reconstruction)

1377

EARTH'S ROTATION

In his *Livre du Ciel et du Monde* (*Book of the Sky and the World*), the French scientist Nicole Oresme argued against any proof of Aristotle's theory of a stationary Earth. He suggested, for example, that it was more likely that Earth rotated on its axis than that the immense celestial spheres circled around it. However, he decided that the arguments on either side were inconclusive and opted to side with Aristotle.

◁ **Nicole Oresme** at his desk

1377–1446

FILIPPO BRUNELLESCHI

A seminal figure of the Italian Renaissance, Brunelleschi was a pioneering architect, engineer, artist, and sculptor. He lived and worked in Florence, designing many of its finest buildings, including the dome of Florence Cathedral.

1377 Ibn Khaldun, Arab historian and pioneering social scientist, suggests that humans developed from monkeys

c. 1415 Filippo Brunelleschi uses mirrors to demonstrate the mathematical technique of linear perspective

1439

▷ Samarkand observatory

1420–29

A CENTRE OF LEARNING IN SAMARKAND

The Timurid sultan Ulugh Beg was only 16 when he became governor of Samarkand, a city in what is now Uzbekistan. As a young man, he was more interested in astronomy and mathematics than statecraft, and he established Samarkand as a major centre of scientific learning, founding a madrasa there, and inviting many Islamic scholars to study in the city. In 1424, he commissioned the construction of an enormous observatory, equipped with the most sophisticated instruments available.

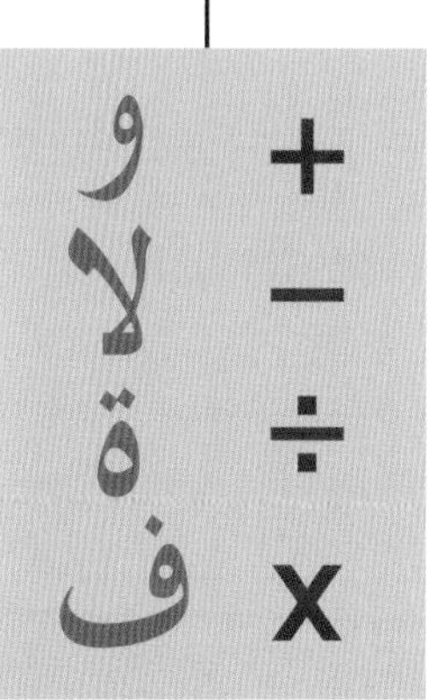

◁ Arabic mathmatical symbols

1430–39

MATHEMATICAL SYMBOLS

While Islamic mathematician Ali al-Qalasadi was not the first to use characters from the Arabic alphabet to represent mathematical operations, he popularized the idea of algebraic symbols through their systematic appearance in his writings.

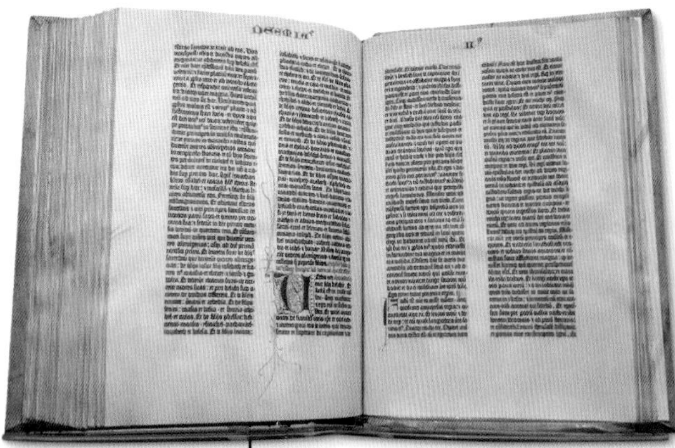

△ Gutenberg Bible

1440–49
PRINTING WITH MOVABLE TYPE

After almost ten years of development, in 1449 Johannes Gutenberg began production of printed texts using a movable-type printing press, the first such machine to be invented in Europe. Within a year he had progressed from simple poems to complete books, and in 1454 he produced a print run of 180 Bibles. His invention had revolutionary implications: mass-production of texts not only meant that ideas and learning could be disseminated quickly, but also challenged the Church's virtual monopoly of learning by removing the need for monastic scribes producing handwritten manuscripts.

> "Like a new star it shall scatter the darkness of ignorance."
>
> *JOHANNES GUTENBERG, ON HIS PRINTING PRESS*

1440

1464 German mathematician Regiomontanus (Johannes Müller) publishes *De Triangulis Omnimodis* (*On Triangles*), a textbook on trigonometry

1453
THE FALL OF CONSTANTINOPLE

In May 1453, the Byzantine capital Constantinople fell to the Ottoman Empire after a 53-day siege. This prompted a large number of the city's Christian inhabitants to flee to Italy, bringing with them a tradition of scholarship stretching back to Classical Greece, and a vast amount of philosophical and scientific learning. This migration influenced the emerging Renaissance movement and reinvigorated the "Scientific Revolution" that went with it.

▷ **The Siege of Constantinople,** miniature by Jean Le Tavernier

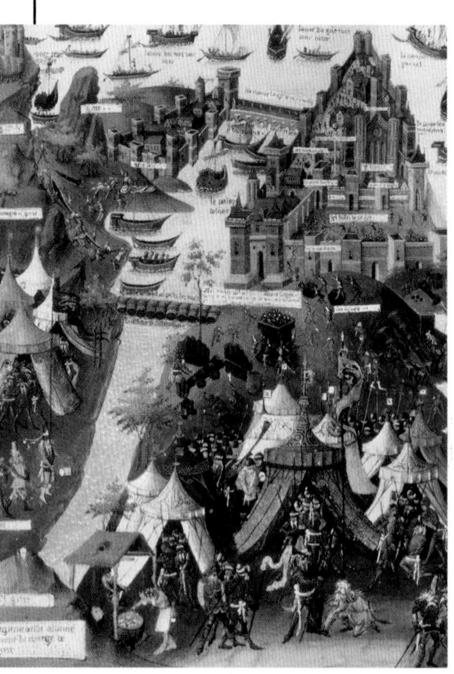

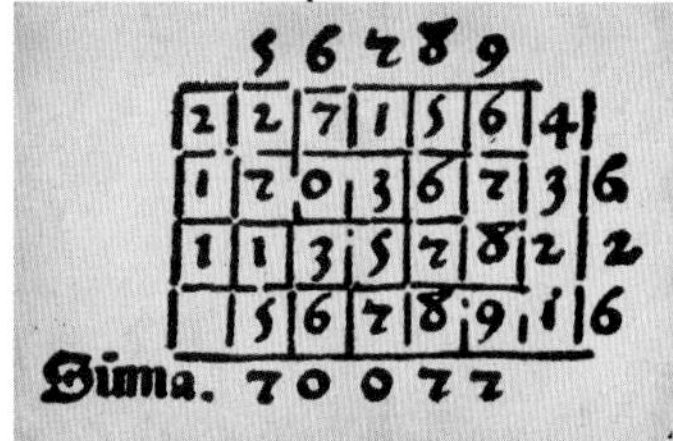

△ **Table from** *The Treviso Arithmetic*

1478
ACCESS TO LEARNING

The publication of *The Treviso Arithmetic* marked the beginning of a wave of printed textbooks on mathematics that was made possible by the invention of the movable-type printing press. Books on other subjects quickly followed, making learning accessible and affordable to a wide readership.

c. 1480s

LEONARDO'S STUDY OF FLIGHT

The notebooks of Leonardo da Vinci contain a wealth of drawings and writings on a wide range of subjects, from sketches for paintings to scientific and technical studies. Particularly striking among the notebooks' pages are the artist's observations of different aspects of flight, which include analyses of the anatomy of flying animals and suggested designs for flying machines.

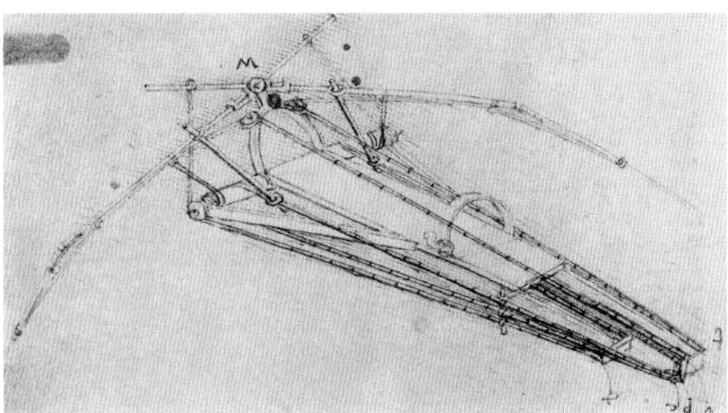

△ **Leonardo sketch** of a flying machine

1452–1519

LEONARDO DA VINCI

A towering figure of the High Renaissance, Leonardo was a true polymath: an accomplished artist, scientist, engineer, and architect. His career began in his native Florence, but took him to Milan, Rome, and later to France, where he died aged 67.

1489 The symbols + and - representing plus and minus operations appear in a treatise by Johannes Widman

c. 1495 Leonardo speculates that fossils are the petrified remains of ancient life-forms

1499

1489 Leonardo begins a series of anatomical drawings based on dissections of animal and human corpses

1490 Leonardo describes capillary action - the flow of liquid in a narrow space without the assistance of gravity

1496 Regiomontanus brings the theories of Ptolemy to a wide readership with the publication of his *Epitome of Ptolemy's Almagest*

△ **Columbus** reaches the Bahamas

1492

COLUMBUS ARRIVES IN THE AMERICAS

Having crossed the Atlantic Ocean in an attempt to discover a westward route to Asia from Spain, Christopher Columbus arrived in the Bahamas, mistaking them for the East Indies. He was unaware that he had stumbled across a continent that was unknown to Europeans as, of course, there was no trace of it on the maps he used. Once the existence of the Americas was realized, it prompted a complete revision of geographers' concepts of the world, as well as a scramble for territory among Europe's colonial powers.

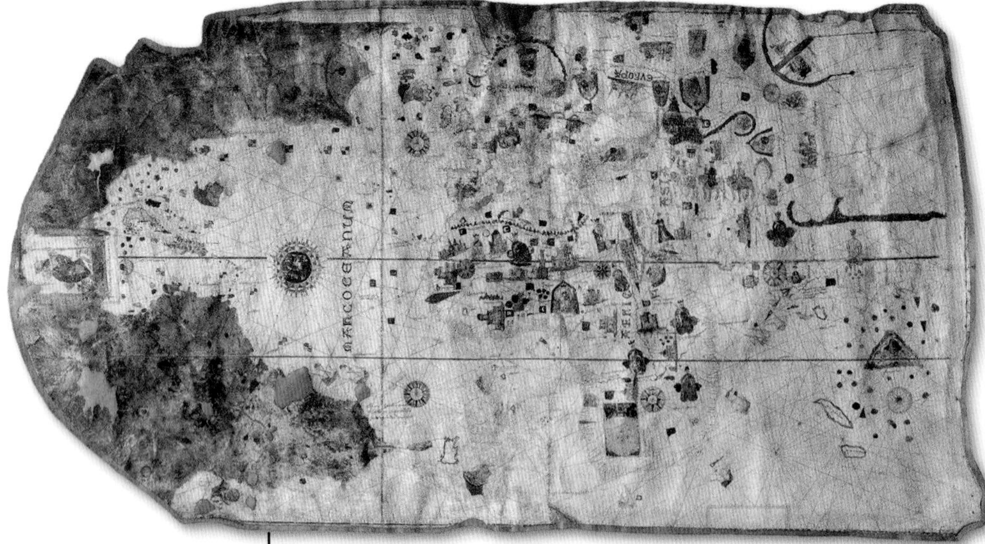

1500

THE NEW WORLD MAPPED

Spanish navigator and cartographer Juan de la Cosa was the master of the *Santa Maria* – the largest of the three ships used in Columbus's crossing of the Atlantic Ocean in 1492. He made seven voyages to the continent over the following 17 years, and in 1500 oversaw the creation of the first world map to include the coast of the Americas (the area painted in green on the far left of the map).

◁ **Juan de la Cosa's** map

De la Cosa accompanied Christopher Columbus on his first three expeditions to the Americas

1500

c. 1500 Indian astronomer Nikalantha Somajayi writes the *Aryabhatiyabhasya*, a commentary on the 6th-century mathematician Aryabhata

1512 The first pocket watch is manufactured by the German clockmaker Peter Henlein in Nuremberg. It can operate for 40 hours without winding

1513 Polish-German astronomer Nicolaus Copernicus writes his *Commentiarolus* with a preliminary account of a heliocentric Solar System

1517 Italian physician Girolamo Fracastoro expresses the view that fossils were originally organic matter

1512

THE THEODOLITE

Martin Waldseemüller, a German cartographer, provided the first description of the theodolite, a surveying instrument (which he called the polimetrum) in the *Margarita Philosophica*. Consisting of a telescope mounted to rotate freely, the theodolite could measure angles between visible points in both the horizontal and vertical planes, making the task of land surveying easier and facilitating the development of triangulation.

△ **Early theodolite**

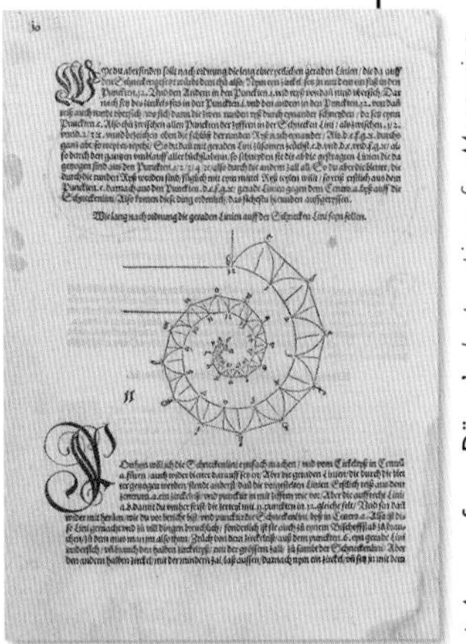

△ **A page from Dürer's** *Instructions for Measuring*

1525

DÜRER'S MATHEMATICS

German artist Albrecht Dürer published one of the first works on applied mathematics and geometry. His *Instructions for Measuring* provided artists and stonemasons with invaluable instructions for constructing regular polygons.

1527
SALTS, SULPHURS, AND MERCURIES
German chemist Paracelsus (Theophrastus von Hohenheim) devised a classification of chemical substances based on a division into salts, sulphurs, and mercuries. Forming part of a mystical philosophy grounded in alchemy, he considered that salts were solid, mercuries volatile, and sulphurs flammable. He believed that changing the proportions of sulphur in a metal could convert it into a different one, offering the possibility of transmuting lead into gold.

▷ **Paracelsus** portrait by Reubens

1537
BALLISTIC SCIENCE
The first modern work on ballistics, the *Nova Scientia*, was written by Italian mathematician Niccolò Tartaglia. He used mathematical principles to show that a cannonball would not travel in a straight line, but in a curved trajectory, and he provided tables of elevation to assist artillery gunners.

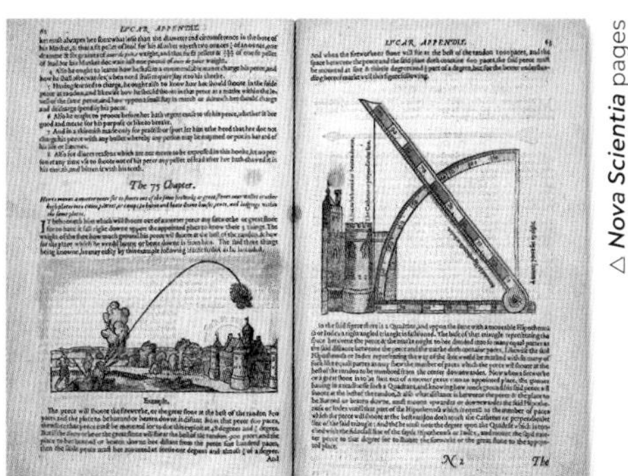

△ ***Nova Scientia*** pages

1539

1526 Gemma Frisius, Flemish mathematician and surveyor, joins the University of Leuven; he goes on to produce the first work describing the full method of triangulation

1530–36
NATURALISTIC DESCRIPTIONS OF PLANTS
A German Carthusian monk, Otto Brunfels published his *Herbarum Vivae Iconis* (*Pictures of Living Plants*), the first modern botanical work, in three volumes. He drew the more than 130 plants illustrated in the work from life, providing a level of detail that set the standard for future botanists and marking a move away from a tradition of medieval herbals that had concentrated on folklore rather than scientific observation.

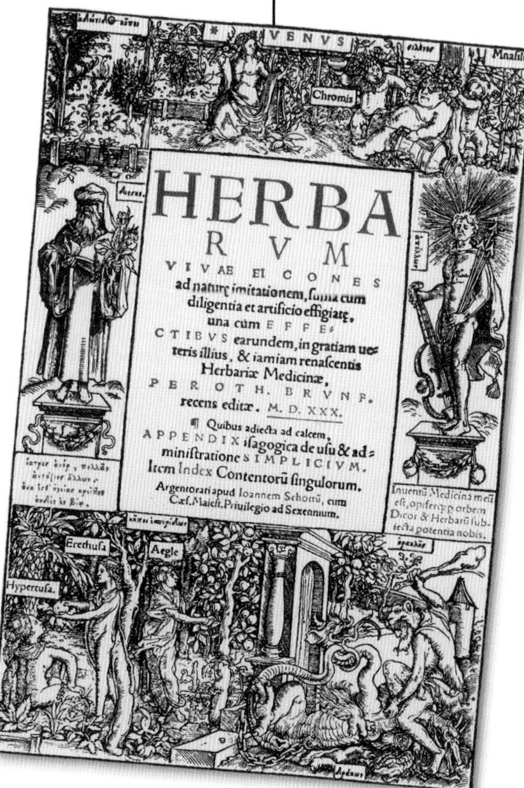

▷ ***Herbarum Vivae Iconis*** title page

▷ 17th-century globe

1541
MERCATOR GLOBE

Completed in 1541 by Dutch cartographer Gerardus Mercator, this large globe was engraved in fine detail. Globes of this period served as calculating devices that allowed navigators to derive times of sunrise and sunset; loxodromes - diagonal lines on the surface of the globe - marked lines of constant bearing.

◁ *De Revolutionibus Orbium Coelestium*

1543
THE COPERNICAN MODEL

German-Polish astronomer Nicolaus Copernicus published a full account of his heliocentric theory in *De Revolutionibus Orbium Coelestium*. His model superseded the system of Ptolemy, the 2nd-century astronomer, which held that Earth was at the centre of the Solar System.

1540

1542 German botanist Leonard Fuchs publishes *De Historia Stirpium* (*History of Plants*), describing around 500 plants with their therapeutic uses

1544 Georg Hartmann, a German instrument maker, describes the phenomenon of magnetic inclination, or magnetic dip

1545 Italian Gerolamo Cardano discovers complex numbers and discusses them in his *Ars Magna*

1473-1543
NICOLAUS COPERNICUS

Born in Toruń, now in Poland, Copernicus studied theology before turning to medicine and astronomy. He employed detailed observations and mathematical calculations to devise his ground-breaking theory of an Earth-centred Solar System.

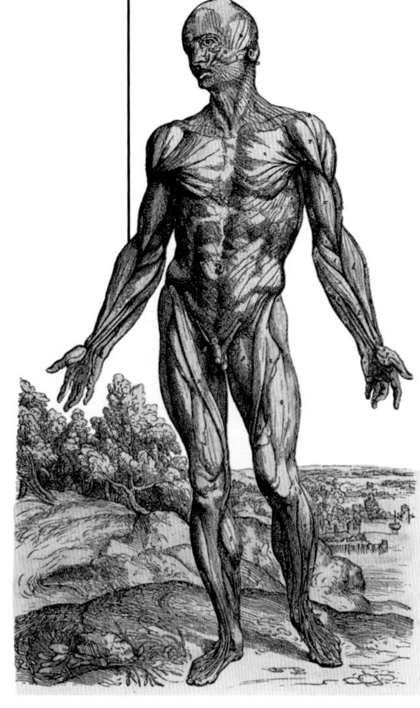

◁ *De Humani Corporis* illustration

1543
THE ART AND SCIENCE OF ANATOMY

The Flemish physician Andreas Vesalius revolutionized the study of anatomy with his publication of *De Humani Corporis Fabrica* (*On the Fabric of the Human Body*). Relying on direct observation obtained through dissection of corpses, his work dispelled many of the errors based on classical understandings of the body, and the high quality of its illustrations assisted generations of medical students.

1546

CLASSIFYING ROCKS AND MINERALS

In his *De Natura Fossilium* (*On the Nature of Fossils*), German scholar Georgius Agricola classified minerals, rocks, and fossils according to their geometric shape, laying the groundwork for modern geological science.

△ *Pentremites spicatus* fossil

1551

ISLAMIC SCIENCE

A gifted astronomer, mathematician, and inventor, the Ottoman-Arab scientist Taqi al-Din Muhammad ibn Ma'ruf al-Shami al-Asadi described a mechanism for a steam turbine driven by a rotating spit, produced improved astronomical tables, and invented a mechanical clock driven by weights.

◁ **Taqi al-Din Muhammad ibn Ma'ruf al-Shami al-Asadi** at work

1546 Girolamo Fracastoro publishes *On Contagion and Contagious Diseases*, setting out an early theory of the mechanism of disease

1551 Swiss naturalist Konrad von Gesner catalogues all the world's known animals in his *Historiae Animalium*, one of the first works of zoology

1552 Bartolomeo Eustachi, an Italian physician, describes the adrenal glands and the workings of the inner ear

1552 Aztec physician Martín de la Cruz writes *The Little Book of the Medicinal Herbs of the Indians*, describing traditional preparations for treating many ailments

1554

HELIOCENTRICITY

In Copernicus's heliocentric model of the Solar System, each planet orbits around the Sun, with Earth the third-innermost planet. The apparent motions of planets in Earth's sky can therefore be explained as effects of our particular location. For example, Venus and Mercury always remain close to the Sun because their orbits always lie Sunward from Earth, whereas Mars, Jupiter, and Saturn follow a general eastward motion around the sky but make backward, or retrograde, loops as Earth "overtakes" them in their orbits. Explaining this through the earlier geocentric, or Earth-centred, model – as Ptolemy did – required the inclusion of complex additional suborbits known as epicycles.

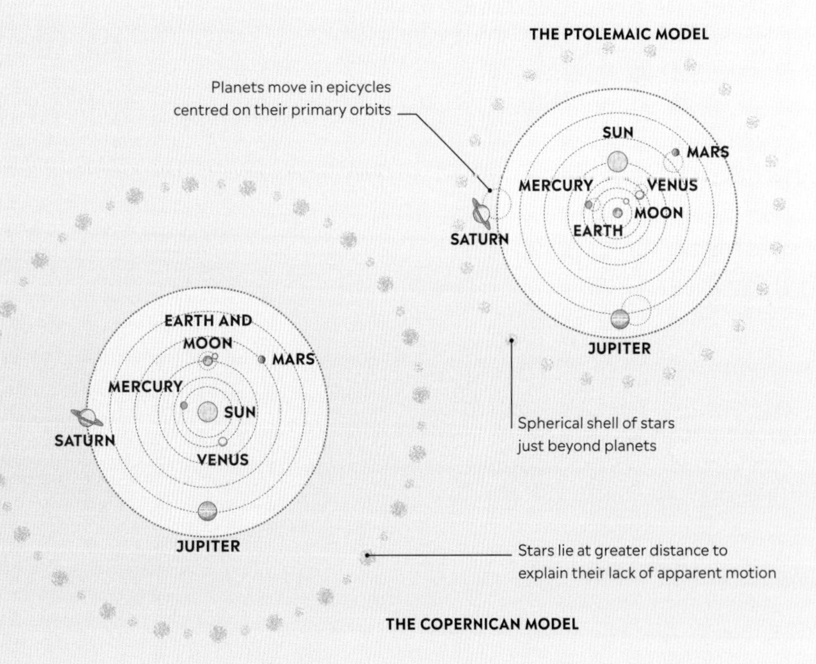

◁ ***De Re Metallica*** illustration

1556
THE STUDY OF METALS

German metallurgist Georgius Agricola published *De Re Metallica* (*On Metals*), the first practical guide to mining engineering. The 12-volume work included sections on the identification of veins of metal-bearing ore, the best means of extraction and the tools required, the testing of ore for quality, and how to process the mined metal.

1564
PROBABILITY

The first mathematical examination of the laws of probability, the *Liber de Ludo Aleae* (*Book on Games of Chance*), was written by Italian mathematician Girolamo Cardano in 1564. A dedicated gambler, he successfully calculated the odds of rolling certain numbers on one or more dice and discovered the law of large numbers, which describes the result of performing the same experiment many times.

△ ***Girolamo Cardano*** depicted on a medal

1555

1560 Giambattista della Porta founds the world's first scientific society, the Academia Secretorum Naturae, in Naples

1562 Gabriele Falloppio, professor of surgery at Padua, Italy, publishes a description of human reproductive organs

Howbeit, for easie alteratiō of *equations*. I will propounde a fewe exāples, bicause the extraction of their rootes, maie the more aptly bee wroughte. And to auoide the tediouse repetition of these woordes : is equalle to : I will sette as I doe often in woorke use, a paire of paralleles, or Gemowe lines of one lengthe, thus: ═══, bicause noe .2. thynges, can be moare equalle. And now marke these nombers.

14.ꝛ.—|—.15.ꝗ═══71.ꝗ.

▷ Recorde's symbol

1557
MATHEMATICAL EQUALITY

In his *Whetsone of Witte*, Welsh physician and mathematician Robert Recorde introduced the equal (or equals) sign as a means of showing equality between two mathematical expressions. The elegance and clarity of his two parallel lines of equal length meant the symbol was soon widely accepted.

1569
MAPPING THE WORLD

Cartographers struggled to portray the Earth on a flat surface until Flemish cartographer Gerardus Mercator solved the problem. By gradually increasing the distance between parallels (lines of latitude) as they approach the poles, but keeping the distance between meridians (lines of longitude) constant, his new projection made lines of constant bearing (or navigational direction) into straight lines. His maps were useful for marine navigation but were later criticized for distorting land-masses towards the poles and the equator.

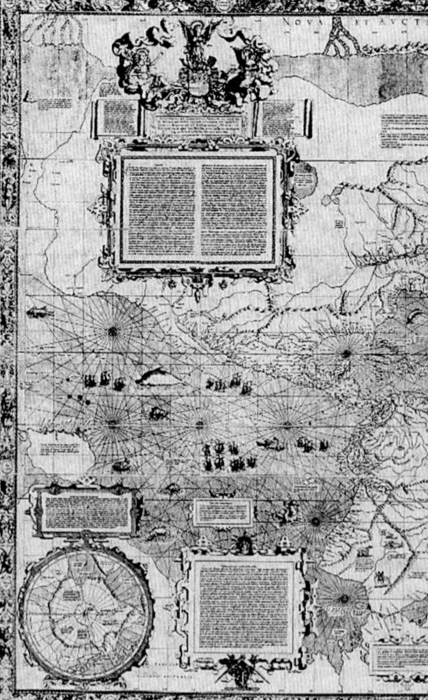

▷ **Mercator projection** world map from 1569

SUPERNOVAE

When a heavyweight star collapses at the end of its life, it triggers an explosion called a supernova. The characteristics of a supernova depend on the nature of the dying star. One form (Type Ia) is created when an already-dead stellar core (a white dwarf) transforms violently into a superdense neutron star. Other types of supernova form when massive stars collapse.

Type II supernova
Massive stars near the ends of their lives develop shells of increasingly heavy elements around their cores. A supernova is triggered when nuclear fusion can no longer support them.

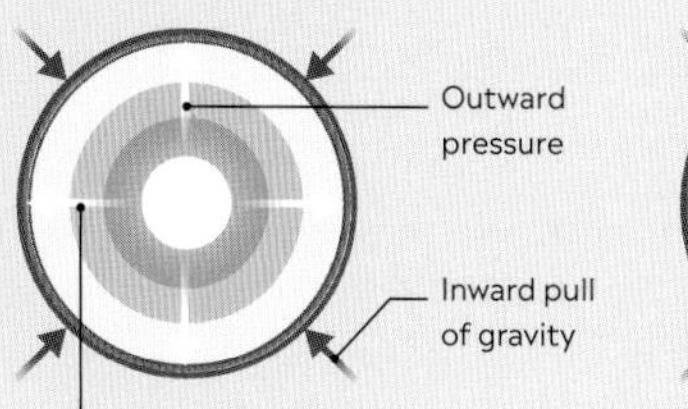

1. While fusion continues, outward pressure from core prevents overlying layers of massive star from collapsing

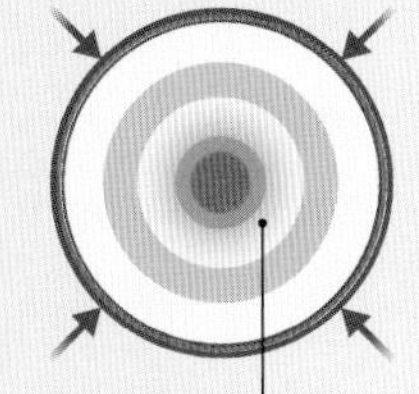

2. Fusion of iron absorbs more energy than it releases, so core fusion abruptly cuts out

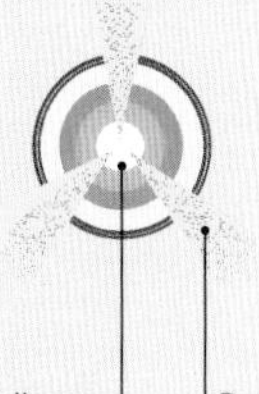

3. Core collapses abruptly into neutron star; upper layers fall in and then rebound

Burst of neutrinos released

4. Powerful shockwave tears through star, triggering supernova explosion with huge wave of nuclear fusion

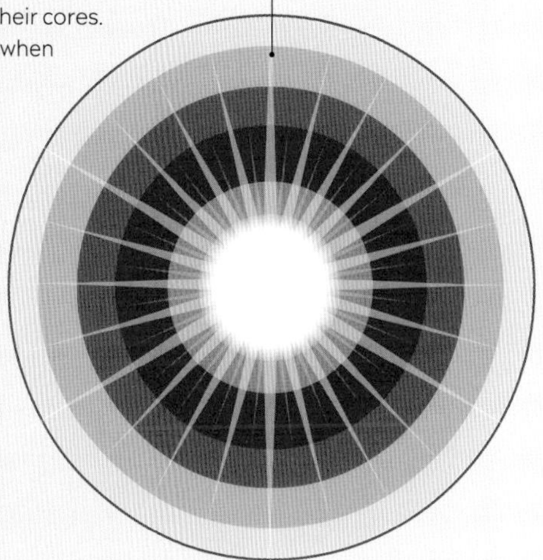

1570 Belgian cartographer Abraham Ortelius publishes the first modern world atlas, the *Theatrum Orbis Terrarum*

1571 On a tour of Mexico, Spanish explorer Francisco Hernández records over 3,000 plants previously unknown to European science, including some in an Aztec botanical garden at Texcoco, and records their medicinal uses

1572 Italian mathematician Rafael Bombelli lays down rules for the use of imaginary numbers

1574

1546-1601
TYCHO BRAHE

Tycho made a vast contribution to astronomy. From his Uraniborg observatory on the island of Hven, he charted the positions of over 770 stars, and determined the path of a comet that lay beyond the Moon.

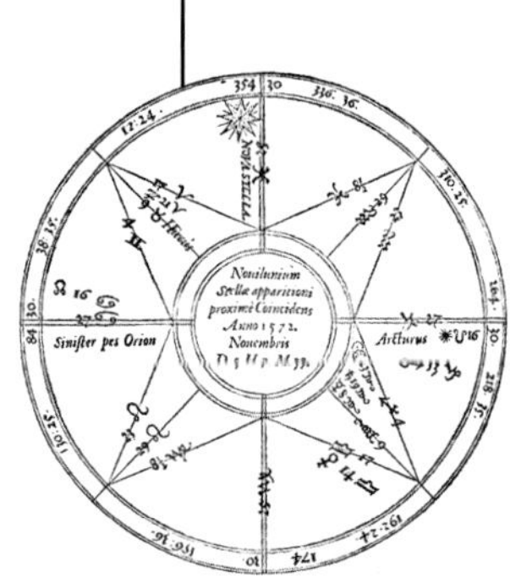

△ *De Stella Nova* illustration

1572
TYCHO'S SUPERNOVA

For two weeks in November 1572, Danish astronomer Tycho Brahe observed a bright object in the constellation of Cassiopeia. It was a supernova, caused by a star exploding, and it made astronomers realize that the Universe was not unchanging.

1578

AZTEC MEDICINE

Compiled by Spanish priest Bernardo de Sahagún with the assistance of Aztec collaborators, the *Florentine Codex* included invaluable information about Aztec mineral and herbal remedies and the devastating impact of smallpox on the people.

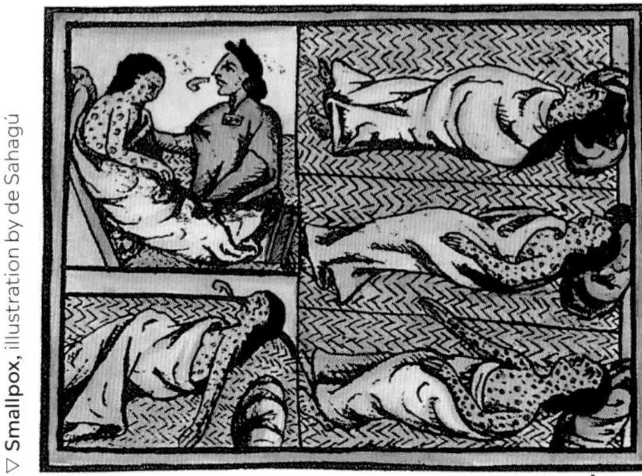

▽ **Smallpox**, illustration by de Sahagú

1575

1577 Taqi al-Din builds an observatory at Istanbul with the latest astronomical instruments, but the Ottoman sultan has it pulled down soon afterwards

1575 In Italy, Francesco Maurolico pioneers the use of mathematical induction, a technique used in mathematical proofs

1577 The Great Comet (now designated C/1577 V1) is observed in Europe by Tycho Brahe and other astronomers

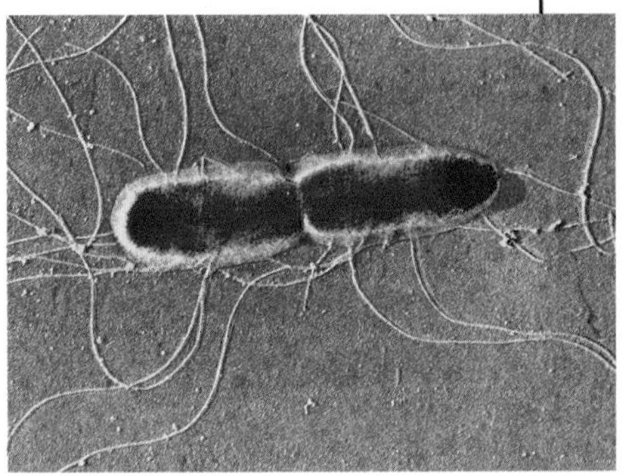

△ **Bacterium** that causes typhoid

1576

TYPHOID DESCRIBED

The Italian physician and mathematician Girolamo Cardano made the first clinical description of typhoid. A disease caused by poor hygiene and contaminated food and water, it is characterized by a fever, acute diarrhoea, and a rash. It spread rapidly in crowded cities and among armies.

◁ *Date palm*

1580

SEXUAL DIFFERENCES IN PLANTS

Venetian physician Prospero Alpini discovered the ability of flowers to self-pollinate after observing the growth of date palms in Egypt. He realized that some trees were male and others female.

1582
THE GREGORIAN CALENDAR

The Gregorian calendar, a reformed calendar that corrected accumulated defects in the old Julian calendar (used since 45 BCE) was adopted by Pope Gregory XIII after extensive research. Eleven days were omitted from October 1582, and the number of leap years per century was reduced from 100 to 97, but at first the calendar was only accepted in Catholic countries.

◁ **Discussion of the calendar reform** before Pope Gregory XIII

1580 Hieronymus Fabricius, a professor of anatomy and surgery in Italy, describes valves in the veins

1584

1581
PENDULUM EXPERIMENTS

Italian astronomer Galileo Galilei began experiments that would eventually establish that a pendulum returns nearly to its release height (showing conservation of energy) and that the square of its period (the time taken for one oscillation) is proportional to its length.

▷ **Galileo** observes the motion of a candelabra

▷ Andrea Cesalpino

1583
BOTANICAL CLASSIFICATION

In his *De Plantis* (*On Plants*), the Italian botanist Andrea Cesalpino developed a scientific method of botanical classification, grouping flowering plants into five broad groups according to the shape of their fruits, seeds, and roots, rather than, as previously done, by their medicinal use or alphabetically.

"The marvellous property of the pendulum… is that it makes all its vibrations… in equal times."

GALILEO GALILEI, LETTER TO GIOVANNI BATTISTA BALIANI, 1639

1564–1642

GALILEO GALILEI

Galileo did important work in many areas of physics, but it was his confirmation of Copernicus's heliocentric Solar System that led to his trial by the Catholic Church when he was forced to recant.

◁ **Galileo** performs his experiment at Pisa

1589–1604

GALILEO'S EXPERIMENTS ON GRAVITY

By dropping objects from a tower, Italian astronomer Galileo Galilei demonstrated that bodies of the same material accelerate towards the ground at the same rate, regardless of their mass. This contradicted the long-held theory proposed by Aristotle that heavier objects should fall faster. It helped establish the modern scientific discipline of dynamics (the study of objects in motion) and further undermined uncritical acceptance of orthodox scientific positions that dated back to classical times.

1585

1586 In *Elements of Hydrostatics,* Flemish mathematician Simon Stevin solves vital questions about the pressure of liquids on the walls of containers

1592 Galileo Galilei invents the thermoscope, a forerunner of the thermometer

1593 Flemish botanist Carolus Clusius, the "father of botany", becomes director of the first botanical gardens in the Netherlands

TYCHONIS
BRAHE DANI,
ASTRONOMIÆ INSTAVRATÆ PROGYMNASMATA.
PRIMA PARS
DE RESTITVTIONE MOTVVM
Solis & Lunæ, Stellarumque inerrantium
tractat.
ET PRÆTEREA DE ADMIRANDA
nova Stella Anno 1572. exorta luculenter agit.
ast Pragæ Bohemiæ absoluta.
PROSTANT
Cum Cæsaris & Regum quorundam privilegiis.

△ Tycho's *A Programme for the Establishment of Astronomy*

1586

COMETARY STUDIES

Tycho Brahe published his *Astronomiae Instauratae Progymnasmata (A Programme for the Establishment of Astronomy)* detailing his observations of comets, including one that appeared in October 1585.

△ ***Natural and Moral History of the Indies*** illustration

1590

LIFE IN THE AMERICAS

Jesuit missionary José de Acosta provided the first comprehensive description of the animal and plant life in the Americas in his *Natural and Moral History of the Indies*. He provided valuable information about the customs of the indigenous people and the physical geography of the area, and was among the first to speculate that indigenous Americans may originally have migrated from Asia.

GRAVITY

The easiest way to understand gravity is to imagine it as an attractive force, pulling objects towards the ground and keeping planets in orbit around the Sun. Every object with mass has a gravitational effect on every other object with mass. The strength of this effect depends on the masses of the two objects and the distance between them.

Mass and acceleration
Although they experience different gravitational forces, light balls and heavy balls fall at the same rate.

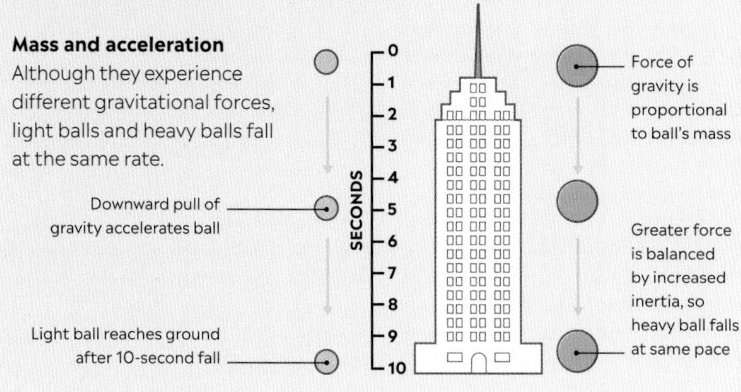

Gravity and mass
For two objects that are a fixed distance (d) apart, the force of gravity (F) is in direct proportion to the product of their masses (m).

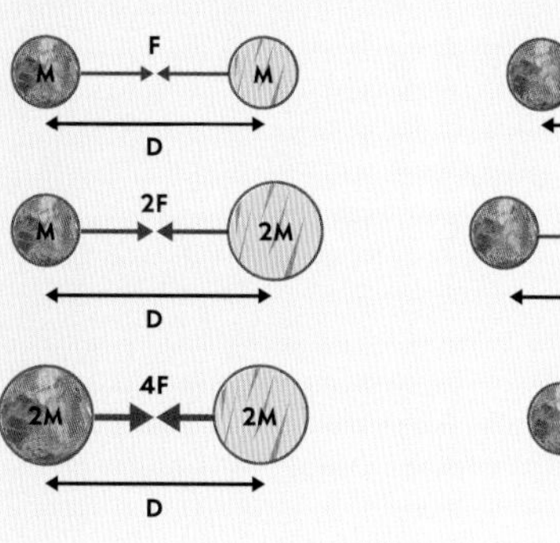

Gravity and distance
For two objects of fixed mass, the force of gravity is inversely proportional to the square of the distance between them.

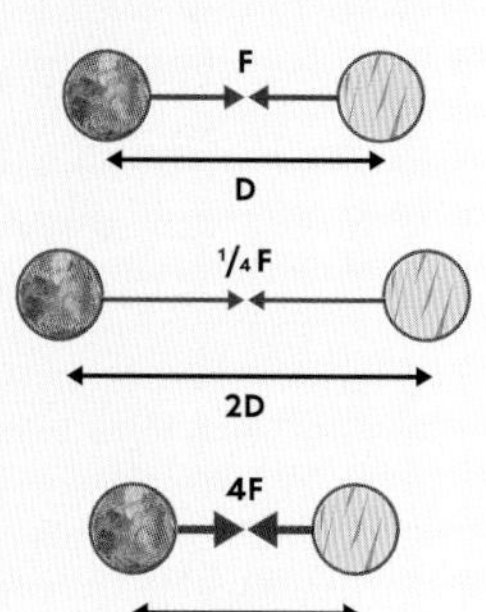

1596 Abraham Ortelius suggests that the continents of Africa and America were once joined together

1597 German metallurgist Andreas Libavius gives the first description of the properties of zinc in his *Alchemia*

1598 Tycho Brahe publishes *Astronomiae Instauratae Mechanica*, which includes a catalogue of over 1000 stars

1599 Italian naturalist Ulisse Aldrovandi begins publishing his three-volume treatise on birds

1599

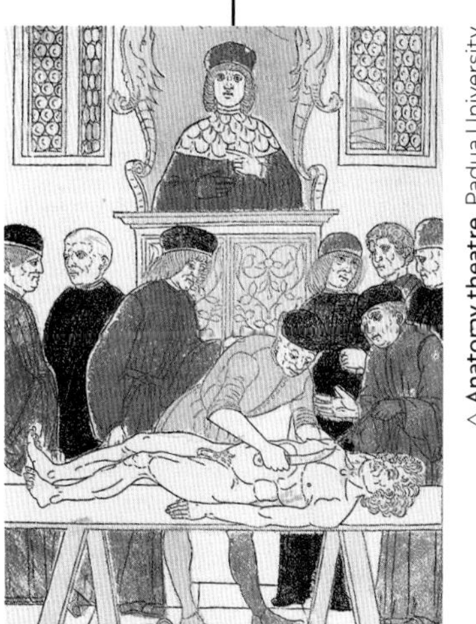

△ **Anatomy theatre,** Padua University

1594
A THEATRE FOR ANATOMY

Hieronymus Fabricius opened the first public theatre for anatomical dissection in Padua, Italy. With capacity for 300 students, the opportunities for direct observation it presented revolutionized medical education.

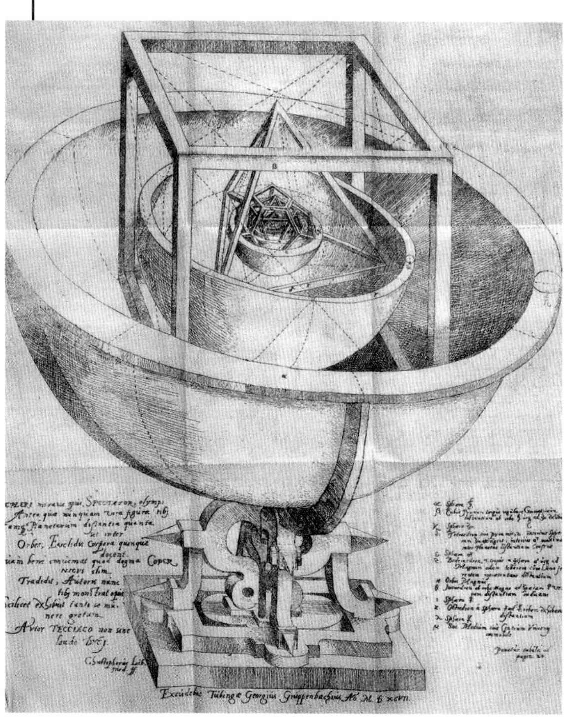

△ ***Mysterium Cosmographicum*** engraving

1596
PLANETARY MOTION

German astronomer Johannes Kepler published the *Mysterium Cosmographicum* explaining the orbits of the planets in geometric terms, with their distance from the Sun linked to ratios derived from Platonic solids (the five regular polyhedra). His findings supported Copernicus's model of a heliocentric Solar System.

▷ *Uranometria* illustration

1603
CELESTIAL ATLAS

German astronomer Johann Bayer published his *Uranometria*, the first comprehensive atlas of the night sky, showing 2,000 stars. It was the first celestial atlas to include the stars around the South Pole. Bayer's method of labelling stars within constellations with Greek letters also brought consistency to previously haphazard naming practices.

1600

1602 English mathematician Thomas Harriot studies the angles of light as it is bent, working out the law of refraction

1604 Johannes Kepler, in his *Astronomiae Pars Optica,* explains the phenomenon of parallax and describes how the eye focuses light

1604 Hieronymus Fabricius publishes work on the circulation of blood in embryos, laying the foundations of embryology

1605 English philosopher Francis Bacon sets out the scientific method in *The Advancement of Learning*

▷ Hans Lippershey

1608
THE TELESCOPE INVENTED

The Dutch instrument-maker Hans Lippershey built the first known optical telescope. A maker of lenses for spectacles, Lippershey realized that placing two convex lenses at either end of a fixed tube would allow objects at a distance to be magnified.

Earth's magnetic field last reversed direction around 42,000 years ago

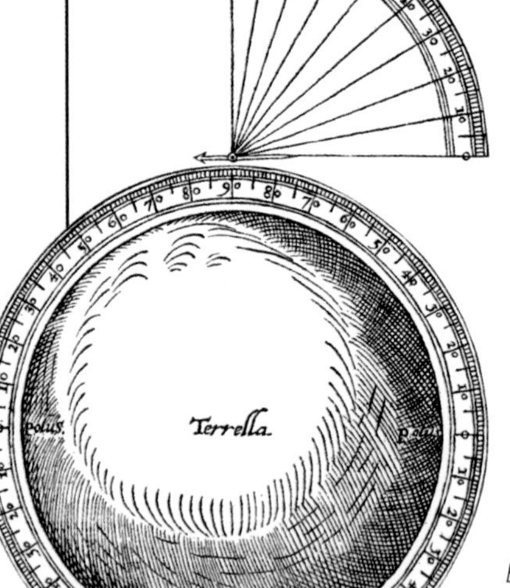

△ *De Magnete* illustration

1600
A MAGNETIC EARTH

In *De Magnete, Magneticisque, et de Magno Magnete Tellure (On the Magnet and Magnetic Bodies, and on the Great Magnet the Earth)*, English scientist William Gilbert concluded that Earth is like a giant magnet, which explains why a compass needle points to the north. He was also the first to use the term magnetic pole.

1609
GALILEO'S TELESCOPE

Hearing of telescopes developed elsewhere, Italian astronomer Galileo Galilei decided to build his own. His first attempts could only magnify objects three times, but he eventually constructed a telescope that could magnify objects 30 times. This made detailed observation of celestial bodies previously invisible to the naked eye possible for the first time.

▷ **Galileo demonstrates** his telescope

1609 Thomas Harriot makes the first drawing of the Moon as observed through a telescope

1609

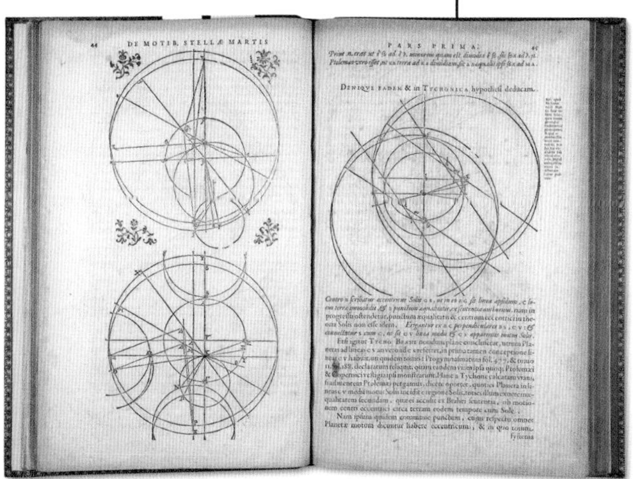

△ *Astronomia Nova* pages

1609
KEPLER'S LAWS

After several years studying the orbit of Mars, Johannes Kepler published *Astronomia Nova* (*New Astronomy*), in which he formulated the first two of his three laws of planetary motion. The first law stated that the planets orbit the Sun in elliptical paths, rather than circular ones, as previously believed. The second law states that planets move faster in their orbits the closer they are to the Sun.

1571–1630
JOHANNES KEPLER

German astronomer Kepler was a key figure in the development of astronomy, his three laws of planetary motion providing a solid theoretical and mathematical basis for Copernicus's heliocentric model.

1610

JUPITER'S MOONS

Using his newly constructed telescope, Galileo noticed several points of light moving near Jupiter. He realized they were satellites of the planet, the first to be discovered apart from Earth's Moon. His description, in *Siderius Nuncius* (*The Starry Messenger*), of what came to be called Io, Ganymede, Europa, and Callisto further reinforced Copernicus's idea that the Solar System did not revolve around Earth.

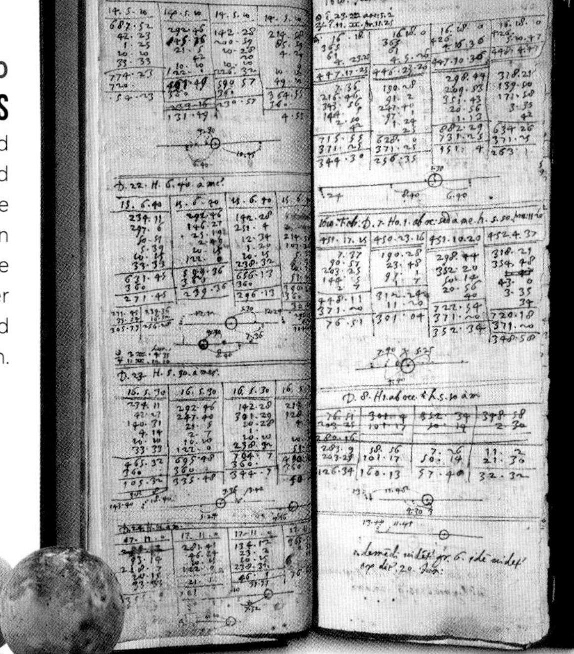

△ *Sidereus Nuncius* with Galileo's notation

▷ Jupiter's moons

1610

1610 Galileo Galilei becomes the first to observe sunspots

1614 Scottish mathematician John Napier devises logarithms as a means of performing complex arithmetical calculations.

1610 French astronomer Nicolas-Claude Fabri de Peiresc becomes the first to observe the Orion Nebula

1611 Johannes Kepler publishes *Dioptrice*, a treatise on optics explaining the workings of the microscope and telescope

1617 Dutch mathematician Willebrod Snell describes a new method to measure Earth's radius using triangulation

1614

PHYSIOLOGICAL EXPERIMENTS

In *De Statica Medicina* (*On the Balance of Medicine*), Italian anatomy professor Santorio Santorio described his experiments on respiration and weight. By measuring fluid and food intake and comparing it to excreted matter, he determined that the body "perspires" or uses up energy, laying the foundation for the study of metabolism.

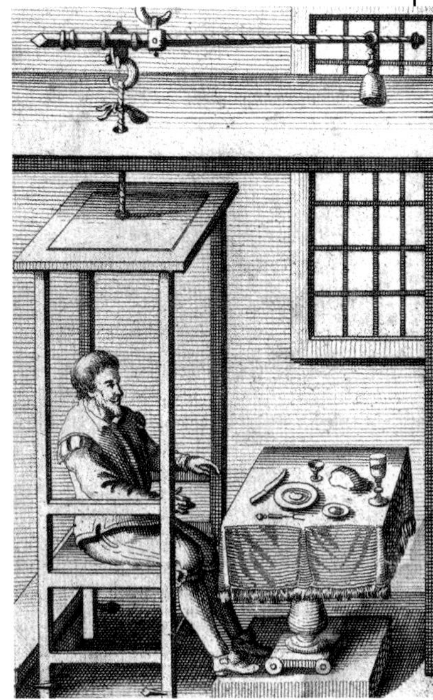

▷ **Santorio** performing an experiment

▷ *Harmonices Mundi* illustration

1619

THIRD LAW OF PLANETARY MOTION

Johannes Kepler published *Harmonices Mundi* (*Harmony of the World*), which contained his third law of planetary motion. This stated that there is a fixed ratio between the period (time taken to orbit) of a planet and its distance from the Sun.

A B C
D E F
▽ **Jan Baptista van Helmont's** apparatus

1620
THE MICROSCOPE

By now, Dutch lens-maker Zacharias Jansen had devised the first compound microscope with one lens to focus on the object and another to magnify its image. His invention made the scientific observation of tiny objects and life forms possible for the first time.

▷ **Zacharias Janssen's** microscope (reproduction)

1620
CARBON DIOXIDE

The Flemish chemist Jan Baptista van Helmont established that there were substances given off during chemical reactions that differed from air. He named these "gases" and determined the gas produced by burning charcoal is the same as that given off by fermenting grape juice, so becoming the first to identify carbon dioxide.

1620 Francis Bacon writes the *Novum Organum*, a definitive restatement of his ideas on scientific method

1624

1621 English scholar Robert Burton writes *The Anatomy of Melancholy* describing various forms of mental disorder

> "Melancholy... is a habit, a serious ailement, a settled humour."
>
> *ROBERT BURTON*, THE ANATOMY OF MELANCHOLY, *1621*

PLANETARY MOTION

Kepler's three laws of planetary motion reflect the effects of the deeper laws of motion and universal gravitation. They apply not only to planets orbiting stars but also to many other situations in which one celestial body orbits another much more massive one under the influence of gravity. Orbits take the shape of a closed oval path called an ellipse, with the larger body at one of two focus points (foci) on either side of the ellipse's centre.

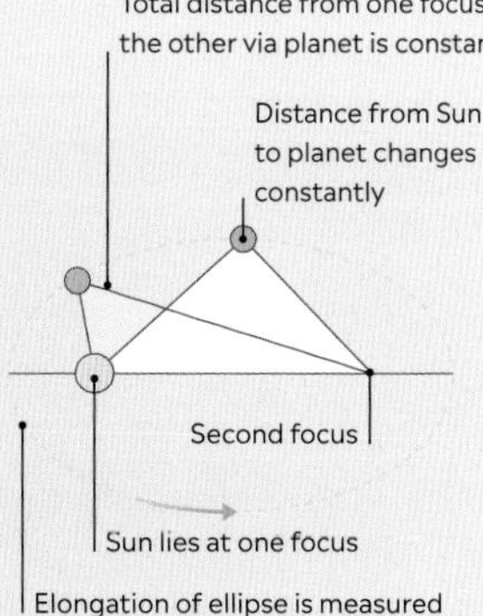
Total distance from one focus to the other via planet is constant
Distance from Sun to planet changes constantly
Second focus
Sun lies at one focus
Elongation of ellipse is measured by its deviation from a circle

First law
All planetary orbits are ellipses, with the Sun at one of the two foci. As a result, the distance between a planet and the Sun is always changing.

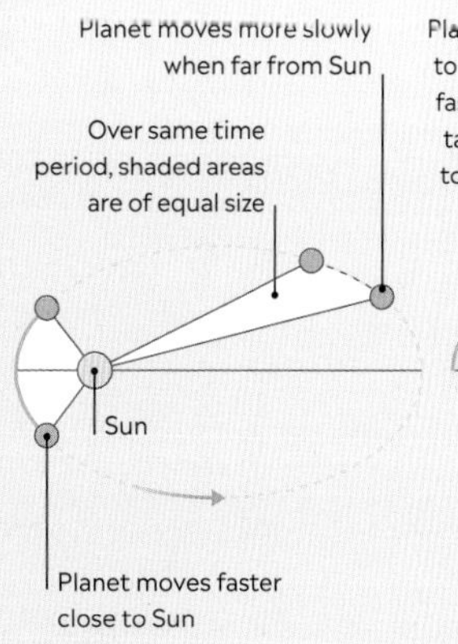
Planet moves more slowly when far from Sun
Over same time period, shaded areas are of equal size
Sun
Planet moves faster close to Sun

Second law
A line segment between a planet and the Sun sweeps out equal areas over an equal period of time.

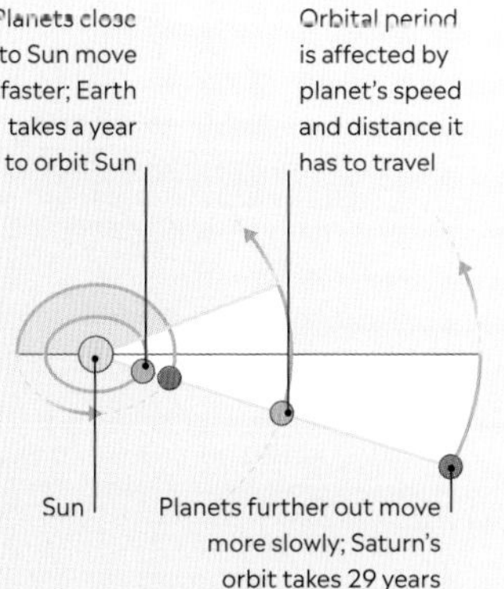
Planets close to Sun move faster; Earth takes a year to orbit Sun
Orbital period is affected by planet's speed and distance it has to travel
Sun
Planets further out move more slowly; Saturn's orbit takes 29 years

Third law
The square of the period of orbit is proportional to the cube of the orbit's semimajor axis (one half of the ellipse's longest dimension).

1627

STUDIES IN FOETAL DEVELOPMENT

The *De Formato Foetu* (*On the Form of the Foetus*) by Flemish anatomist Adriaan van den Spiegel is published. It describes the development of the human foetus in the womb, contributes to the understanding of its morphology, and establishes embryology as a scientific discipline.

▷ ***De Formato Foetu,*** illustration

1630

TREE GROWTH

Jan Baptista van Helmont planted a willow tree to which he added only rainwater for five years. Finding it had gained 74 kg (164 lb) in weight, while the soil had lost only a minuscule amount, he concluded the increase was caused by some chemical process. His experiment was a first step towards the discovery of photosynthesis.

◁ **Jan Baptista van Helmont**

1625

1627 The Rudolphine Tables, a catalogue of nearly 1,500 stars, is completed and published by Johannes Kepler

1626 Italian physician Santorio adds a graduated scale to a thermoscope, creating the first thermometer

1578-1657

WILLIAM HARVEY

British physician Harvey performed practical experiments to devise his theory of circulation: he cut the veins of fish and snakes and tied ligatures to human arms to block the circulation so he could observe where the trapped blood caused swelling.

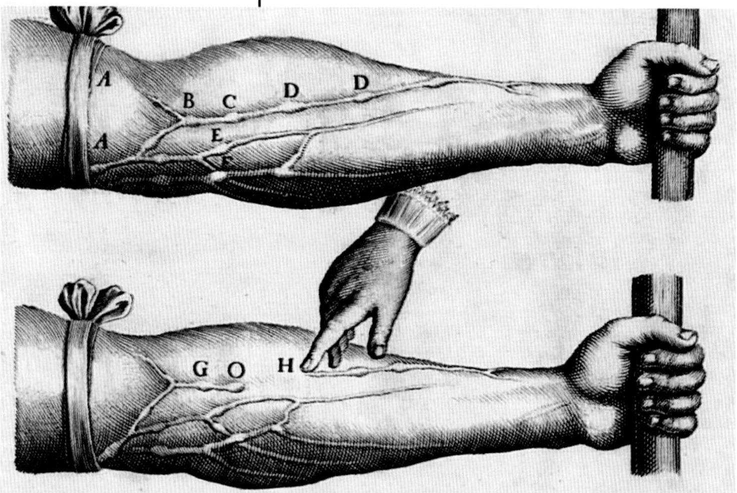

△ ***De Motu Cordis*** illustration

1628

THE CIRCULATION OF BLOOD

William Harvey published his *De Motu Cordis* (*On the Heart's Motion*), resolving the age-old problem of how blood circulates around the body. Instead of the lungs playing the key role, as Greek authors thought, Harvey found that the heart pumps blood out through the veins and it then returns from the body's peripheries through the arterial system.

PHOTOSYNTHESIS

While animals get their food by eating other organisms or their products, plants make their own food using a process called photosynthesis. The word photosynthesis comes from photo, meaning "light", and synthesis, meaning "to make". In the chemical process, sunlight, water, and carbon dioxide are harnessed to produce glucose (a simple sugar) and oxygen. The plant uses the glucose to release energy and form substances such as proteins and cellulose. The oxygen is released into the atmosphere.

Green plants
Photosynthesis takes place in tiny structures called chloroplasts that are found in plant leaves. Chloroplasts, located near the leaf's surface, contain a green pigment called chlorophyll.

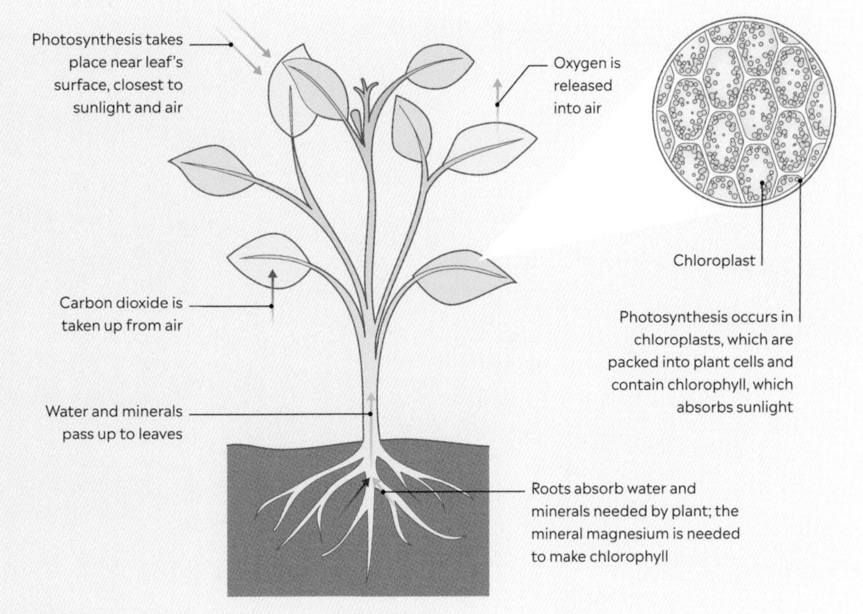
Photosynthesis takes place near leaf's surface, closest to sunlight and air
Oxygen is released into air
Chloroplast
Photosynthesis occurs in chloroplasts, which are packed into plant cells and contain chlorophyll, which absorbs sunlight
Carbon dioxide is taken up from air
Water and minerals pass up to leaves
Roots absorb water and minerals needed by plant; the mineral magnesium is needed to make chlorophyll

1632 Italian surgeon Marco Severino publishes first textbook on surgical pathology

1632 English clergyman William Oughtred introduces the multiplication sign in *The Key of the Mathematicks*

1634

1631

ACCURATE MEASUREMENT

French mathematician Paul Vernier created an instrument for taking accurate measurements of very small objects. It consisted of two sliding scales, with the markings on the secondary scale allowing for fine adjustments to measured readings.

△ **Vernier calipers** in brass

1632

GALILEO'S TRIAL

Galileo's publication of *Dialogue Concerning the Two Chief World Systems* in which he defended the Copernican heliocentric system led to his trial by the Catholic Church's Inquisition on charges of heresy. He was sentenced to house arrest and all his works were banned.

△ **Galileo** at the Papal Inquisition

The dissection of Thomas Parr
William Harvey was the first person to suggest that blood flows continuously to and from the heart. In this painting from c.1900, he is shown in 1635 dissecting a man believed at the time to be 152.

HUMAN CIRCULATORY SYSTEM

The circulatory system delivers oxygen and nutrients around the human body and removes waste products. The entire process relies on a series of tubes and valves, as well as a continuous pump - the heart. The human heart beats around 100,000 times a day, pumping about 5 litres (9 UK pints) of blood around the body.

The discovery that blood circulates through the body was made by 17th-century physician William Harvey. Before that, going back as far as Roman times, doctors thought blood was made in the liver and then used up by the muscles.

The human circulatory system is described as a double system because blood passes through the heart twice on each circuit of the body. Arteries carry blood away from the heart towards an organ, while veins carry blood back towards the heart. The pulmonary artery carries blood from the heart to the lungs, where it becomes oxygenated and is then returned to the heart by the pulmonary vein. The aorta is the main artery that carries blood from the heart to the rest of the body. To reach the body's extremities, arteries branch into smaller tubes called capillaries.

Blood is pumped out of the heart by two muscular chambers called ventricles. It leaves via the aorta, pumped by the ventricle on the left side of the heart. The two main veins carrying blood back to the heart are the superior vena cava and the inferior vena cava. The superior vena cava carries blood from the head, neck, upper chest, and arms. The inferior vena cava is the largest vein in humans, and it carries blood back to the heart from the lower and middle parts of the body.

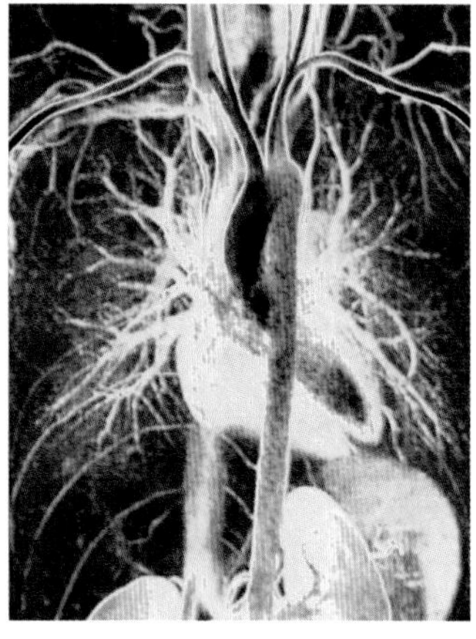

Medical imaging and circulation
Imaging technology allows doctors to study blood vessels and circulation in a living patient. This image was captured using a technique called magnetic resonance angiography.

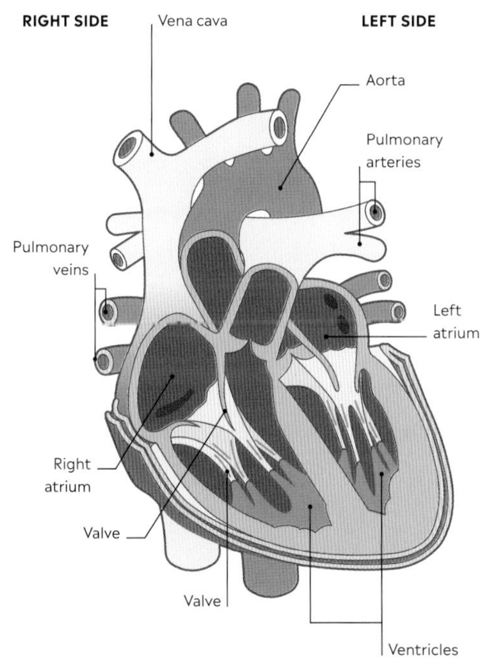

INSIDE THE HEART

The heart is a muscular pump. The right ventricle pumps blood without oxygen to the lungs. The left ventricle has a thicker muscular layer around it, enabling it to pump blood with oxygen (oxygenated blood) around the whole body. Valves control the flow around the heart, preventing the blood from flowing backwards.

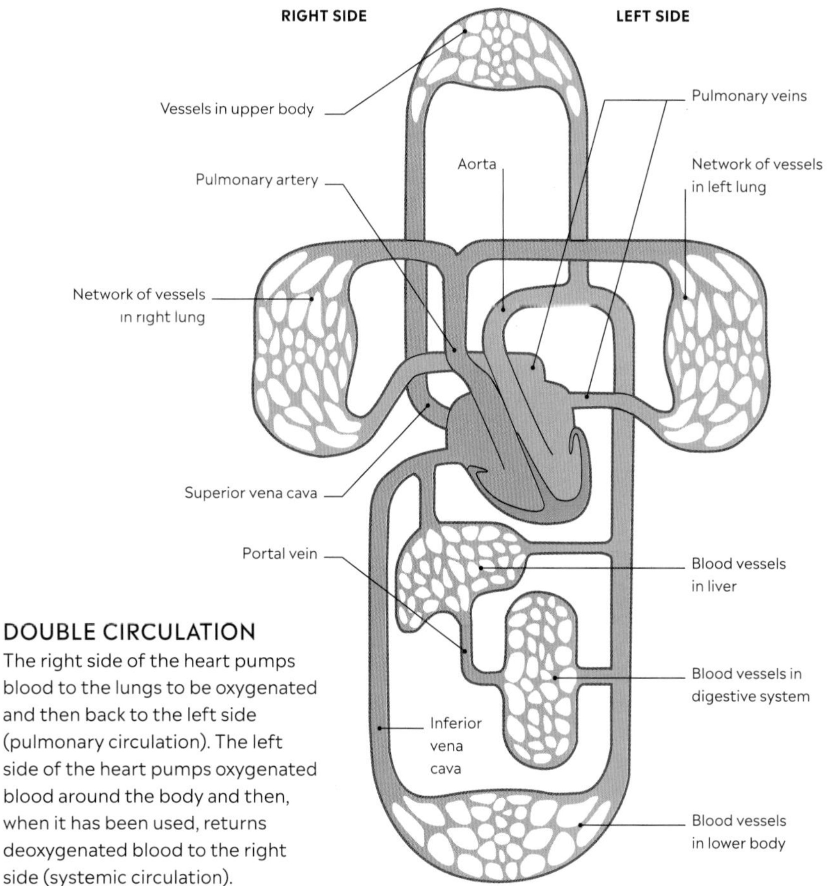

DOUBLE CIRCULATION

The right side of the heart pumps blood to the lungs to be oxygenated and then back to the left side (pulmonary circulation). The left side of the heart pumps oxygenated blood around the body and then, when it has been used, returns deoxygenated blood to the right side (systemic circulation).

1637

CHINESE ENCYCLOPEDIA

Chinese scholar Song Yingxing published the *Tiangong Kaiwu*, an encyclopedia that covered a wide range of practical applications of science in agriculture, metallurgy, papermaking, transportation, hydraulics, and the production of gunpowder.

△ **Song's encyclopedia** illustration

1642

EARLY CALCULATOR

The French mathematician Blaise Pascal invented a mathematical calculating machine, the Pascaline, or Arithmetic Machine. Originally intended to help with tax calculations, it used a system of toothed gears that could carry forward numbers into adjacent columns to perform addition and subtraction operations.

▷ **Pascal's** adding machine

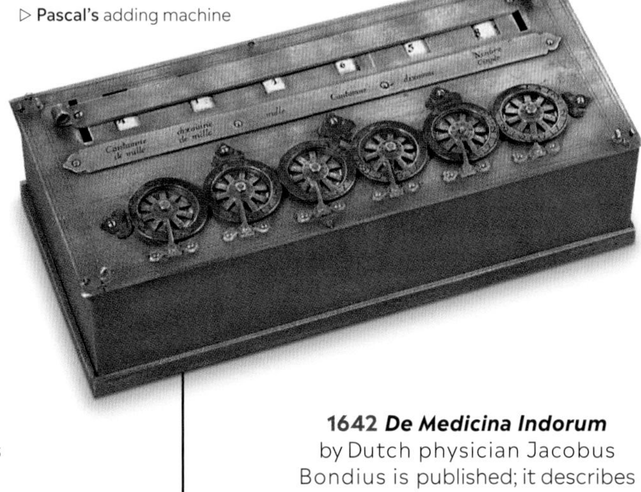

1635

1637 French mathematician Pierre de Fermat scribbles his theorem in the margin of an old textbook

1642 *De Medicina Indorum* by Dutch physician Jacobus Bondius is published; it describes a range of tropical diseases

▷ René Descartes

1637

CARTESIAN CO-ORDINATES

In an appendix to his *Discourses on Method*, French philosopher and mathematician René Descartes introduced a system of mapping points on a plane using horizontal (x-) and vertical (y-) co-ordinates. His system has been used by mathematicians in creating graphs ever since.

1639

PREDICTING A TRANSIT OF VENUS

English astronomer Jeremiah Horrocks predicted and observed the transit of Venus, a rare astronomical event in which Venus is visible passing across the disc of the Sun. The event occurs in pairs eight years apart roughly every 120 years and was used by astronomers to help calculate the size of the Solar System.

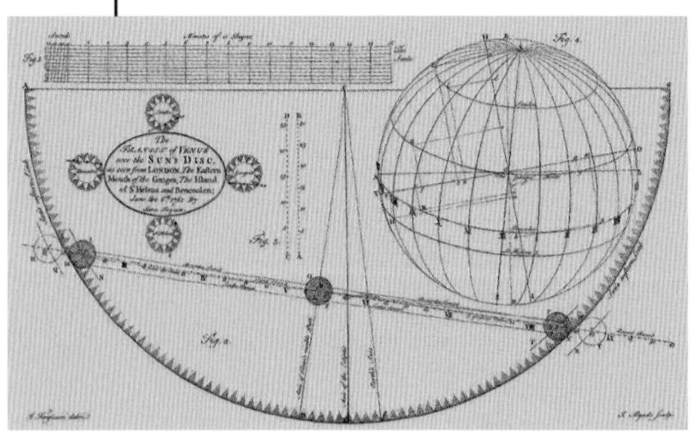

△ **Diagram** of the transit of Venus from Horrocks' observation

There have been seven transits of Venus since the invention of the telescope

▽ **Torricell** demonstrates his barometer

1644
THE BAROMETER

An assistant to Galileo, the Italian physicist Evangelista Torricelli made the first barometer by creating a vacuum inside a sealed glass tube partially filled with mercury. He realized that the rise and fall of the mercury inside it was caused by what he called an "ocean of air", allowing him to measure atmospheric pressure. His experiment also made him the first person to create a sustained vacuum.

1644 Italian astronomer Giovanni Odierna publishes the first book on microscopic life, which includes a description of a fly's eye

1646 Blaise Pascal discovers that the atmosphere is less dense at altitude through experiments done by his brother-in-law Florin Périer

1649 French scientist Pierre Gassendi proposes that the properties of matter are determined by the shapes of atoms

1649

1647
MOON MAP

Motivated by a theory that longitude at sea could be calculated using the Moon's shadow cast during an eclipse, Polish astronomer Johannes Hevelius spent five years of star-gazing at his home-made observatory in Gdańsk. The result of his observations was the first atlas of the Moon. Its 40 beautifully engraved plates show the Moon at various of its phases.

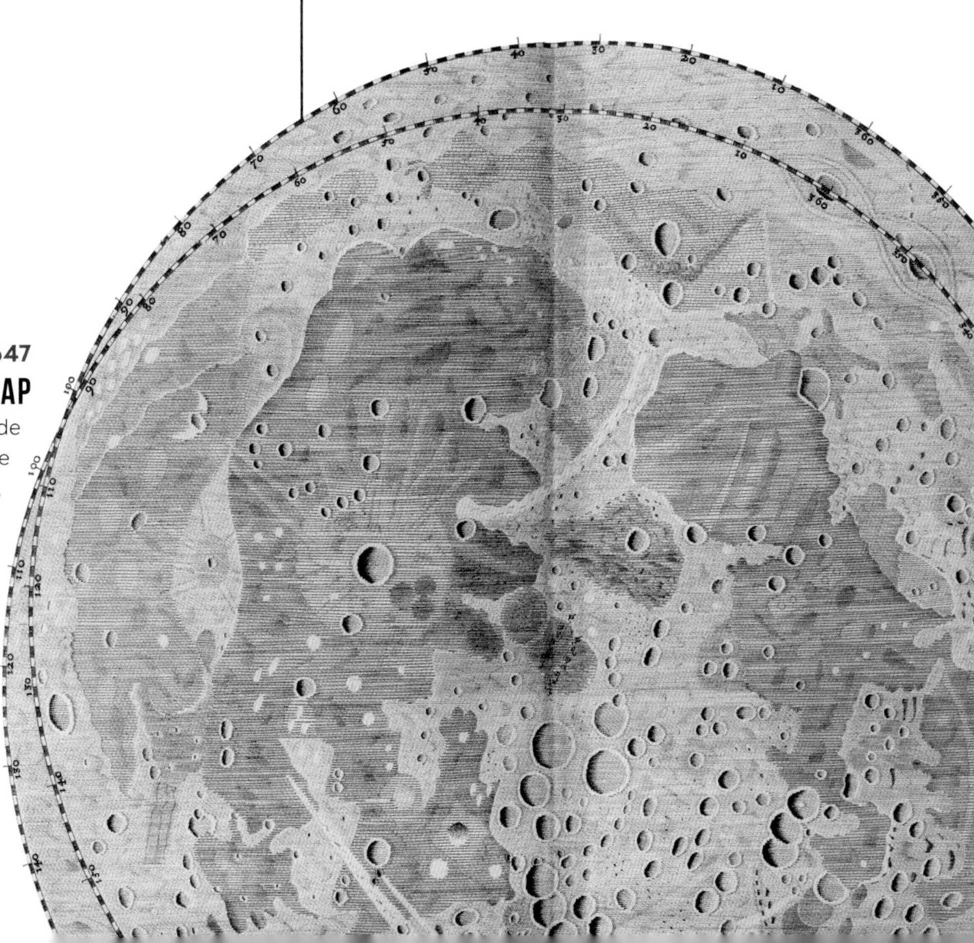
▷ Moon map by Hevelius

△ **Otto von Guericke** demonstrates vacuums

1650

1652 Danish physician Thomas Bartholin gives the first full description of the human lymphatic system

1653 Blaise Pascal, in his *Treatise on the Equilibrium of Liquids,* proposes Pascal's Law that a liquid's pressure in a small closed system is equal in all directions

▽ **Illustrations** from *The English Physician*

> "Truly my own body being sickly, brought me easily into a capacity, to know that health was the greatest of all earthly blessings."
>
> *WILLIAM CULPEPER,* THE ENGLISH PHYSICIAN, *1652*

1652

MEDICINAL PLANTS

The English botanist Nicolas Culpeper published *The English Physician* (later known as *The Complete Herbal*), which attempted to make medicine more accessible to poorer people by writing in English rather than Latin and listing readily accessible herbal remedies, such as mugwort for easing labour pains.

1658
COMPARATIVE ANATOMY
Dutch biologist Jan Swammerdam was the first scientist to describe red blood cells, after he observed them with a microscope in the leg of a frog. His other work yielded information on the metamorphosis of caterpillars into butterflies, and the discovery that the chief bee in a hive is a queen (rather than a male as previously believed).

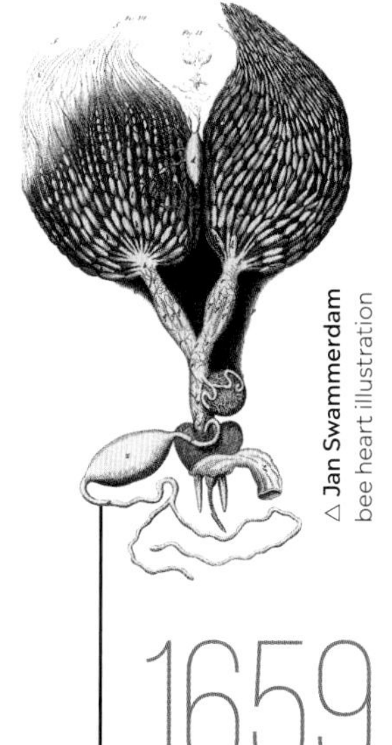
△ **Jan Swammerdam** bee heart illustration

1654
VACUUM POWER
German scientist Otto von Guericke demonstrated the power of a vacuum by pumping all the air out of a sphere formed of two copper bowls, known as Magdeburg spheres. The atmospheric pressure on the vacuum was so strong that even two teams of horses could not pull them apart.

1655 English mathematician John Wallis devises a symbol for infinity and develops a way to find the tangential lines to a curve

1655 English scientist Robert Hooke suggests lunar craters are caused by giant bubbles of volcanic lava bursting

1659

1656 Christiaan Huygens builds the first pendulum clock, unprecedented in its level of accuracy

▽ **Blaise Pascal**

1653
PASCAL'S TRIANGLE
Blaise Pascal published his *Treatise on the Arithmetical Triangle* as part of his research into the probability of throwing numbers at dice. The work's diagram of a triangle in which each number on successive layers is created by adding the two numbers above; it forms one of the bases of probability theory.

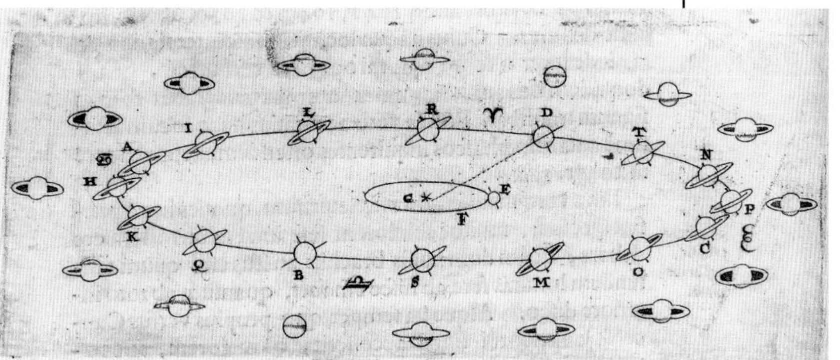
△ **Study of Saturn** by Christiaan Huygens

1655
STUDIES OF SATURN
The Dutch scientist Christiaan Huygens used a refracting telescope to discover Titan, the first of Saturn's Moons to be identified. He also theorized that Saturn was surrounded by a thin, flat ring – an idea he confirmed four years later by observing the planet's rings.

1661
BLOOD CAPILLARIES
Seeing a network of fine tubes on the surface of a frog's lung, Italian biologist Marcello Malpighi discovered capillaries. He speculated that these connected the arteries and lungs, allowing blood to flow back to the heart.

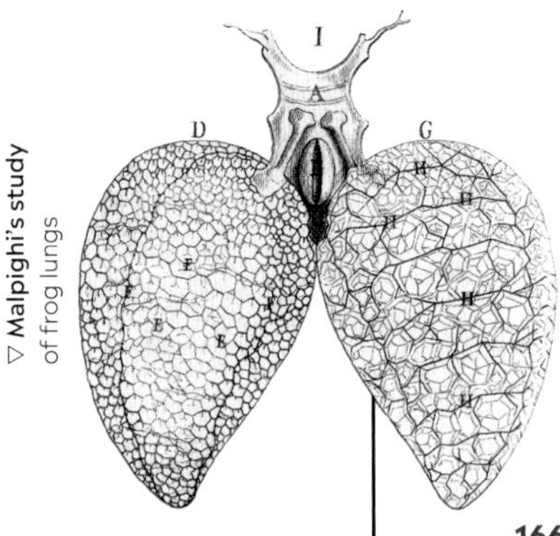

▽ **Malpighi's study** of frog lungs

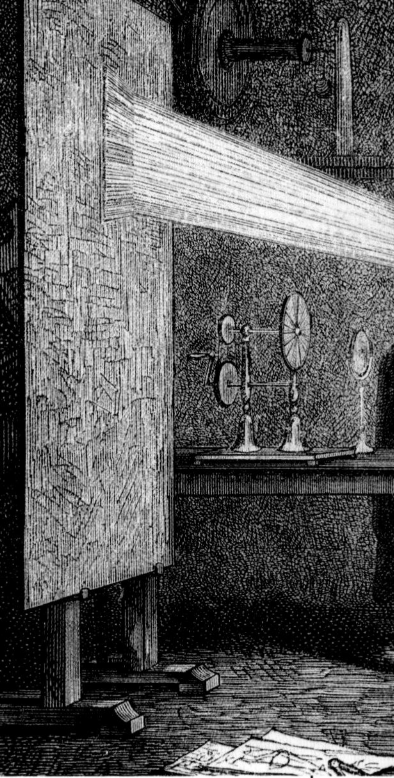

1666
ANALYSING WHITE LIGHT
English physicist and mathematician Isaac Newton used a prism to refract white light, which split into rays of coloured light. He recombined these rays into white light by refracting them through another prism. Newton had discovered that white light is composed of a spectrum of colours, which always appear in the same order.

1660

1660 The Royal Society is founded in London, Britain's oldest scientific institution

1661 Anglo-Irish natural philosopher Robert Boyle defines an element as a substance that cannot be broken down further

1627–91
ROBERT BOYLE
Born in Lismore, Ireland, Boyle did important work in physics, hydrostatics, and Earth science, but is best known as a founders of modern chemistry and a champion of experimental work. Like some others of his time, he believed in alchemy.

▷ **Boyle's air pump** experimental apparatus

1662
BOYLE'S LAW
Robert Boyle used an air pump to help discover the law which states that at constant temperature the pressure of a gas is inversely proportional to its volume, meaning that the pressure of a gas will decrease as its volume increases.

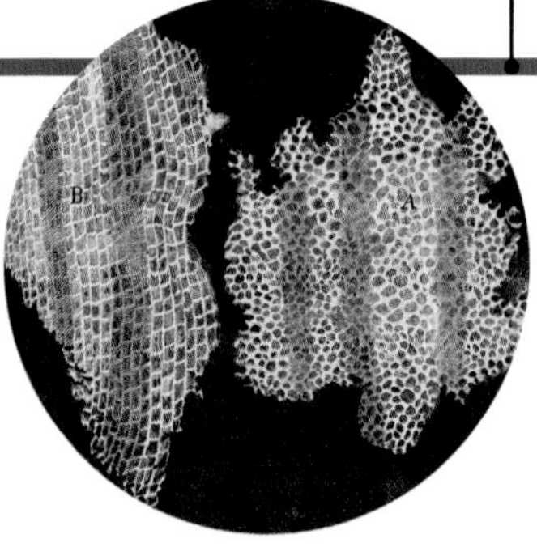

△ **Robert Hooke's** study of cork

1665
DISCOVERY OF THE CELL
Using an improved microscope with three lenses, Robert Hooke observed a honeycomb-like lattice on the surface of cork bark. He had discovered plant cells, marking the start of cell biology. He published his findings in his *Micrographia,* filled with extraordinarily detailed illustrations.

△ **Isaac Newton** perfoming his refraction experiment

1669
STRATIGRAPHY

Danish geologist Nicholas Steno realized that so-called "tongue stones" resembled the teeth of modern sharks and were in fact fossilized ancient sharks' teeth. He also proposed that such fossils became buried in the ground through a process of layers of rock being laid down in sequence, so establishing the principles of stratigraphy.

△ **Fossilized** shark teeth

1666–7 The first attempts at animal to human blood transfusion are made by English physician Richard Lower and French physician Jean-Baptiste Denis

1669

1666 Giovanni Cassini, working in Italy, observes that Mars has a polar cap

1668 English chemist John Mayow describes respiration, arguing that the lungs are designed to bring air into contact with the blood, and the heart to pump blood to them

1668 Francesco Redi disproves spontaneous generation theory (the idea that living organisms regularly arise from non-living matter) through experiments with flies

CELLS, TISSUES, AND ORGANS

The building blocks of all living things, cells can either work alone – for example, blood cells – or be organized into tissues. Tissues are composed of cells of similar structure and function that work together as one. Tissues join up to form organs, like the heart or the brain, which in turn form systems, such as the circulatory or nervous systems.

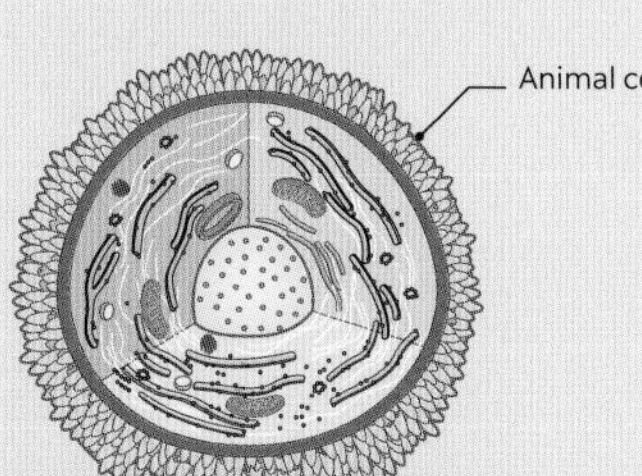

Cells
There are hundreds of different cell types, each with a unique function determined by genes and location.

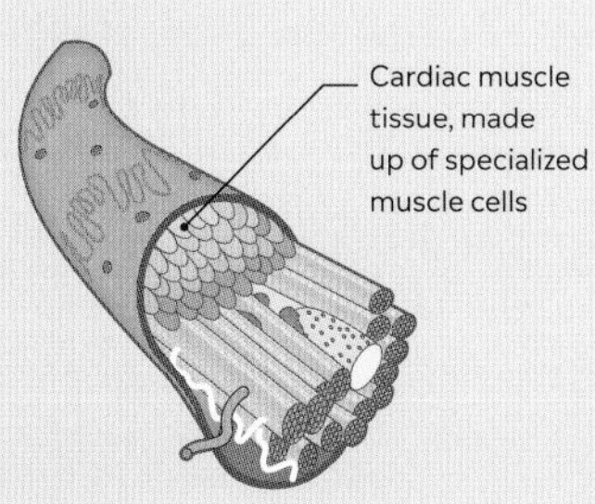

Tissues
Cells of the same type join up to form tissues such as skin, bone, or cardiac muscle (above).

Organ systems
Different organs combine to form systems. The heart and blood vessels, for example, join together to form the circulatory system.

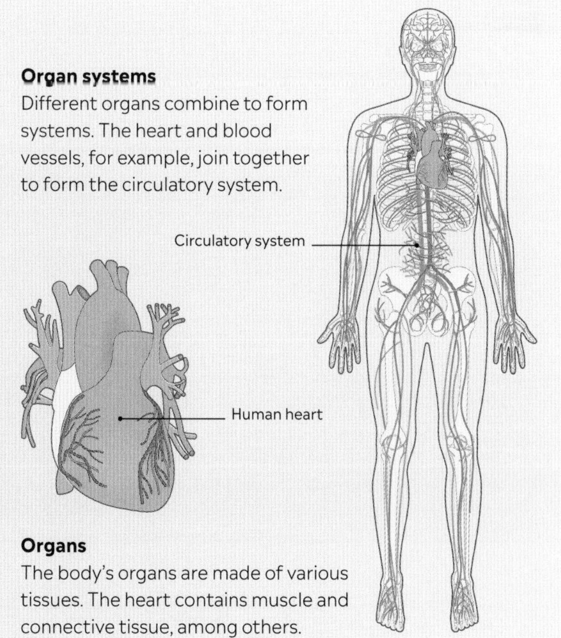

Organs
The body's organs are made of various tissues. The heart contains muscle and connective tissue, among others.

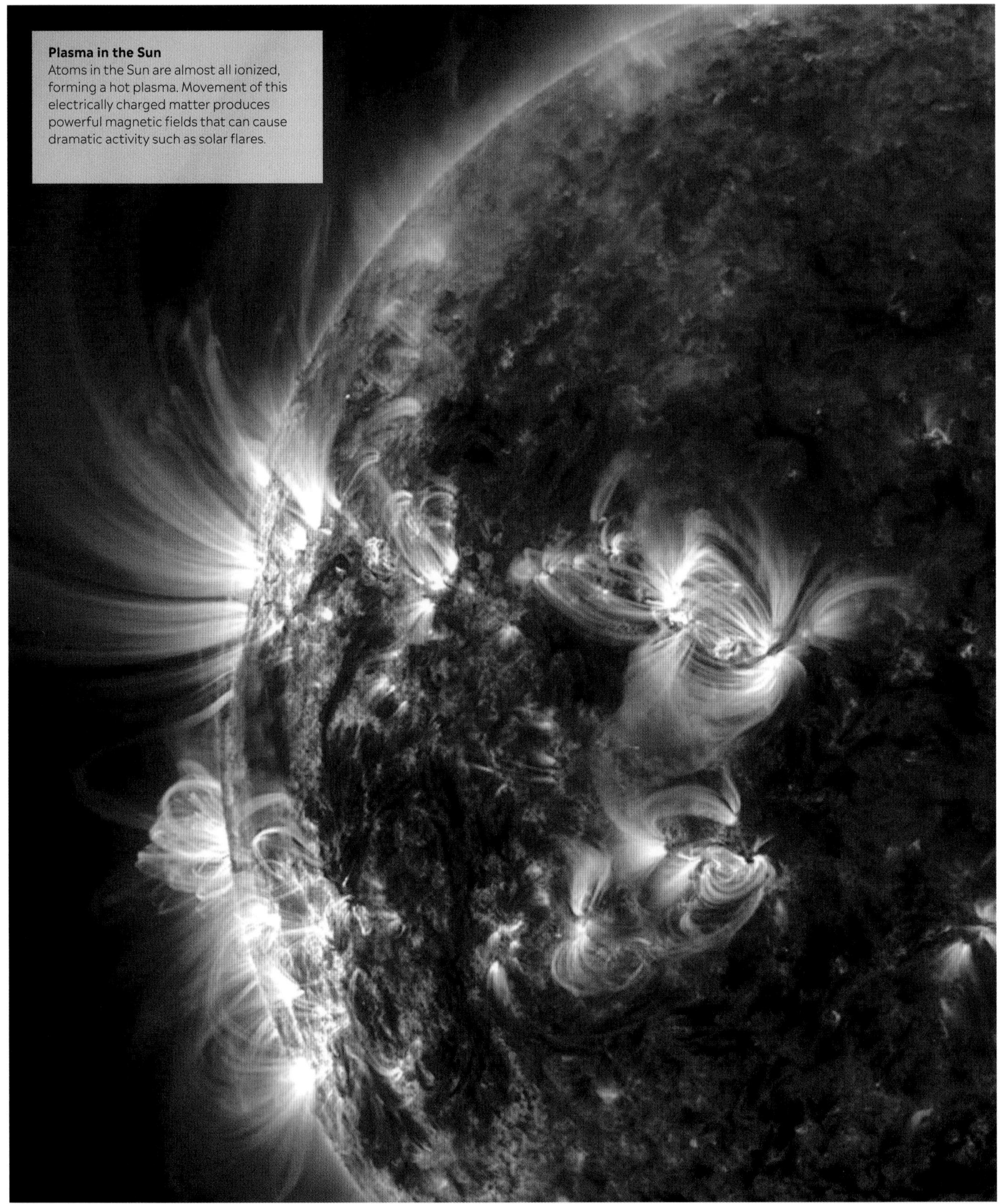

Plasma in the Sun
Atoms in the Sun are almost all ionized, forming a hot plasma. Movement of this electrically charged matter produces powerful magnetic fields that can cause dramatic activity such as solar flares.

THE STATES OF MATTER

Unusual metal
Melting at -39°C (-38°F), mercury is the only metal to be liquid at room temperature. This makes it useful in many devices, such as thermometers, mercury switches, and barometers.

Matter can exist in different forms, known as states of matter. It is the arrangement of particles in a substance that determines its state and many of its properties. The three "classical" states of matter are solid, liquid, and gas.

In a solid, particles are packed close together and fixed in position so they cannot move freely. Solid matter keeps its size and shape, assuming that temperature and pressure remain the same. In a liquid, particles are packed more loosely than in a solid, and they are not fixed in place. This means a liquid tends to keep its size but change its shape to fit its container. In a gas, particles are neither close together nor fixed, moving energetically about in space. A gas can change both its size and shape to fit its container.

Matter can change between these states through a phase transition. A phase transition can be caused by changes in pressure and temperature (*see below*).

A fourth state of matter, plasma, is less familiar in everyday life but is the most common state in the Universe. It is usually formed when a gas is heated to extreme temperatures, tearing electrons away from their atoms and causing the substance to become ionized (electrically charged). The Sun and other stars are made from plasma, and on Earth plasma is generated by lightning and neon lights. In addition to solids, liquids, gases, and plasmas, there are many other states of matter, such as Bose-Einstein condensates and quark-gluon plasmas, which do not exist in nature.

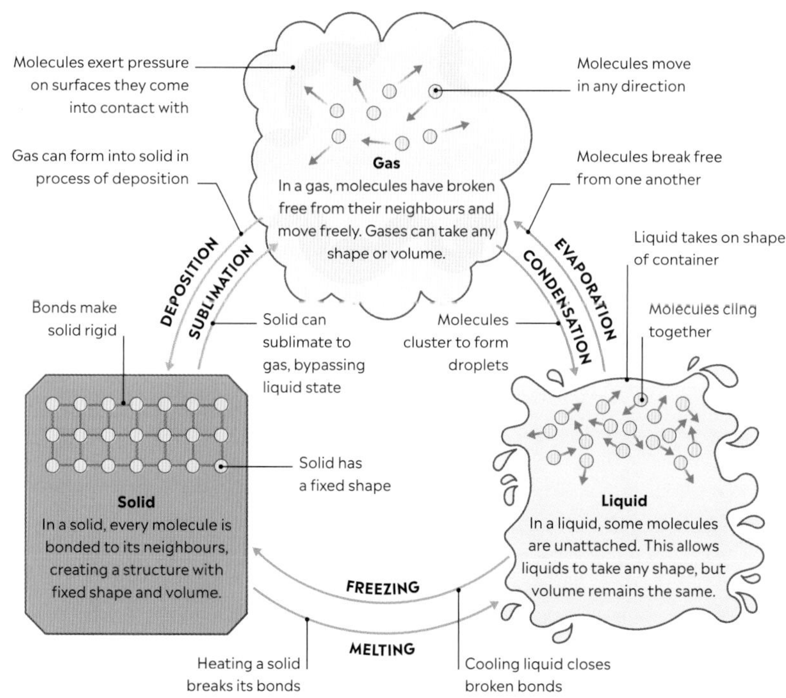

CHANGES OF STATE

States change through a phase transition - that is, a transformation in structure and properties. At low temperatures, matter is solid. Heating it causes it to melt and become a liquid, which when boiled becomes a gas. This occurs at different temperatures for different substances. For example, water boils at 100°C (212°F), while nitrogen boils at -196°C (-321°F).

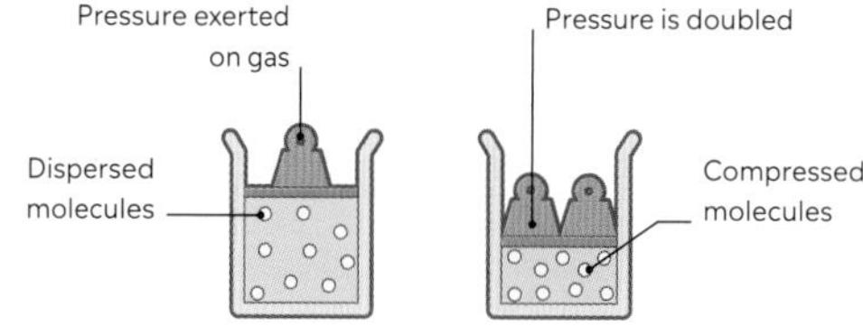

BOYLE'S LAW

The pressure of a gas is inversely proportional to its volume. This means that if pressure is doubled, volume is halved, as long as temperature is kept the same.

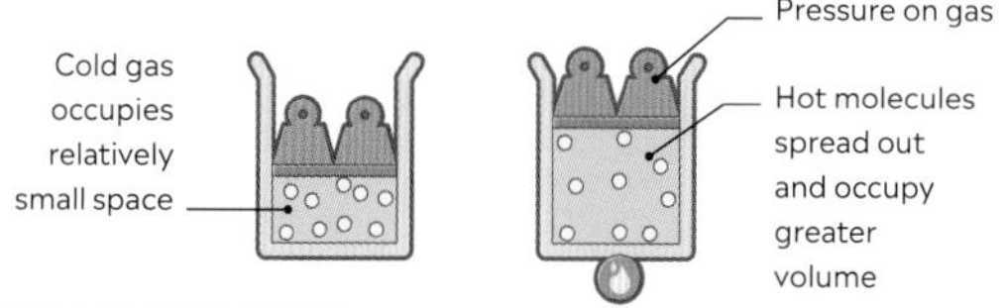

CHARLES'S LAW

Applying heat to a gas increases its volume proportionately. Its molecules move faster and require more space if pressure is to remain the same.

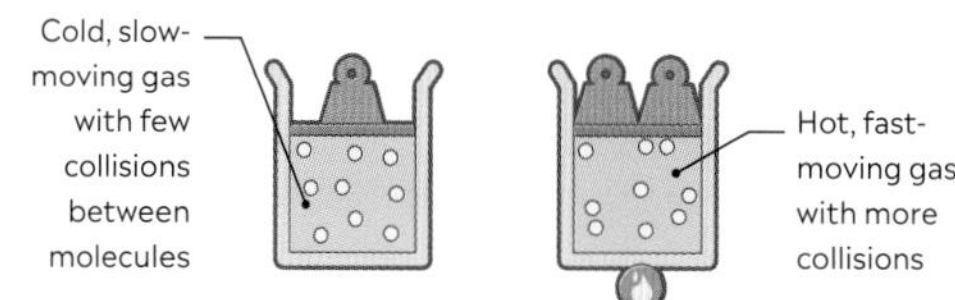

GAY-LUSSAC'S LAW

The pressure of a gas is proportional to its temperature. This means that if the temperature of a gas doubles (measured using the Kelvin scale), pressure also doubles.

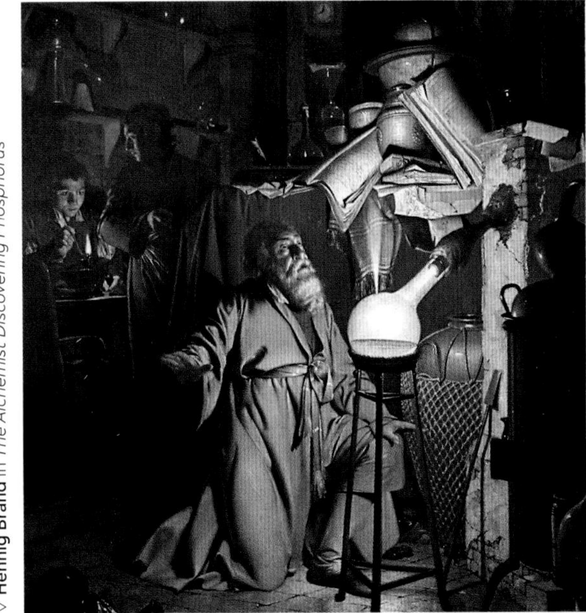

▽ **Hennig Brand** in *The Alchemist Discovering Phosphorus*

1670

PHOSPHORUS ISOLATED

German chemist Hennig Brand became the first person since ancient times to discover a new element. He was evaporating urine while trying to discover an alchemical process for producing gold when he found a light, waxy substance that glowed in the dark, which he named phosphorus ("light bearer").

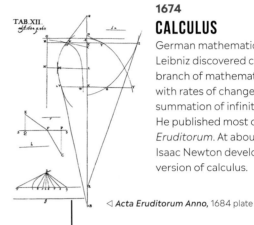

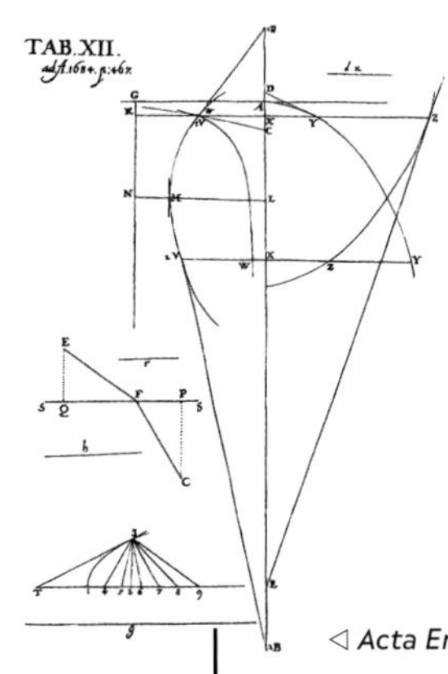

◁ ***Acta Eruditorum Anno,*** 1684 plate

1674

CALCULUS

German mathematician Gottfried Leibniz discovered calculus, the branch of mathematics dealing with rates of change and the summation of infinitely small factors. He published most of his work in *Acta Eruditorum*. At about the same time, Isaac Newton developed his own version of calculus.

1670

1670 Robert Boyle discovers hydrogen by pouring acid onto iron

MICROORGANISMS

Also known as microbes, microorganisms are organisms that are too small to see with the naked eye. Most of the life forms on Earth - including bacteria, viruses, and many plants, animals, fungi, and algae - can only be seen with a microscope. Some microorganisms spread disease, while others are beneficial to our existence. All surfaces of the human body - from the skin, to the lining of the gut - are alive with millions of microorganisms, many of which keep us safe from disease.

The size scale

Microorganisms range in size from a thousandth of a metre (millimetre, mm) for mites, to a millionth of a metre (micrometre or micron, μm) for bacteria, and 10 times smaller than that for viruses.

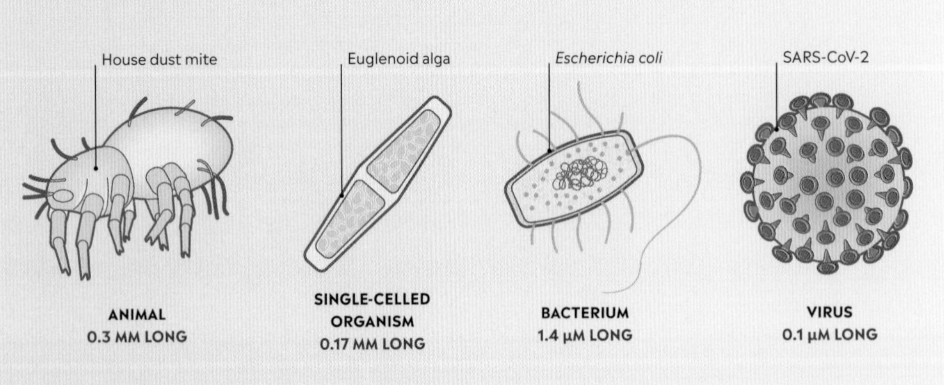

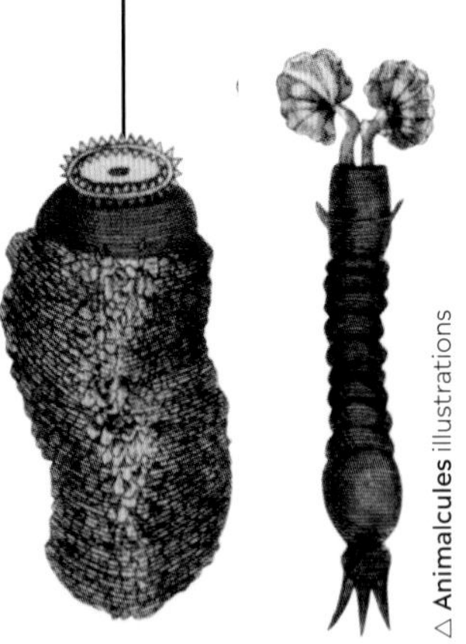

△ **Animalcules** illustrations

1675

ANIMALCULES

Having built a microscope with a magnification factor of x275, Dutch scientist Antonie van Leeuwenhoek was able to see tiny animals, and was the first to discover protozoa, which he called "animalcules".

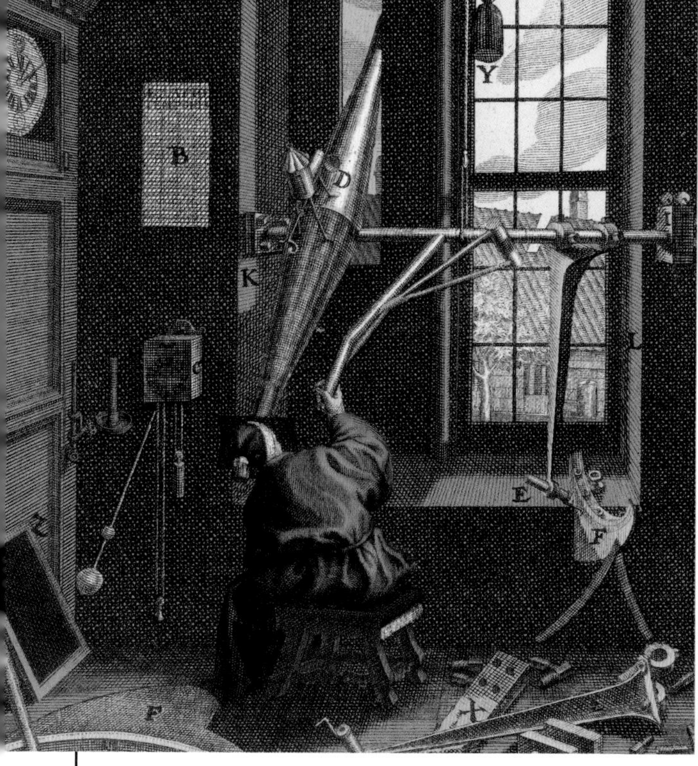

△ **Ole Romer** in his observatory

239 BCE was the first recorded sighting of Halley's Comet (in China)

1676

THE SPEED OF LIGHT

Scientists considered the speed of light to be infinite until Danish astronomer Ole Römer measured it for the first time. While observing eclipses of the satellite Io by Jupiter, he found the period between the eclipses shortened when Earth was closer to Jupiter. He realized this was because the light from the eclipse had less far to travel, allowing him to calculate its speed.

1678 Christiaan Huygens proposes that light is made up of waves that vibrate up and down as the light travels

1679 Gottfried Leibniz describes a system of binary arithmetic

1680 Italian physiologist Giovanni Borelli founds biomechanics with his study of muscle contraction

1682 English astronomer Edmond Halley observes the comet whose return he will later predict

1684 Giovanni Cassini discovers two moons of Jupiter (Dione and Tethys), having already found two others

1684

1678

HOOKE'S LAW

Robert Hooke observed that elastic materials such as metals changed shape when subjected to force. He then devised his famous law (outlined in *De Potentia Restitutiva*), which states that the amount of this stretch is proportional to the amount of force applied. This insight eventually led to the development of coiled springs..

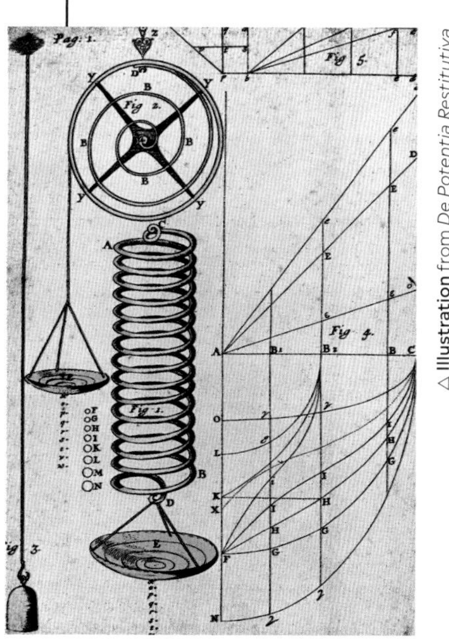

△ **Illustration** from *De Potentia Restitutiva*

1656–1742

EDMOND HALLEY

A major figure in astronomy, Halley made the first star chart of southern constellations, devised a method for observing the transit of Venus, and mapped the orbits of 24 comets, including the one named after him (whose return in 1758 he predicted).

1685

1686
THE BIOLOGICAL SPECIES
English naturalist John Ray produced the first scientific definition of "species" to describe a group of animals or plants that have the same characteristics and can reproduce together. This later formed the basis of taxonomy, the scientific classification of species.

◁ **John Ray**

1686 Edmond Halley suggests surface winds form because of a pattern of atmospheric circulation caused by the Sun's heat

1687
LAWS OF MOTION
Isaac Newton published his *Principia Mathematica*, in which he put forward his theory of gravity and the laws of motion, including that a body's speed will remain constant unless another force acts on it, and that bodies exert equal and opposite forces on each other.

△ **Isaac Newton's** notebook

"Every body perseveres in its state of rest, or of uniform motion in a right line, unless it is compelled to change that state by forces impressed thereon."

ISAAC NEWTON'S FIRST LAW OF MOTION, FROM PRINCIPIA MATHEMATICA, *1687*

1689 English physician Walter Harris publishes one of the first books on paediatric medicine

1690 Johannes Hevelius publishes his final star chart; it includes many constellation names, such as Lynx, still used today

1694
SEXUAL REPRODUCTION IN PLANTS
In *De Sexu Plantarum*, German botanist Rudolph Camerarius set out his discovery of the process of sexual reproduction in plants. By studying mulberry bushes, he identified the male (anthers) and female (pistils) organs of flowering plants and described the function of pollen in plant reproduction.

▷ **Mulberry** plant

1696
SCIENCE AND RELIGION

English theologian and natural scientist William Whiston published *A New Theory of the Earth* in which he sought to give scientific explanations for events in the Bible, such as Noah's flood (which he believed was caused by a comet passing close to the Earth). It marked a further move away from religious explanations for natural phenomena and an early attempt to scrutinize the Bible scientifically.

◁ ***Noah's Ark on Mount Ararat*** by Simon de Myle

1697 German chemist Georg Stahl theorizes that a substance called phlogiston is released into the air when something is burnt

1697 Swiss mathematician Johann Bernoulli solves a problem regarding the trajectory of objects; his solution has implications for ballistics

1698 Christiaan Huygens' *Cosmothereos* is published posthumously; it theorizes that life might exist on other planets

1699

1698
THE MINER'S FRIEND ENGINE

The English military engineer Thomas Savery created the first successful steam engine to extract water from flooded mineshafts. The water was channelled into a closed vessel, where steam under pressure forced it up and out of the shaft.

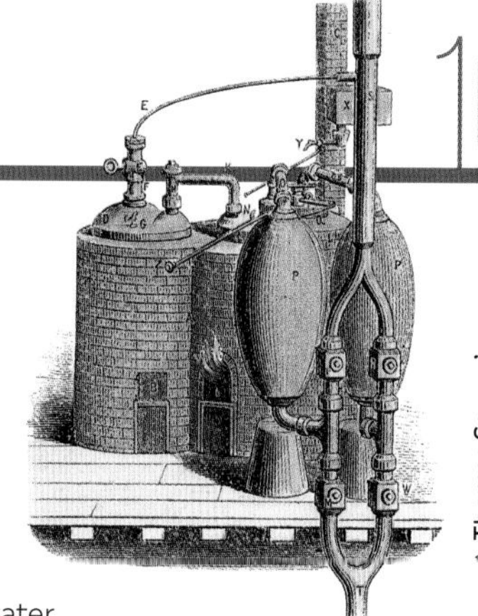

△ **Thomas Savery's** steam pump

1643–1727
ISAAC NEWTON

English mathematician Isaac Newton's pursuit of a mechanical framework for the Universe led him to advances in optics and calculus, and the formulation of his theory of gravity, which underpinned the science of physics for over 250 years.

1699
COMPARATIVE PHYSIOLOGY

While dissecting a chimpanzee, English physician Edward Tyson found that its brain and other internal organs were very similar to those of humans. His dissections of other animals, including porpoises and rattlesnakes, made him the founder of comparative physiology.

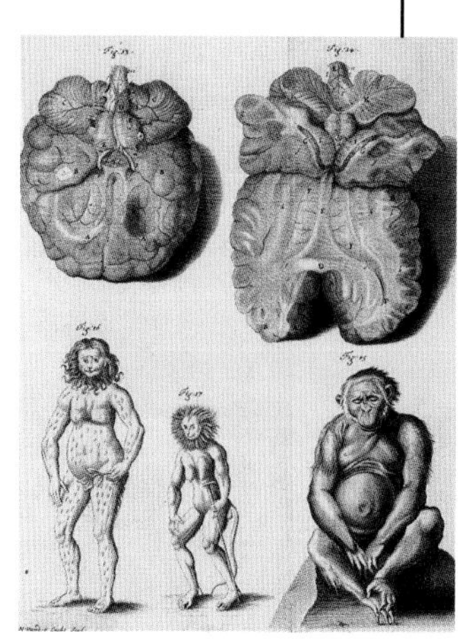

▷ **Edward Tyson's** *Pongo pygmaeus* (orangutan) illustration

THE LAWS OF MOTION

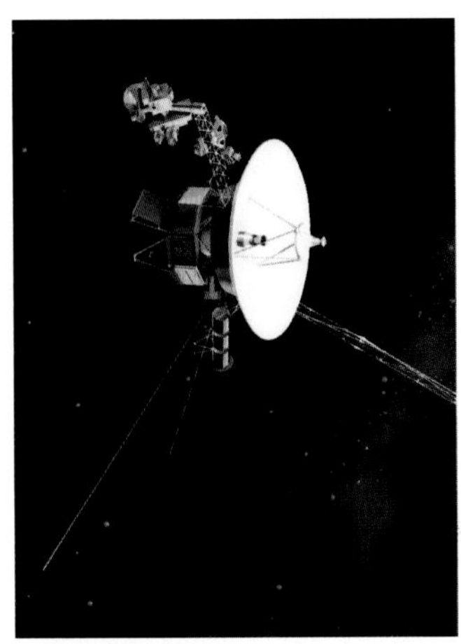

Unopposed motion
Voyager 1 travels in nearly empty space, where there are essentially no forces acting on it. This allows it to continue moving indefinitely.

Sir Isaac Newton's laws of motion are the foundations of classical mechanics. The three laws explain the relationship between an object's motion and the forces that act on it, allowing us to predict accurately how objects from cannonballs to rockets move under the influence of forces.

The first law of motion introduces the concept of inertia - that is, the tendency of an object to continue in its existing state of motion unless acted on by an unbalanced force. This directly contradicts the historic idea that an object continues to move only if a force continues to act on it.

The second law describes the relationship between the force applied to an object, its mass, and the resultant acceleration: the heavier the object, the less it accelerates under a given force.

According to the third law of motion, for every action, there is an equal and opposite reaction. For example, the Sun's force on Earth is matched by the force of Earth on the Sun (although the latter produces a much smaller effect, due to the much larger mass of the Sun).

These three laws can be built upon to provide further insights into physical systems, such as applying them to systems containing large numbers of particles to explain how gases behave. More recent developments in physics - such as relativity (*see pp.186–87 and 197*) and quantum mechanics (*see pp.180–81*) - have emerged to describe the behaviour of objects that are very small, very fast, or very massive, but Newton's three laws of motion remain a good approximation for everyday phenomena.

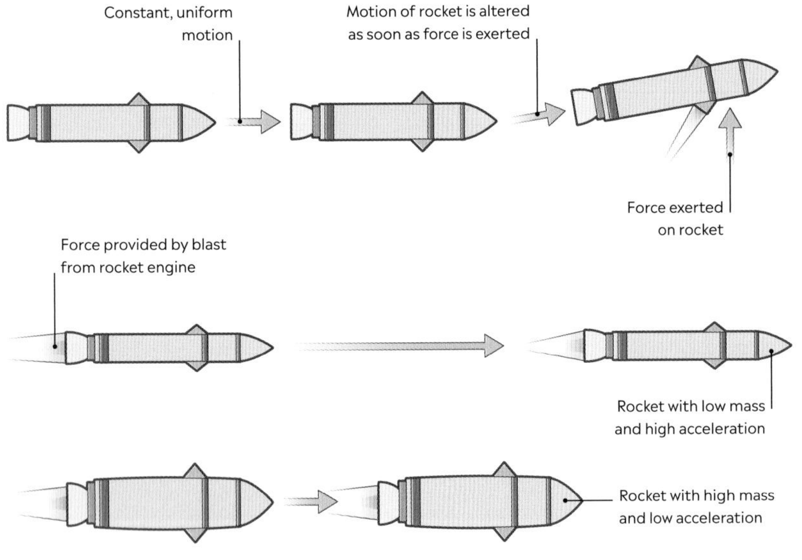

FIRST LAW OF MOTION

Newton's first law of motion states that an object at rest remains at rest, and an object in motion remains in motion with constant speed and direction unless acted on by external forces. In other words, an object (such as a rocket) will not accelerate unless an unbalanced force (such as that provided by a booster) acts on it.

SECOND LAW OF MOTION

The second law says that the acceleration (a) of an object depends on the mass (m) of the object and the force (F) applied to it. This can be written with the formula: $F = ma$. The consequence of this is that an object with a lower mass accelerates more under a given force than an object with a higher mass.

Action, in form of backwards blast generated by combustion of fuel

Reaction takes form of forwards movement

THIRD LAW OF MOTION

The third law says that whenever an object exerts a force on another object, that second object exerts an equal and opposite force on the first. In other words, every action has an equal and opposite reaction. For example, the forwards thrust of a rocket can be considered the reaction to the backward blast of combusted fuel.

Action and reaction
As a real-world example of what is stated in the third law of motion, when rowers push water backwards with their oars (action), the water exerts a forwards push on the boat, propelling it forwards (reaction).

WAVE–PARTICLE DUALITY

In some experiments, waves act like particles, and particles act like waves. This makes it impossible to describe objects at the quantum scale (the scale of atoms and even smaller objects) as wholly one or the other. Instead, light, for example, is considered to have characteristics of both waves and particles. Known as wave-particle duality, this is an important concept in quantum physics.

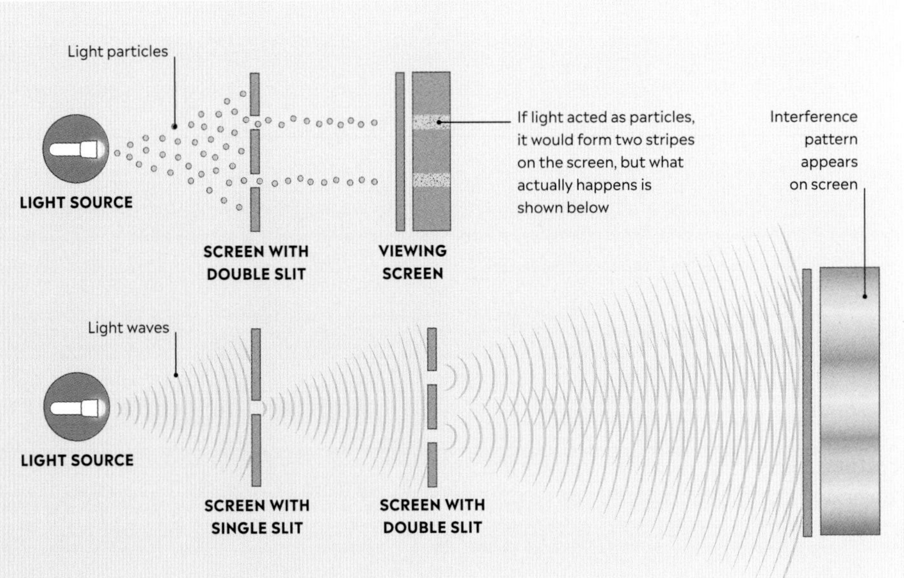

Wave-like behaviour
Light that is split into two beams by a pair of slits in a screen diffracts and interferes with itself, producing a distinct pattern of dark and light fringes on a viewing screen.

1700

1704 The Italian anatomist Antonio Valsalva publishes the first detailed account of the physiology of the human ear

1705 Edmond Halley predicts the return in 1758 of what will become known as Halley's Comet

1706 William Jones, a Welsh mathematician, uses the Greek letter π to represent the ratio of a circle's circumference to its diameter

1707 Denis Papin, a French engineer, invents the high-pressure boiler

▷ Georg Ernst Stahl

1703

PHLOGISTON THEORY

Building on earlier work (*see p.81*), German chemist Georg Ernst Stahl set about trying to provide experimental proof of the existence of phlogiston (a supposed element thought to be involved in combustion). Although wrong about phlogiston, Stahl's work encouraged experimentation that ultimately led to the discovery of oxygen.

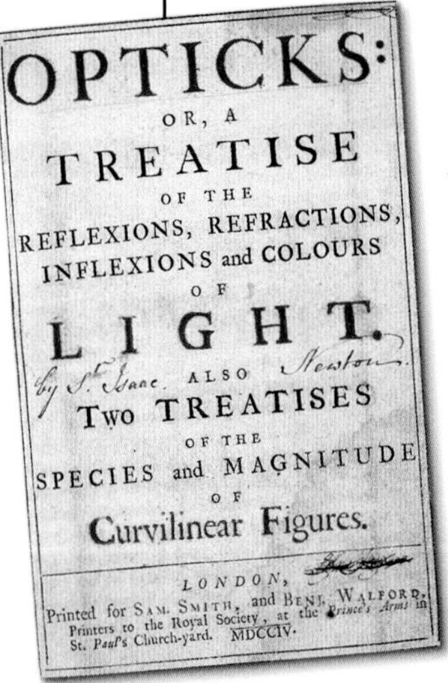
OPTICKS: OR, A TREATISE OF THE REFLEXIONS, REFRACTIONS, INFLEXIONS and COLOURS OF LIGHT. ALSO Two TREATISES OF THE SPECIES and MAGNITUDE OF Curvilinear Figures.

LONDON, Printed for SAM. SMITH, and BENJ. WALFORD, Printers to the Royal Society, at the Prince's Arms in St. Paul's Church-yard. MDCCIV.

◁ ***Opticks*** by Isaac Newton, title page

1704

NEWTON'S OPTICKS

Isaac Newton published his *Opticks*, his magnum opus on the nature of light. A broad-ranging work, it detailed his experiments with refraction using prisms and lenses and with diffraction using sheets of glass, but also explored the nature of heat, electrical phenomena, and the possible cause of gravity.

> "Are not rays of light very small bodies emitted from shining substances?"
>
> *ISAAC NEWTON*, OPTICKS, *1704*

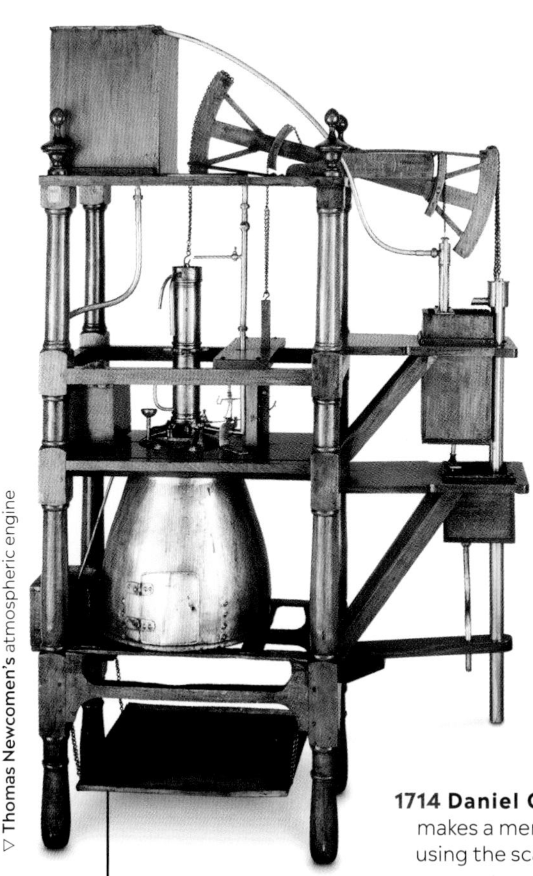

▽ **Thomas Newcomen's** atmospheric engine

1712
THE NEWCOMEN ENGINE

English inventor Thomas Newcomen built his "atmospheric engine", the first truly practical steam engine, vastly improving on the performance achieved by Savery's earlier engine (*see p.81*). The steam created a vacuum that pulled down a piston attached to one end of a rocking beam; a chain attached to the other end was put to use pumping water out of flooded mines.

1714
THE SYRINGE

A French military surgeon, Dominique Anel invented a fine-pointed suction syringe to extract dirt and infection from soldiers' wounds. His invention remained in use for centuries, particularly in the treatment of diseases affecting the tear duct.

◁ **Dominique Anel's** lachrymal syringe

1714 Daniel Gabriel Fahrenheit makes a mercury thermometer using the scale that will later be named after him

1719

c. 1713 The sequence known as the Bernoulli numbers is discovered by the Swiss mathematician Jacob Bernoulli

1715
SMALLPOX INOCULATION

News of a technique used in Turkey to combat smallpox, known at the time as "variolation", reached Europe. English aristocrat Lady Mary Wortley Montagu, who had lived in Constantinople, was influential in spreading its use after demonstrating the technique on her own daughter.

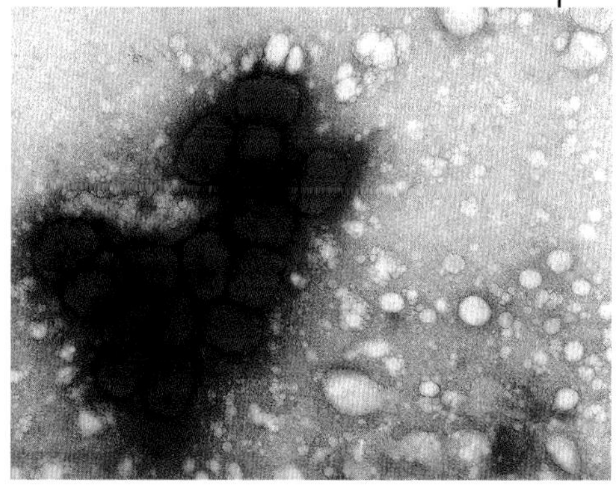

△ **Smallpox** viruses

1718
THE PROPER MOTION OF STARS

By comparing his own observations with those of the ancients, the English astronomer Edmond Halley – then employed as a professor at England's Oxford University – demonstrated the proper motion of the "fixed" stars as seen from Earth. His measurements revealed that the stars Arcturus, Sirius, and Aldebaran had all moved more than half a degree from the positions noted by the Greek astronomer Hipparchus almost two millennia earlier.

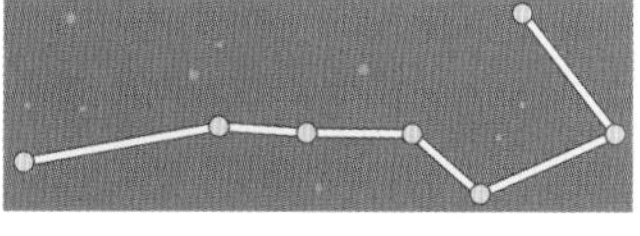

100,000 BCE

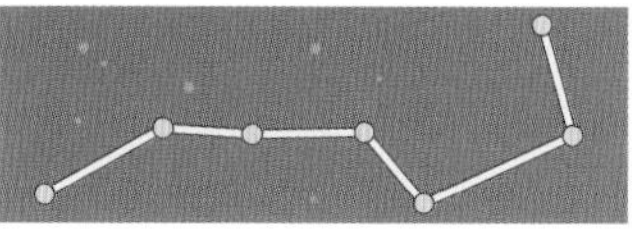

2000 CE

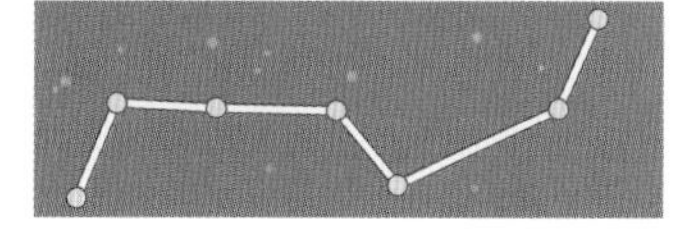

100,000 CE

△ **The changing** shape of the Plough in Ursa Major

1725
A NEW STAR CATALOGUE

John Flamsteed's *Historia Coelestis*, published after his death, catalogued the positions of nearly 3,000 stars, becoming for many years a standard reference source. Flamsteed had served as Britain's Astronomer Royal for over 40 years, having personally laid the foundation stone of the Greenwich Observatory in London in 1675.

▷ **John Flamsteed** at Greenwich Observatory

1720

1721 Jan Palfijn, a Flemish surgeon, introduces forceps as a means of facilitating birth

1721 English clockmaker George Graham invents the mercury pendulum for clocks, designed to stop the pendulum rod expanding and contracting with changes in temperature

1723 German physicist Jacob Leupold publishes the first of nine volumes of his *Theatrum Machinarum Generale*, the first systematic treatise on mechanical engineering

1724 Emperor Peter the Great founds the Academy of Sciences at St Petersburg in Russia and invites foreign scholars to improve the state of knowledge in the country

▷ **Gravesande's** ring experiment

"Plants very probably draw through their leaves some part of their nourishment from the air."

STEPHEN HALES, VEGETABLE STATICKS, *1727*

1720
THERMAL EXPANSION

Willem 's Gravesande published *Physices Elementa Mathematica, Experimentis Confirmata* (*Mathematical Elements of Natural Philosophy Confirmed by Experiments*), providing experimental proof of the laws of Newtonian mechanics. One of Gravesande's demonstrations, designed to prove heat expansion, featured a metal ball that, when heated, distended to no longer fit through the ring.

1728
INDIAN OBSERVATORIES

Construction of the Jantar Mantar observatories began in Rajasthan, on the orders of Jai Singh II, ruler of the kingdom of Jaipur in north-western India. In all, 19 separate structures were built, including the world's largest sundial. Designed to aid stargazing with the naked eye, the instruments served to predict eclipses and other astronomical events.

▷ **Jantar Mantar** observatory

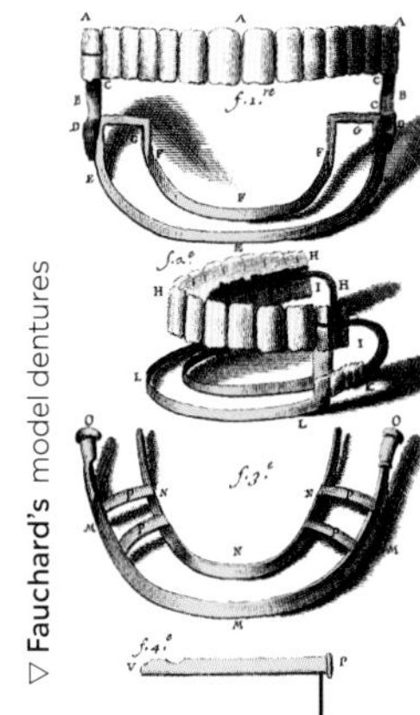
▷ **Fauchard's** model dentures

1728
EARLY DENTISTRY

French physician Pierre Fauchard published *Le Chirurgien Dentiste*, a formative text on dentistry that included the first description of the use of orthodontic braces. He identified 103 separate oral problems, suggesting ways to treat them, and also suggested limiting sugar intake to lessen the risk of tooth decay.

1729

1728 **English astronomer James Bradley** explains the periodic shifts in the position of fixed stars in terms of the aberration of light

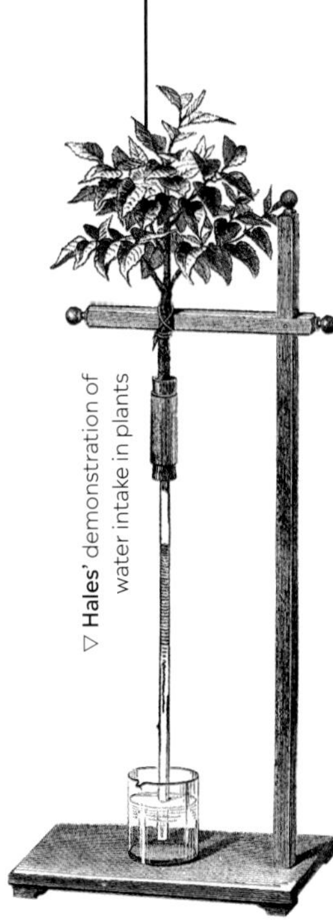
▷ **Hales'** demonstration of water intake in plants

1727
STUDIES IN PHYSIOLOGY

English scientist Stephen Hales published his *Vegetable Staticks*, a pioneering work recording experiments in plant and animal physiology. He studied the circulation of water in plants, including transpiration, by which liquid moving through a plant is lost by evaporation from its leaves.

△ **Gray's** flying boy

1729
STATIC ELECTRICITY

In a series of pioneering experiments, English scientist Stephen Gray showed that some bodies conduct electricity while others do not. His famous "flying boy" experiment revealed how electricity can flow through materials he termed "conductors" but not others ("insulators").

1730

COBALT METAL ISOLATED

The Swedish chemist Georg Brandt discovered cobalt - the first metal to be identified by a named individual. He did so by distinguishing it from bismuth, known to be a metal since ancient times and often found alongside cobalt. Having separated the two substances, Brandt subjected each to chemical and other tests, proving along the way that it is cobalt rather than bismuth that can give glass a blue colour.

△ **Cobalt** metal

ELEMENTA CHEMIAE,
QUAE
ANNIVERSARIO LABORE DOCUIT,
IN PUBLICIS, PRIVATISQUE,
SCHOLIS,
HERMANNUS BOERHAAVE.
TOMUS SECUNDUS.
QUI CONTINET OPERATIONES CHEMICAS.

LUGDUNI BATAVORUM,
Apud ISAAGUM SEVERINUM.
M. D. CCXXXII.

1732

ELEMENTS OF CHEMISTRY

Published this year, Hermann Boerhaave's *Elementa Chemiae* (*Elements of Chemistry*) was to become a standard chemistry textbook for the rest of the century. A professor of botany and medicine at Leiden University in the Netherlands, Boerhaave also made major contributions to physiology and medical education in general and encouraged direct investigation rather than reliance on classical texts.

◁ ***Elements of Chemistry*** title page

1730

1732 English inventor John Kay invents the flying shuttle loom, a first step in the mechanization of textile manufacture that helps to drive the Industrial Revolution

1735 English meterologist George Hadley describes the Hadley cell - the pattern of wind circulation that creates the trade winds

1730 French scientist René de Réaumur constructs an alcohol thermometer with a graduated scale from 0° (marking the freezing point of water) to 80° (boiling point)

1735

THE MARINE CHRONOMETER

Responding to a prize offered by the British Parliament 21 years earlier, John Harrison unveiled the first marine chronometer, now known as H1, in 1735. The device was designed to compensate for changes in temperature and to run without the need for lubrication, employing two interconnecting swinging balances to counteract a ship's swaying. By providing accurate time-keeping, it enabled mariners more accurately to work out longitude (their east-west position) on the world's oceans.

▷ **H1 chronometer** by John Harrison

ATMOSPHERIC CIRCULATION

Earth's atmosphere is constantly circulated around the globe as wind. The Sun heats tropical air in low-pressure zones, and this is then circulated towards the cold poles. However, because of Earth's shape and rotation, as well as the thickness of the atmosphere, the circulation is in the form of giant "cells" – in the tropics, the temperate zone, and at the poles.

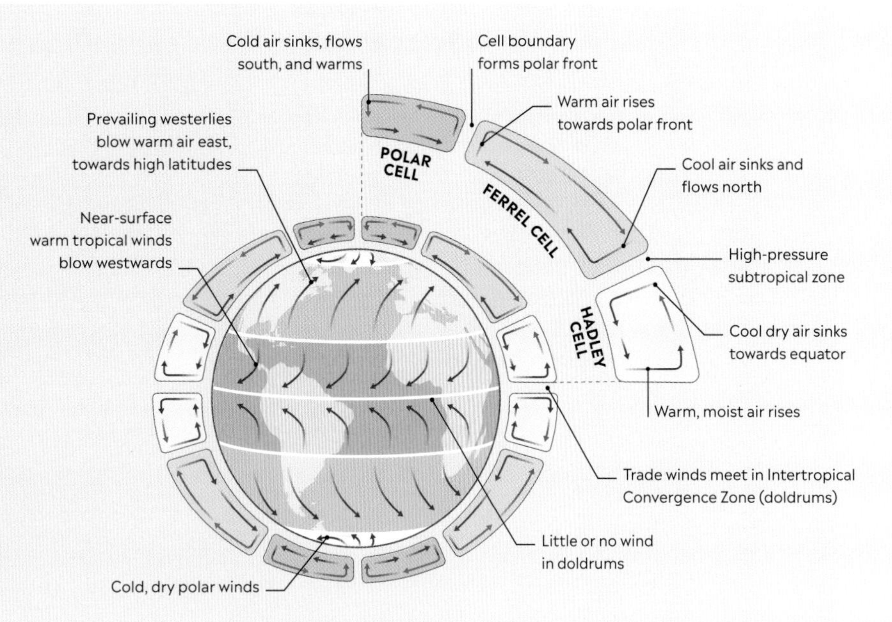

Cold air sinks, flows south, and warms
Cell boundary forms polar front
Warm air rises towards polar front
Prevailing westerlies blow warm air east, towards high latitudes
POLAR CELL
FERREL CELL
Cool air sinks and flows north
Near-surface warm tropical winds blow westwards
High-pressure subtropical zone
HADLEY CELL
Cool dry air sinks towards equator
Warm, moist air rises
Trade winds meet in Intertropical Convergence Zone (doldrums)
Little or no wind in doldrums
Cold, dry polar winds

Circulation cells
Warm tropical air rises, cools, sinks, and is recycled, while cold polar air flows outwards, is warmed, and is also recycled. In between, the temperate cell has warm air flowing polewards and rising when it meets cold polar air.

1736 Claudius Aymand, a French surgeon, performs the first successful operation for appendicitis

1736 American physician William Douglass describes scarlet fever

1739

1736 Swiss mathematician Leonhard Euler publishes his *Mechanica*, generally considered to be the first systematic textbook of mechanics, analysing the mathematics governing movement

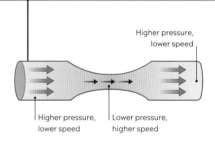

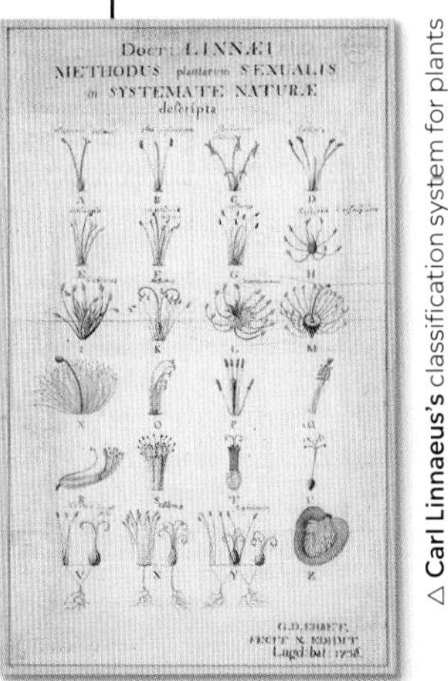

Doct: LINNÆI
METHODUS plantarum SEXUALIS
in SYSTEMATE NATURÆ
descripta

△ Carl Linnaeus's classification system for plants

1735
NATURAL CLASSIFICATION

In *Systema Naturae* (*System of Nature*), the Swedish naturalist Carl Linnaeus introduced a new classification system for organisms. He divided the natural world into three kingdoms – animal, vegetable, and mineral – and established the convention, still used today, of identifying each species by two Latin names.

1700–82
DANIEL BERNOULLI

Born into a Swiss family of mathematicians, Bernoulli worked as a professor, teaching mathematics, botany, physiology, and physics. Active in many fields of study, he made advances in mechanics, astronomy, and maritime science.

Higher pressure, lower speed

Higher pressure, lower speed

Lower pressure, higher speed

Speed and pressure
As the diameter of a vessel through which a fluid passes narrows, the speed of flow increases at the same time as pressure decreases, allowing the same volume of fluid to circulate unhindered.

1738
BERNOULLI'S PRINCIPLE

In his work *Hydrodynamica*, Daniel Bernoulli investigated the forces exerted by fluids and propounded what became known as the Bernoulli principle that a fluid's potential energy, measured in terms of static pressure (the force it exerts on a body at rest), decreases as its speed increases.

1743

EARTH'S SHAPE

A former child prodigy, French mathematician Alexis-Claude Clairaut was just 30 years old when he published his *Théorie de la Figure de la Terre, Tirée des Principes de l'Hydrostatique* (*Theory of the Shape of the Earth, based on the Principles of Hydrostatics*). The work demonstrated how to compute gravitational forces at any latitude. The theory he propounded, which became known as Clairaut's theorem, confirmed Isaac Newton's view that Earth is an oblate spheroid and flattens toward the poles .

△ Alexis-Claude Clairaut

1740

1740 King Frederick the Great revives the Prussian Academy of Science in Berlin as part of his domestic reforms

1740 English inventor Benjamin Huntsman introduces a new, improved "crucible" process for smelting steel

1743 Benjamin Franklin and others come together to found the American Philosophical Society in Philadelphia, America's first scientific society

1744 French cartographer César-François Cassini oversees the triangulation of France, the first national geographic survey

1742

THE CENTIGRADE THERMOMETER

Swiss astronomer Anders Celsius invented the centigrade thermometer. He initially chose 0° as boiling point and 100° for freezing, but French physicist Jean Pierre Christin switched the two figures round in the following year, giving us the system we use today.

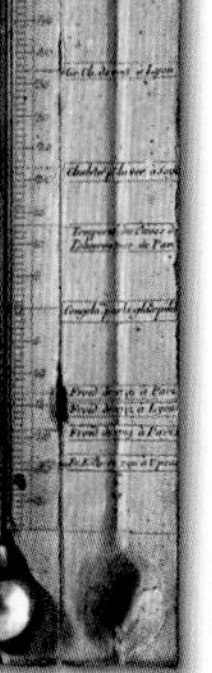

△ **Celsius** thermometer

1743

D'ALEMBERT'S PRINCIPLE

In his *Traité de Dynamique* (*Treatise on Dynamics*), French philosopher Jean d'Alembert spelled out what would become known as d'Alembert's principle. This stated that actions and reactions in a clos ed system of moving bodies are in equilibrium. The book applied this rule to the solution of problems in mechanics.

△ Jean d'Alembert

George Washington, Thomas Jefferson, and Alexander Hamilton were all early members of the American Philosophical Society

1706–90

BENJAMIN FRANKLIN

Born in Boston, Mass., US, Franklin was a polymath whose interests spanned literature, politics, diplomacy, science, and invention. His practical creations included the lightning rod, bifocal spectacles, and the metal-lined fireplace known as the Franklin stove.

1746

STORING ELECTRICAL CHARGE

Working separately, German physicist Ewald Georg von Kleist and Dutch scientist Pieter van Musschenbroek invented the Leyden jar – the first practical way to store static electricity. Von Kleist's receptacle took the form of a medicine bottle filled with alcohol and stoppered with a cork skewered by a metal nail.

▷ Leyden jar

1749

1745 Swiss biologist Charles Bonnet publishes his *Traité d'Insectologie* (*Treatise on Insectology*), describing his observations of parthenogenetic reproduction and the metamorphosis of aphids

1746 French mineralogist Jean-Étienne Guettard draws up the first geological map of France

1749

A NEW NATURAL HISTORY

The Comte de Buffon, a French polymath, published the first of 36 volumes of his *Histoire Naturelle* (*Natural History*), a summation of the current state of knowledge on animals and minerals; a planned section on plant life was to remain unwritten. In time, Buffon's exquisite literary style made the work one of the most widely read scientific texts of its era.

▷ **Buffon's** *Histoire Naturelle*

1750

THE SOUTHERN SKY

Under the leadership of the French astronomer Nicolas de Lacaille, an expedition travelled to the Cape of Good Hope to study the southern hemisphere sky. Keeping watch every night from a specially constructed observatory, Lacaille noted some 10,000 stars over the following two years, giving many of them the names by which they are still known today. He eventually identified and named 14 new constellations.

◁ **The galaxy M83,** discovered by Lacaille

1750

1751 Robert Whytt, a Scottish pioneer of neurology, publishes *On the Vital and other Involuntary Motions of Animals*, distinguishing involuntary from consciously willed actions

1752 The Murder Act allows medical schools in England to dissect the bodies of executed murderers in the interests of anatomical research

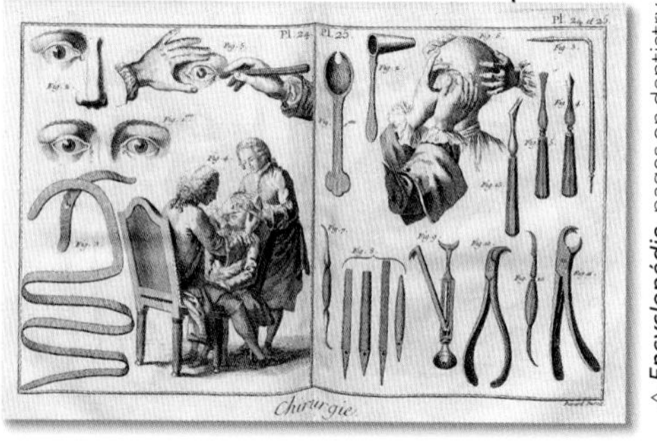

△ ***Encyclopédie,*** pages on dentistry

1751

ENLIGHTENMENT ENCYCLOPEDIA

The first volume of the *Encyclopédie, ou Dictionnaire Raisonné des Science, des Arts, et des Métiers*, generally known as the *Encyclopédie*, was published in France. Edited by Jean d'Alembert and Denis Diderot, the work eventually ran to 28 volumes, and came to be seen as the quintessential scientific text of the 18th-century Age of Enlightenment.

1752

CONDUCTING LIGHTNING

Seeking to show that lightning is a form of electrical discharge, Benjamin Franklin described an experiment involving flying a kite in a storm. Aware of the dangers, he recommended carefully insulating oneself against the risk of electrocution, and suggested that the electricity flowing down the kite string should be diverted into a Leyden jar for storage. The experiment inspired the invention of lightning rods to protect buildings from storm damage.

▷ **Benjamin Franklin** performing his experiment

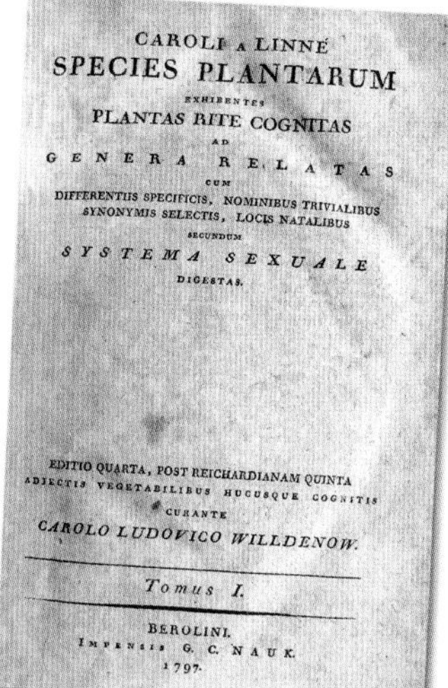
CAROLI A LINNÉ
SPECIES PLANTARUM
EXHIBENTES
PLANTAS RITE COGNITAS
AD
GENERA RELATAS
CUM
DIFFERENTIIS SPECIFICIS, NOMINIBUS TRIVIALIBUS
SYNONYMIS SELECTIS, LOCIS NATALIBUS
SECUNDUM
SYSTEMA SEXUALE
DIGESTAS.
EDITIO QUARTA, POST REICHARDIANAM QUINTA
ADJECTIS VEGETABILIBUS HUCUSQUE COGNITIS
CURANTE
CAROLO LUDOVICO WILLDENOW.
Tomus I.
BEROLINI,
IMPENSIS G. C. NAUK.
1797.

△ *Species plantarum* front cover

1753

LINNAEAN CLASSIFICATION

With the publication of *Species Plantarum* (*Plant Species*), Carl Linnaeus completed the work he had begun two decades earlier to catalogue, classify, and name all known components of the natural world. Listing 5,940 different plants, the book was the first to apply consistently the Linnaean system of nomenclature, giving each plant a two-word Latin name to identify first the genus and then the individual species.

1756

ISOLATING CARBON DIOXIDE

After experimenting with heated limestone, the Scottish chemist Joseph Black announced the isolation of carbon dioxide, which he termed "fixed air". He later demonstrated that the gas was also produced by animal respiration.

◁ Joseph Black

1754 Dorothea Erxleben is the first woman to graduate in medicine, completing her studies at the University of Halle, Germany

1755 Italian chemist Sebastian Menghini studies the effects of camphor on animals

1756 Mineralogist Johann Lehmann studies rock strata in Germany's Harz Mountains and the Erzgebirge range, stimulating the use of stratigraphy in geology

1758 The reappearance of Halley's Comet confirms the accuracy of Edmond Halley's prediction, made 53 years earlier

1759 John Harrison completes his H4 marine timekeeper, which incorporates most of the features necessary to calculate longitude at sea

1759

CLASSIFYING LIFE

Living organisms are classified according to their characteristics. This gives us five kingdoms: plants, animals, fungi, protists (including the amoeba), and prokaryotes (including bacteria). The five-kingdom system is widely used, but there are many organisms that do not fit perfectly into any kingdom. More recently, and based on modern molecular evidence, a system of three domains or superkingdoms has been suggested. These domains are bacteria, archaea (similar but distinct from bacteria), and eukarya (which includes plants, animals, and fungi).

Taxonomic ranks

Every organism is ranked according to species, genus, family, order, class, phylum, kingdom, and domain. Human beings belong to the genus *Homo* and the species *Homo sapiens*.

DOMAIN Tigers – and all animals, plants, fungi, and protists – belong to the domain eukarya. All members have a membrane-bound cell nucleus.

KINGDOM The animal kingdom contains multicellular organisms that obtain their energy by eating food. Most animals have nerves and muscles.

PHYLUM Tigers and other members of the phylum Chordata have a strengthening rod, or spine, that runs the length of their bodies.

CLASS Tigers are mammals, meaning they are warm-blooded, have hair, and suckle their young. Most mammals give birth to live young.

ORDER Members of the order Carnivora have teeth that are specialized for biting and shearing. Tigers and other carnivores live primarily on meat.

FAMILY Tigers belong to the family Felidae, which includes carnivores with short skulls and well-developed claws. Most members also have retractable claws.

GENUS *Panthera* was originally a genus of spotted big cats. Today, the unifying characteristic is a flattish, evenly convex skull.

SPECIES The tiger (*Panthera tigris*) is the only striped species belonging to the genus *Panthera*.

▷ Johann Heinrich Lambert

1760
PHOTOMETRIA PUBLISHED

Switzerland's Johann Heinrich Lambert published *Photometria*, establishing a system for measuring the perceived brightness of light. He introduced the term "albedo" to describe the fraction of light that a surface reflects; the property of a surface reflecting equally in all directions is still known as Lambertian reflectance.

1761
VENUSIAN ATMOSPHERE

Encouraged by the French astronomer Joseph-Nicolas Delisle, scientists from many countries travelled to different parts of the globe to observe the transit of Venus across the Sun. In the course of the study, Russia's Mikhail Vasilievich Lomonosov, working near St Petersburg, established that the planet has an atmosphere.

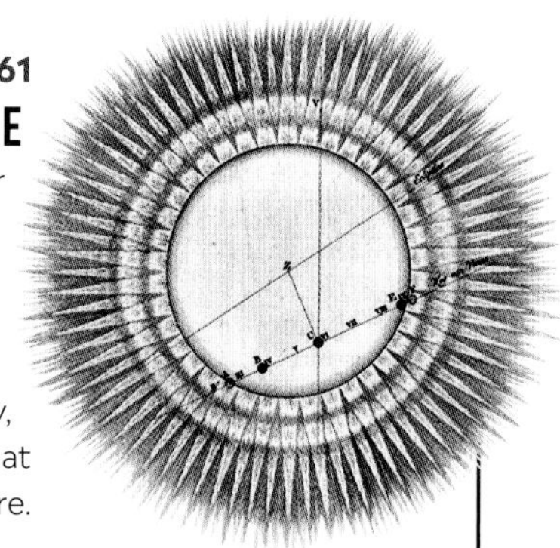

△ **Transit of Venus** drawing

1760

1761 Italian anatomist Giovanni Morgagni publishes *On the Seats and Causes of Disease*, a milestone in pathological anatomy that challenges the notion of disease as an imbalance of humours

1761 Joseph Black discovers latent heat, the energy released or absorbed by a substance as it changes state

1760
STUDIES OF EARTHQUAKES

Working at Cambridge University, John Michell published his *Conjectures Concerning the Cause and Observations upon the Phaenomena of Earthquakes*, inspired in part by the catastrophic convulsion that had devastated Portugal's capital Lisbon in 1755. Basing his observations on his knowledge of geological strata in England, he was able to identify the epicentre of the Lisbon earthquake, as well as spelling out more generally a pioneering view of the role played by faults in Earth's crust in seismological activity.

▷ **Lisbon** earthquake

1711–65

MIKHAIL VASILIEVICH LOMONOSOV

Born in the far north of Russia, Lomonosov was a Renaissance man whose interests stretched beyond astronomy and geology to include physics, chemistry (where he helped discredit the phlogiston theory), history, and Russian grammar.

1764

TEXTILE INDUSTRY

English weaver and carpenter James Hargreaves invented the spinning jenny, a multi-spindle spinning frame that enables a single worker to operate eight or more spools at a time. The device sped up the process of production, marking a key step in the industrialization of cloth manufacturing.

▷ **Hargreaves'** spinning jenny

1762 English industrialist John Roebuck patents a method of making cast iron malleable "by the action of a hollow pit-coal fire"

1762 Astronomer James Bradley dies; he has catalogued the measured locations of 60,000 stars

1762 Lomonosov publishes his pioneering work on geology, *On the Strata of the Earth*

1764

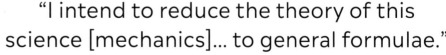

1764 Robert Whytt, publishes *On Nervous, Hypochondriac, or Hysteric Diseases*

▷ Joseph-Louis Lagrange

1762

CALCULUS OF VARIATIONS

Born in Italy but later naturalized French, the mathematician Joseph-Louis Lagrange published the results of his work on the calculus of variations – a term coined six years earlier by Swiss mathematician Leonhard Euler. Building on each other's work, the two jointly developed a new approach to mechanics, giving their name to the Euler-Lagrange equations.

> "I intend to reduce the theory of this science [mechanics]... to general formulae."
>
> *JOSEPH-LOUIS LAGRANGE,* MÉCANIQUE ANALYTIQUE, *1788*

1736–1819

JAMES WATT

Trained as an instrument maker, James Watt was employed by the University of Glasgow in his native Scotland. While seeking to repair a model Newcomen device (*see p.85*), he had the idea for the improved steam engine that was to bring him worldwide fame.

1766

IDENTIFICATION OF HYDROGEN

Aged 35, the English scientist Henry Cavendish published his first paper on what he termed "factitious airs". One gas that he identified as "inflammable air", experimentally produced by the action of acids on metal filings, was in fact hydrogen, which he correctly surmised to be proportioned two to one with oxygen in water.

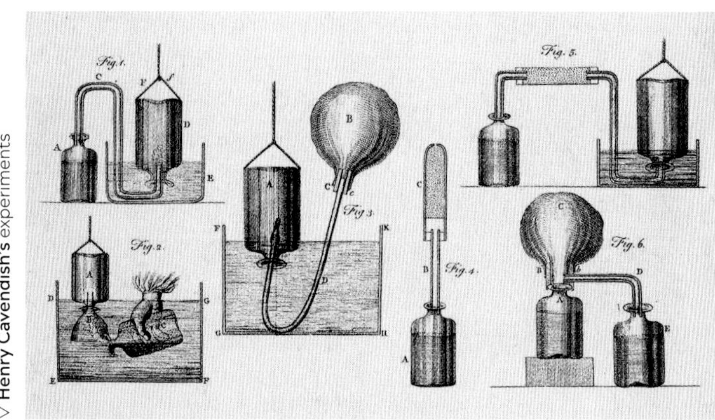

▽ **Henry Cavendish's** experiments

1765

1765 Italian biologist Lazzaro Spallanzani rebuts the theory of spontaneous generation in his first published scientific work

1765 Leonhard Euler publishes *Theoria Motus Corporum Solidorum seu Rigidorum (Theory of the Motion of Solid and Rigid Bodies)*

1765

WATT'S ENGINE

Aware of flaws in the steam engine designed by Thomas Newcomen 50 years earlier, Scottish engineer James Watt conceived a model in which steam condensed in a secondary cylinder apart from the piston, thereby avoiding heat loss and conserving energy. Introduced commercially 11 years later, Watt's machine and its successors became the main power source for British factories.

◁ **James Watt's** steam engine (replica)

△ **Cugnot's** steam-powered vehicle

1769
THE FIRST AUTOMOBILE

The earliest automobile was a three-wheeled steam-powered vehicle designed by French military engineer Joseph Cugnot to transport cannons. A larger version of the vehicle, also three-wheeled and with the front wheel supporting the steam boiler, was trialled in the following year. Travelling at less than 4 kilometres (2½ miles) per hour, the vehicle proved unstable and was quickly abandoned.

1766 Swiss inventor Horace-Bénédict de Saussure invents an early electrometer – an instrument for measuring electrical charge

1767 The first edition of *The Nautical Almanac* is published under the supervision of British astrononmer Nevil Maskelyne

1768 French chemist Antoine Baumé invents the graduated hydrometer, introducing the Baumé scale

1769

1767
PROBING ELECTRICITY

In *The History and Present State of Electricity*, English scientist Joseph Priestley combined a review of existing knowledge on electricity with thoughts for future research. His own experiments showed carbon could conduct an electrical charge, disproving the idea that only water and metals were conductors.

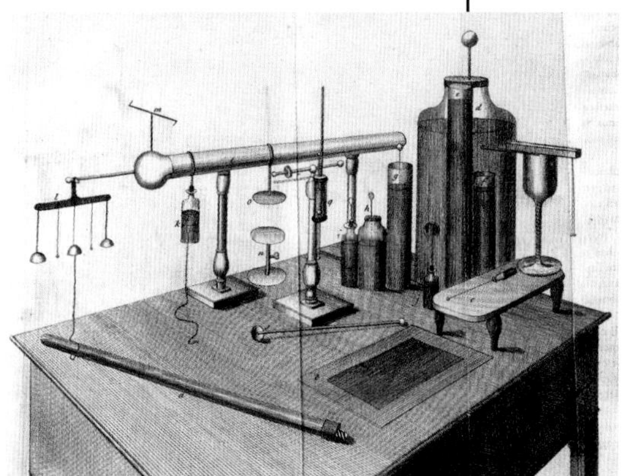
△ **Illustration** from Priestley's book

△ ***Banksia ericifolia***

1769
HEADING SOUTH

On an expedition led by Captain James Cook to the South Pacific that arrived in Tahiti in 1769, English botanist Joseph Banks identified several hundred new plant species.

The Nautical Almanac tabulated lunar distances to help mariners determine longitude by observing the Moon

1771
GLOBAL VOYAGE

French admiral Louis-Antoine de Bougainville published his *Voyage Around the World*, describing his circumnavigation of the globe in 1766–69. Naturalists on board his vessels collected many samples, including plants of a genus later named after the admiral.

▷ *Bougainvillea* flowers

1772
LAGRANGIAN POINTS

Discovered by Leonhard Euler and Italian Joseph Louis Lagrange, Lagrangian points are the five points (L1–L5) where the gravitational forces between bodies orbiting each other balance. They exist throughout the Solar System where a massive body orbits a larger one.

Lagrangian points in the Earth-Moon-Sun system
Smaller objects sent to Langrangian points tend to keep their position relative to the two larger bodies, making them useful locations for satellites.

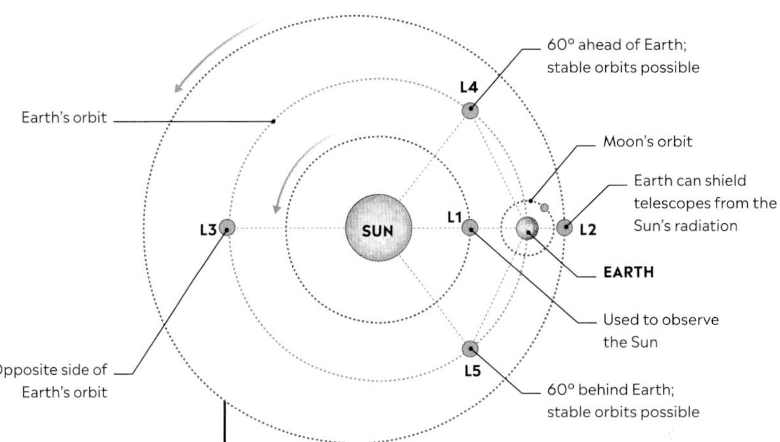

1770

1770 Swiss mathematician Leonhard Euler publishes his *Elements of Algebra*, his well-written and accessible mathematics textbook

1771 Scottish surgeon John Hunter's *Natural History of the Human Teeth* lays the foundations for a scientific approach to dentistry

1772 Carl Wilhelm Scheele, a Swedish-German chemist, discovers oxygen, which he terms "fire air"; he publishes his results five years later

1774 German geologist Abraham Werner publishes the first modern mineralogy textbook, *On the External Characteristics of Fossils*

1730–1817
CHARLES MESSIER

Fascinated by star-gazing from youth, Messier was employed from the age of 21 by France's leading astronomer, Joseph-Nicolas Delisle, to record his observations of the heavens. His catalogue is used to this day to identify celestial phenomena.

1774
MESSIER CATALOGUE

French astronomer Charles Messier published the first part of a catalogue of the celestial phenomena he and his assistant had observed, designating the 45 entries M1 to M45. Over the course of the next decade, the list grew to 103 objects, which were subsequently identified as galaxies, nebulae, and star clusters.

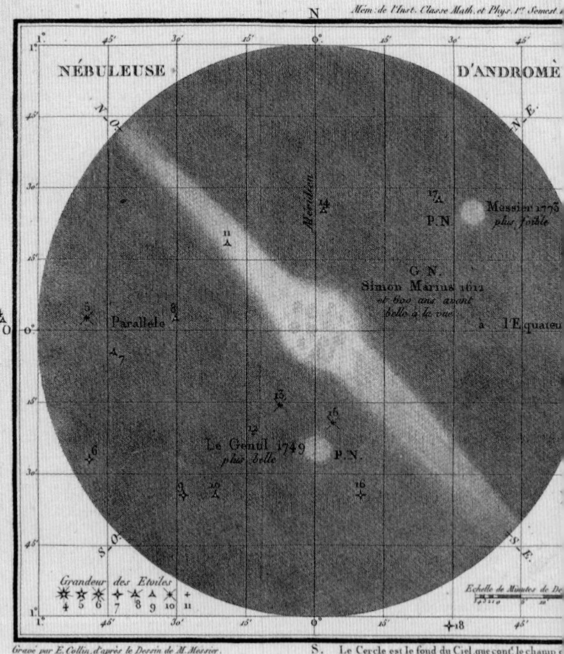

▷ **Sketch by Charles Messier** of the M31 Andromeda nebula

THE CARBON CYCLE

Life depends on carbon for food, and humans also depend on carbon for clothing, homes, transport, and energy. Carbon is cycled through living organisms into nonliving matter and is stored in rocks and the atmosphere. Through the process of photosynthesis, plants extract carbon from the atmosphere. Plants are then eaten by animals, which, through breathing, release some carbon into the atmosphere, and some is buried in soil and rock when they die. Human burning of rock carbon in the form of fossil fuels returns carbon to the atmosphere.

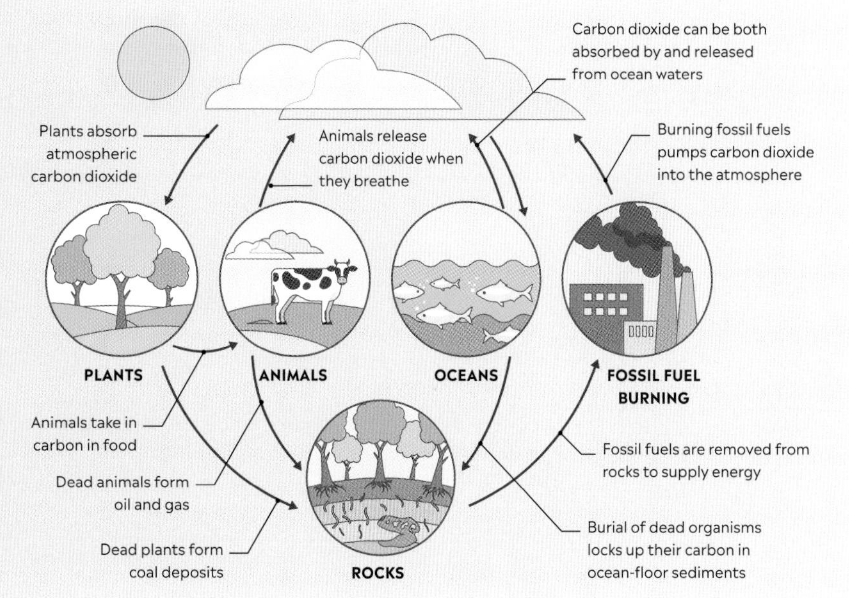
Carbon dioxide can be both absorbed by and released from ocean waters
Plants absorb atmospheric carbon dioxide
Animals release carbon dioxide when they breathe
Burning fossil fuels pumps carbon dioxide into the atmosphere
PLANTS
ANIMALS
OCEANS
FOSSIL FUEL BURNING
Animals take in carbon in food
Dead animals form oil and gas
Dead plants form coal deposits
ROCKS
Fossil fuels are removed from rocks to supply energy
Burial of dead organisms locks up their carbon in ocean-floor sediments

1774 French chemist Antoine Lavoisier notes that animal respiration and combustion both produce carbon dioxide; he links the phenomena to the carbon cycle

1774 In France, Nicolas Desmarest publishes the results of his decade-long study of the geology of the Auvergne region, recognizing its volcanic origins

1774

▽ **Franz Mesmer** at work

1774

MESMERISM

Franz Mesmer, an Austrian doctor practising in Vienna, began to develop his theory of "animal magnetism", which he viewed as a natural force inside all living things. He experimented briefly with the use of actual magnets to treat patients suffering from hysteria, but soon developed alternative techniques, including procedures later identified as hypnosis.

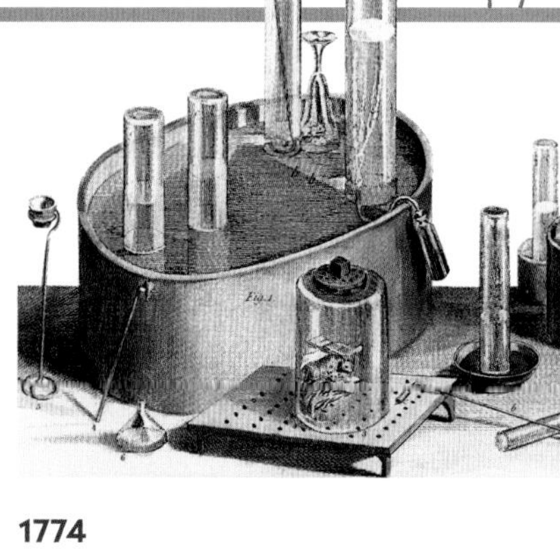
△ **Joseph Priestly's** apparatus

1774

OXYGEN ISOLATED BY PRIESTLEY

English scientist Joseph Priestley isolated oxygen, which he called "dephlogisticated air". He was the first to publish news of the discovery, though Carl Wilhelm Scheele had independently prepared oxygen at least three years earlier. In the first part of his *Experiments and Observations on Different Kinds of Air*, later expanded to six volumes, Priestley also announced the discovery of other water-soluble gases, including ammonia, sulphur dioxide, and hydrogen chloride.

> "A responsive influence exists between the heavenly bodies, the earth, and animated bodies."
>
> *FRANZ MESMER,* PROPOSITIONS CONCERNING ANIMAL MAGNETISM, *1779*

"It was not very difficult . . . to perceive that the active herb could be no other than the Foxglove."

WILLIAM WITHERING, AN ACCOUNT OF THE FOXGLOVE AND SOME OF ITS MEDICAL USES, *1785*

1776

DRUG EXTRACTION AND ISOLATION

While working as a physician at Birmingham General Hospital in England, William Withering divided his time between medicine and botanical studies. Combining both interests, he noticed the significance of foxgloves in traditional herbal remedies, paving the way for the use of their active ingredient, digitalis, in the treatment of heart conditions.

◁ ***Digitalis purpurea,*** or foxglove

1776 The first of James Watt's steam engines are installed in commercial enterprises, speeding the onset of the Industrial Revolution

1775 American inventor David Bushnell builds the *Turtle*, the world's first submarine, while enrolled at Yale University

▽ ***Cossus* moth**, a genus included in Fabricius's classification

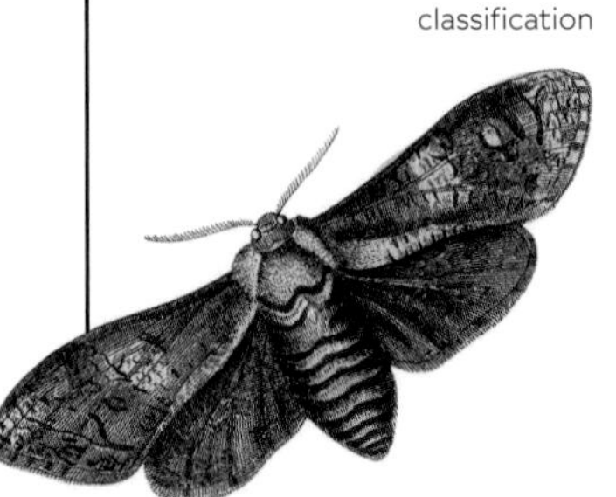

1775

ENTOMOLOGICAL SYSTEM

Danish zoologist Johan Christian Fabricius published his *Systema Entomologiae* (*Entomological System*), the first of a series of works in which he clarified and expanded the classification of insects as originally undertaken by Carl Linnaeus. He eventually identified almost 10,000 species, far more than the 3,000 known to, and listed by, Linnaeus.

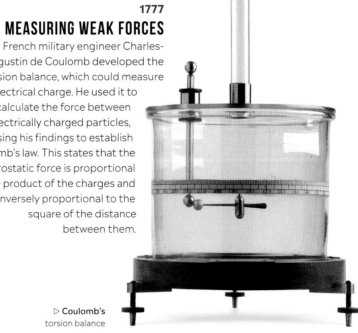

1777

MEASURING WEAK FORCES

French military engineer Charles-Augustin de Coulomb developed the torsion balance, which could measure electrical charge. He used it to calculate the force between electrically charged particles, using his findings to establish Coulomb's law. This states that the electrostatic force is proportional to the product of the charges and inversely proportional to the square of the distance between them.

▷ **Coulomb's** torsion balance

1778
OXYGEN

Partly inspired by the work of Joseph Priestley, French chemist Antoine Lavoisier published his *Easter Memoir*, detailing his research on the nature of air. Having established experimentally that Priestley's "dephlogisticated air" made up only about one-sixth of the atmosphere we breathe and that the other chemical property involved was unable to support respiration or combustion on its own, he coined the word "oxygen" to describe what he termed "nothing else than the healthiest and purest part".

◁ **Respiration experiment** by Lavoisier

1778 French aristocrat the Comte de Buffon speculates on the origins of the Solar System in *Les Époques de la Nature* (*The Epochs of Nature*)

1779 Italian priest Lazzaro Spallanzani experiments with artificial insemination, using a dog as a research tool

1779 In *Experiments and Observations on Animal Heat,* Irish chemist Adair Crawford shows that respiration brings about chemical changes affecting air's heat capacity

1779 Samuel Crompton creates the spinning mule, which industrializes textile production in England

1779

1743–1794
ANTOINE LAVOISIER

A French nobleman, Lavoisier was a central figure in the 18th-century chemical revolution, notable for naming oxygen and uncovering its role in combustion. His work was cut short by the French Revolution, when he was charged with fraud and guillotined.

1779
PRINCIPLES OF PHOTOSYNTHESIS

Working in England, the Dutch biologist Jan Ingenhousz expanded on Joseph Priestley's work to show that plants give off bubbles of oxygen when exposed to light and emit carbon dioxide when in the dark. He thereby demonstrated that light plays an essential part in the way in which green plants create and store chemical energy, uncovering the basic principle of photosynthesis.

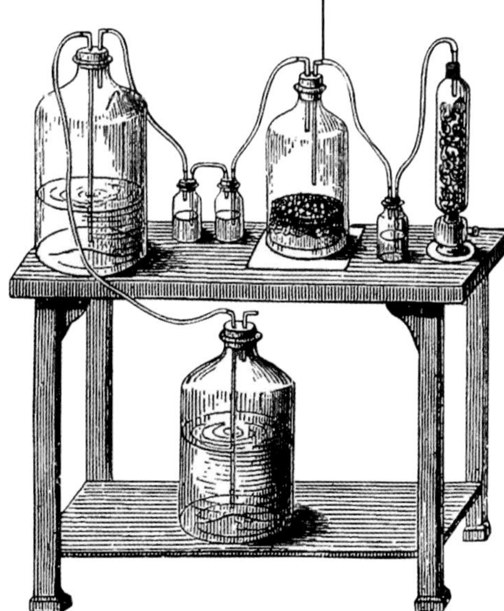

△ **Jan Ingenhousz's** experiment

1738–1822
WILLIAM HERSCHEL
Born in Hanover, Germany, Herschel emigrated to England at the age of 19. After starting his career as a musician, he turned to astronomy in 1773. The star catalogue he published in 1820 from his detailed notes listed some 5,000 objects.

1782
FIRST MANNED FLIGHT
Following experiments with parachutes and small-scale models, the French brothers Joseph-Michel and Jacques-Étienne Montgolfier constructed a balloon powered by hot air. After a successful test flight carrying a sheep, a duck, and a rooster, Étienne made a tethered ascent. Soon after, two aristocrats, Jean-François Pilâtre de Rozier and the Marquis d'Arlandes, made the first free, piloted flight; it lasted 25 minutes.

▷ **Montgolfier brothers'** balloon

1780

1780 In France, Antoine Lavoisier and Claude Berthollet begin a new era in chemical discovery by combusting (burning) organic compounds and analysing the products to detemine their chemical composition

1780 Italian polymath Felice Fontana discovers the water-gas shift reaction, enabling the production of hydrogen from carbon monoxide and water vapour

1780 Swiss scientist Ami Argand invents the oil lamp that will be named after him

▽ **Galvani's** experiment

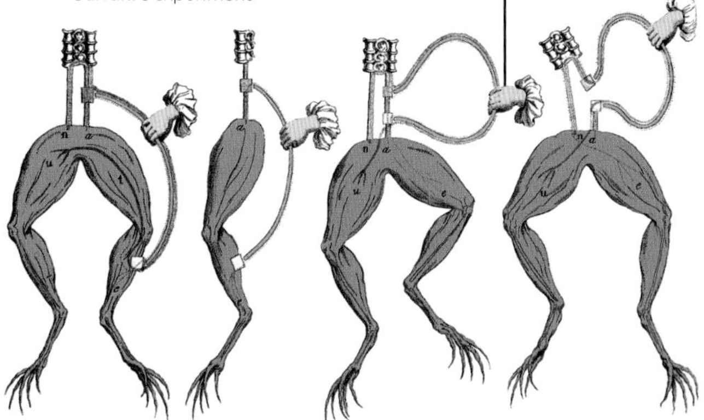

1780
NERVE IMPULSES
The Italian physicist Luigi Galvani and his wife Lucia Galeazzi found that the muscles of dead frogs twitched when struck by an electric spark. News of the results, attributed by Galvani to "animal electricity", reached the novelist Mary Shelley, who was influenced by them in writing her novel *Frankenstein*.

△ **Herschel's telescope** (replica)

1781
DISCOVERY OF URANUS
Using a telescope of his own making, the astronomer William Herschel discovered an unfamiliar object in the Gemini constellation. Further observation revealed this to be a previously unrecorded planet, the first to be discovered since antiquity. Eventually it would be given the name Uranus, after the ancient Greek sky god.

"Light is influenced by gravity in the same way as massive objects."

JOHN MICHELL IN A LETTER TO HENRY CAVENDISH, 1783

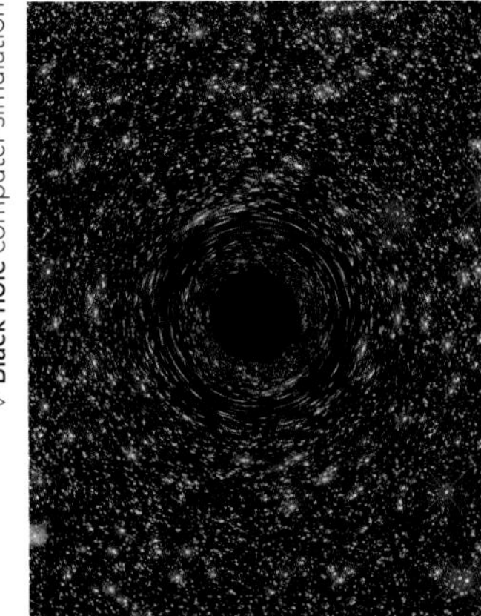

▷ **Black hole** computer simulation

1783

DARK STARS

John Michell, an English clergyman with wide scientific interests, used concepts of Newtonian physics to posit the idea of black holes (which he called "dark stars"). In a paper read to London's Royal Society, he suggested that some stars might exert such a strong gravitational pull that light might not be able to escape from them (*see below*).

1784

CRYSTAL STRUCTURE

In France, René-Just Haüy, a priest and member of the French Academy of Sciences, published his *Essai d'une Théorie sur la Structure des Crystaux* (*Essay on a Theory of the Structure of Crystals*). The book was a pioneering work in the relatively new field of crystallography and demonstrated that crystals are made up of orderly arrangements of specially shaped constituent molecules.

▷ **René-Just Haüy's** wooden crystal structure model

1784

1784 French scientists Antoine Lavoisier and Pierre-Simon Laplace measure the amount of oxygen consumed and carbon dioxide and heat produced in respiration and combustion

1784 In *Experiments on Air*, English scientist Henry Cavendish describes how he produced water by exploding two parts of "inflammable air" (hydrogen) with one part of "dephlogisticated air" (oxygen)

BLACK HOLES

A black hole originates when a star far more massive than the Sun runs out of fuel. Its core then collapses with enough violence to overcome the repulsive forces between subatomic particles that give matter its structure. The result is a superdense, infinitely small point called a singularity, which can grow by pulling in more material from its surroundings. When such available material is plentiful – such as during the formation of galaxies – black holes can undergo runaway growth to become supermassive objects with the mass of many millions of Suns.

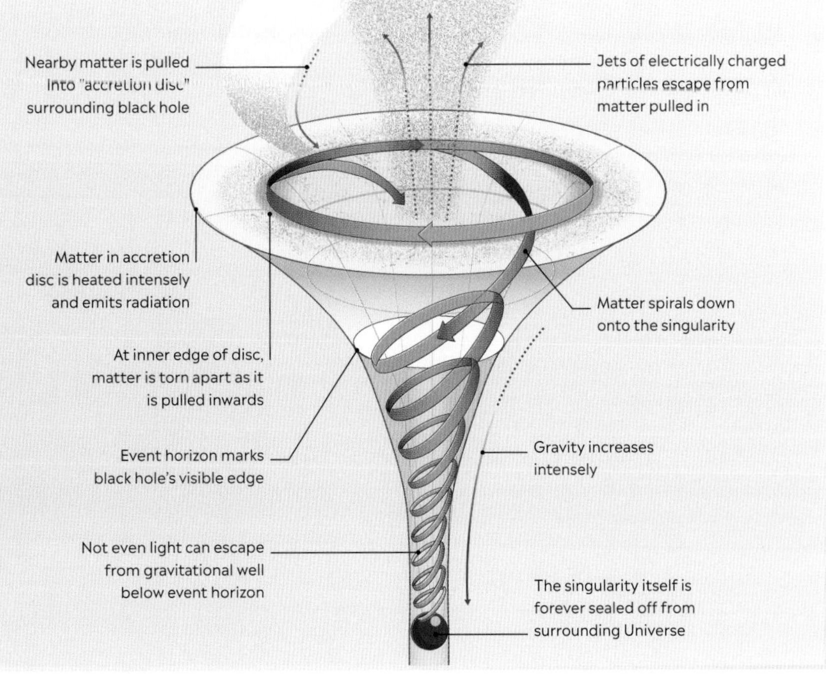

Nearby matter is pulled into "accretion disc" surrounding black hole
Jets of electrically charged particles escape from matter pulled in
Matter in accretion disc is heated intensely and emits radiation
Matter spirals down onto the singularity
At inner edge of disc, matter is torn apart as it is pulled inwards
Event horizon marks black hole's visible edge
Gravity increases intensely
Not even light can escape from gravitational well below event horizon
The singularity itself is forever sealed off from surrounding Universe

The structure of a black hole

In Einstein's general theory of relativity, a singularity warps the shape of space-time around it, creating a steep-sided "gravitational well" of distortion from which not even light can escape.

1785

ATTRACTION AND REPULSION

In his *First Memoir on Electricity and Magnetism*, Charles-Augustin de Coulomb elaborated the principles of Coulomb's law, which he had established in 1777 (*see p.100*). He published six more memoirs by 1789, further exploring the laws of attraction and repulsion between electric charges and magnetic poles.

▷ **Charles-Augustin de Coulomb**

1787

MOUNTAIN STUDIES

Driven by the desire to study geological and meteorological conditions in extreme situations, Genevan-born academic and inventor Horace de Saussure became one of the first climbers to ascend Mont Blanc, the highest mountain in western Europe. Saussure carried barometers, thermometers, and other instruments of his own devising to the tops of the mountains he climbed, taking measurements on the way.

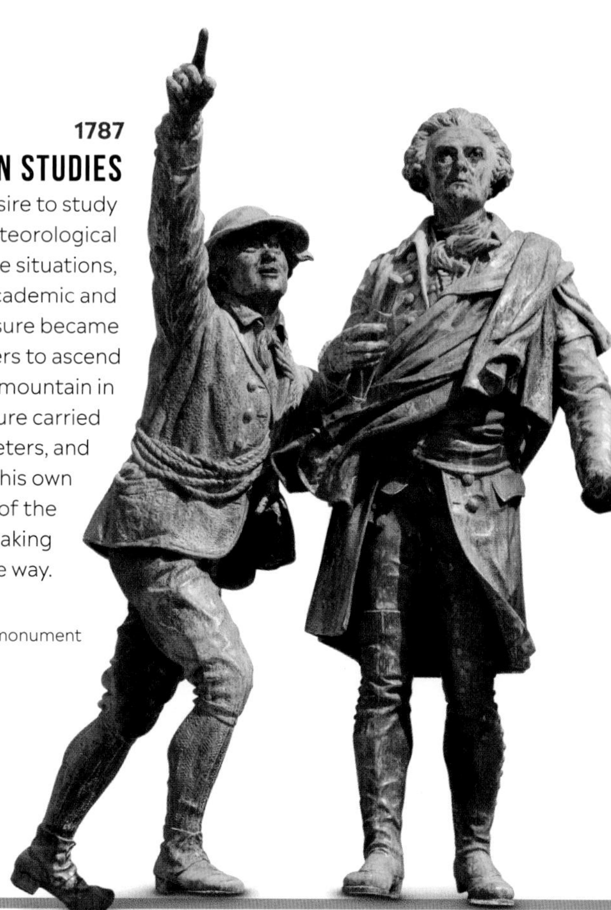

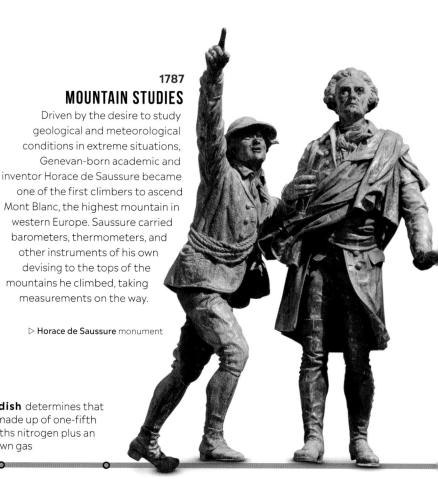

▷ **Horace de Saussure** monument

1785

1785 Henry Cavendish determines that the atmosphere is made up of one-fifth oxygen and four-fifths nitrogen plus an unknown gas

1785 French chemist Claude Louis Berthollet discovers the bleaching properties of chlorine gas

1786 William Herschel publishes the first results of his deep-sky telescope observations, listing 1,000 star clusters. He adds a further 1,500 objects to his catalogue over the next 16 years.

THE ROCK CYCLE

Earth's rock cycle sees rock materials taken from the surface, reworked into the interior of the planet, and then brought back to the surface. The processes of surface weathering and erosion break down the rocks into their mineral parts, and these, carried by wind and water, are deposited in sedimentary layers. Over time, burial lithifies the sediments into sedimentary rock, and increased heat and pressure transform them into metamorphic rocks. Greater heat also melts rocks, and the magma either cools into intrusive igneous rock or erupts as extrusive igneous rock.

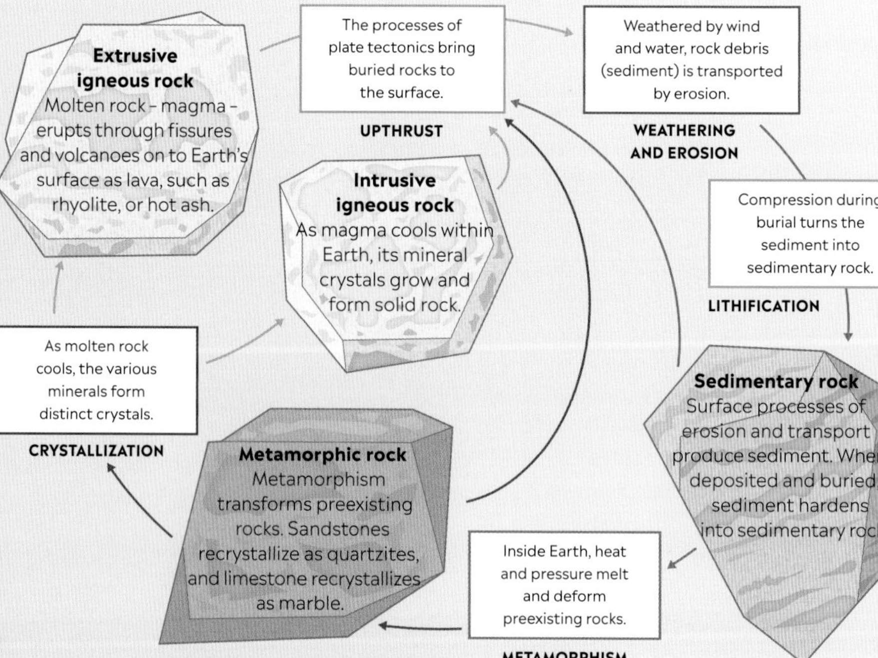
Extrusive igneous rock
Molten rock – magma – erupts through fissures and volcanoes on to Earth's surface as lava, such as rhyolite, or hot ash.
The processes of plate tectonics bring buried rocks to the surface.
UPTHRUST
Weathered by wind and water, rock debris (sediment) is transported by erosion.
WEATHERING AND EROSION
Intrusive igneous rock
As magma cools within Earth, its mineral crystals grow and form solid rock.
Compression during burial turns the sediment into sedimentary rock.
LITHIFICATION
As molten rock cools, the various minerals form distinct crystals.
CRYSTALLIZATION
Sedimentary rock
Surface processes of erosion and transport produce sediment. When deposited and buried, sediment hardens into sedimentary rock.
Metamorphic rock
Metamorphism transforms preexisting rocks. Sandstones recrystallize as quartzites, and limestone recrystallizes as marble.
Inside Earth, heat and pressure melt and deform preexisting rocks.
METAMORPHISM

1789
THEORY OF COMBUSTION

In his *Traité Élémentaire de Chimie* (*Elementary Treatise on Chemistry*), often seen as the first modern textbook on the subject, Antoine Lavoisier championed a new theory of combustion, maintaining that oxygen supports combustion and respiration. His work was a rejection of the existing phlogiston theory.

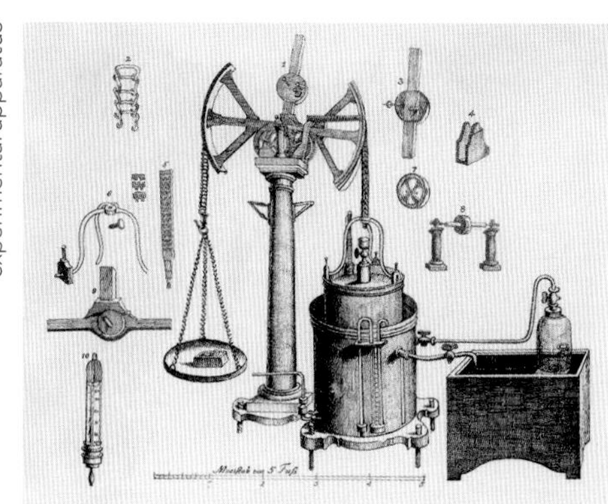

▷ **Antoine Lavoisier's** experimental apparatus

"I have looked further into space than ever human being did before me."

ATTRIBUTED TO WILLIAM HERSCHEL, c. 1813

1789
PLANT GENERA

French botanist Antoine Laurent de Jussieu published his *Genera Plantarum* (*Plant Genera*) followed by many other significant works on classification that built on the legacy of Carl Linnaeus by recognizing the importance of divisions, classes, and orders in plant taxonomy.

▷ **Plate from Jussieu's** *Dictionary of Natural Science*

1788 Joseph-Louis Lagrange publishes his *Mécanique Analytique* (*Analytical Mechanics*), a summation of 16 years' work on simplifying the formulae of classical and Newtonian mechanics

1789

1789 German chemist Martin Klaproth identifies the chemical element uranium, naming it after the recently discovered planet Uranus; he discovers zirconium in the same year

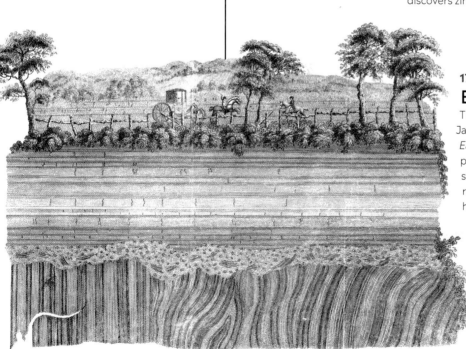

1788
EARTH'S HISTORY

The farmer, manufacturer, and geologist James Hutton published his *Theory of the Earth*, spelling out his view that Earth's past history is embodied in the current state of its rocks. From close study of rock formations in his native Scotland, he came to the conclusion that most had been formed not by a single act of creation but instead by the coming together of different materials in primeval times at the bottom of the ocean. At the time, his ideas were proclaimed atheistic, but they gradually gained acceptance, becoming the foundation of the science of geology.

△ **Hutton's unconformity** at Jedburgh

NERVE SIGNALS

Nerve signals are electrical impulses passed between neurons by chemicals called neurotransmitters. Neurons comprise long fibres (axons) with branched endings (dendrites) that contact thousands of other neurons. Signals are triggered when receptor neurons detect a stimulus (for example, touch). They pass between neurons until reaching effector neurons that produce a response (for example, muscle contraction). If a person steps on a pin, the signal travels from the foot to their brain and finally to the leg muscles, in a fraction of a second.

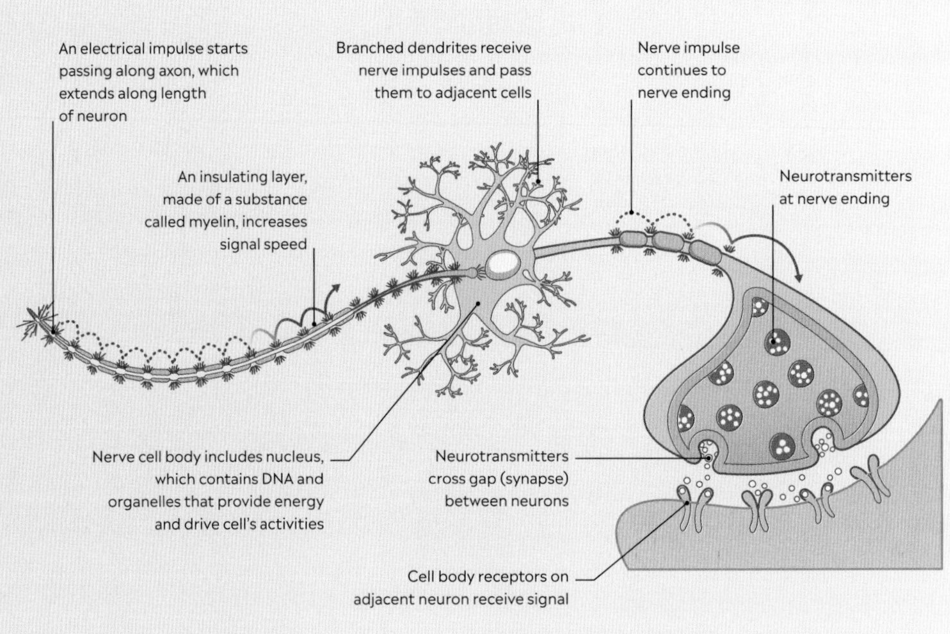
An electrical impulse starts passing along axon, which extends along length of neuron
Branched dendrites receive nerve impulses and pass them to adjacent cells
Nerve impulse continues to nerve ending
An insulating layer, made of a substance called myelin, increases signal speed
Neurotransmitters at nerve ending
Nerve cell body includes nucleus, which contains DNA and organelles that provide energy and drive cell's activities
Neurotransmitters cross gap (synapse) between neurons
Cell body receptors on adjacent neuron receive signal

1790

1791 Italian scientist Luigi Galvani determines that a form of electricity is present in living tissues and involved in nerve conduction and muscle contraction

1792 German chemist Jeremias Richter coins the term"“stoichiometry” to describe the principle of fixed chemical reactions

1790 The US Congress passes a Patent Act, the country's first

△ **Metric weights**

1791

METRIC SYSTEM

In the wake of the French Revolution, the French Academy of Sciences set up a commission designed to establish a rational classification of weights and measures. The result was the metric system, with the basic units of metres and kilograms divided or multiplied in decimal units of 10. The metre itself was calculated to be one ten-millionth of the distance from the equator to the North Pole.

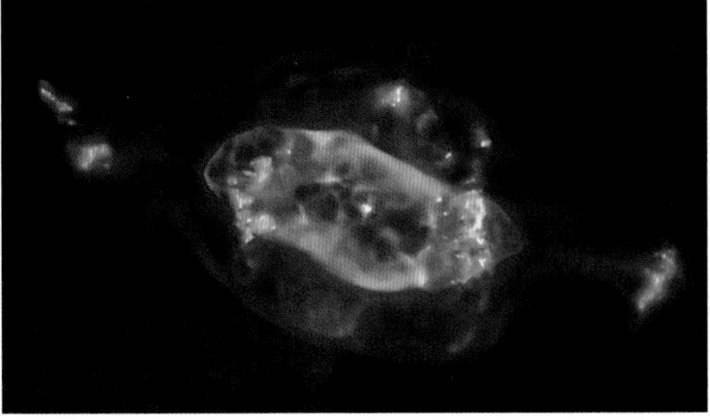

c. 1790

PLANETARY NEBULAE

William Herschel coined the term planetary nebulae to describe a group of circular nebulae he had observed. Although the name continues to be used, it is actually a misnomer, as the round shape comes from gases emitted by stars reaching the ends of their lives and has nothing to do with planets.

◁ **Saturn Nebula,** an early discovery by Herschel

1794

PROGRESS IN MENTAL ILLNESS

The French physician Philippe Pinel's *Memoir on Madness* advocated a more humane treatment of persons judged mentally ill and made the case that mental disorders can be treated and cured. Instead of the bleeding and purging treatments that had been the norm, Pinel suggested careful observation and lengthy conversation with patients to seek to understand the individual circumstances of each case.

▷ **Philippe Pinel** releasing mental patients from chains

1793

INSECT POLLINATION

German schoolmaster and naturalist Christian Sprengel published *The Secret of Nature in the Form and Fertilisation of Flowers*, a pioneering text studying the vital role played by insects in plant pollination; previously they had generally been regarded as nectar thieves.

△ **Bee pollinating** a flower

1794

1794 The pioneering chemist Antoine Lavoisier is guillotined, a victim of the French Revolution's Reign of Terror

1794 French chemist Joseph Louis Proust propounds the law of definite proportions, by which chemical compounds always combine in constant ratios

1794 English physicist John Dalton explores the condition of colour blindness, from which he himself suffers

1794

THE LAWS OF ORGANIC LIFE

The English thinker Erasmus Darwin published the first part of *Zoonomia or The Laws of Organic Life*. While primarily concerned with the human body, the book suggested in passing that all life on Earth might have sprung from what Darwin termed "the great first cause". This has been viewed as an early iteration of the idea of evolution, which Darwin's grandson Charles would later expound in detail.

◁ **Erasmus Darwin**

> "In the great length of time, since the earth began to exist ... warm-blooded animals have arisen from one living filament."

ERASMUS DARWIN, ZOONOMIA, *1794*

1749-1834
GILBERT BLANE

Scottish doctor Blane was physician to the Fleet. His naval health reforms, which included the provision of lemon juice, soap, medical necessities, and adequate ventilation, along with the introduction of hospital ships, changed the lives of servicemen.

1796
COWPOX VACCINE

Having scraped pus from cowpox blisters on the hands of a milkmaid, English physician Edward Jenner used the material experimentally to inoculate the young son of his gardener, hoping thereby to give the boy immunity against smallpox, a similar but more deadly disease. He published the results of his experiment two years later in his *Inquiry into the Causes and Effects of the Variolae Vaccinae*, coining the term for cowpox from which the word "vaccine" would subsequently be derived.

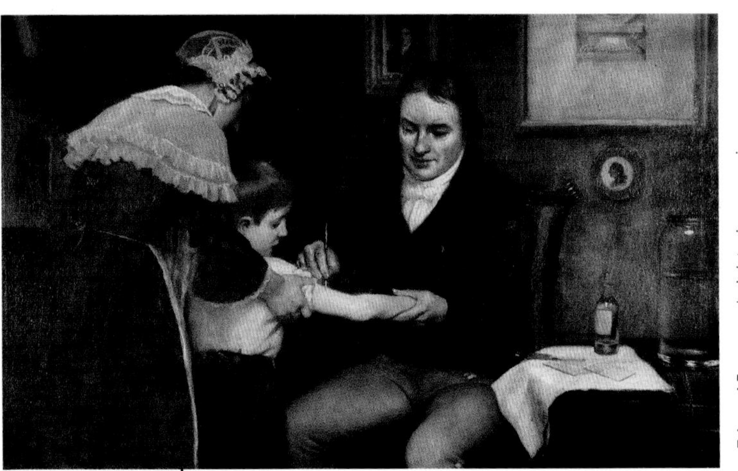

△ **Edward Jenner** administering a vaccine

1795

1795 Gilbert Blane introduces the mandatory provision of lemon juice to prevent scurvy in the British navy; the disease soon disappears from ships

1796 Aged just 19, German Carl Friedrich Gauss makes major breakthroughs in mathematics, including advances in modular arithmetic and the study of prime numbers

1797 German astronomer Heinrich Olbers publishes the results of his work on calculating the orbits of comets

1796
COMPARATIVE ANATOMY

Since described as the founding father of palaeontology, the French zoologist Georges Cuvier published the first results of his studies of animal skeletons. His work established that African and Indian elephants belonged to different species and that mammoth bones belonged to neither so must have come from an extinct line. He also identified the mastodon as a separate entity and named the *Megatherium*, an extinct giant sloth found in South America.

▷ **Indian elephant**

"It is not knowledge, but the act of learning, not possession, but the act of getting there, which grants the greatest enjoyment."

CARL FRIEDRICH GAUSS, LETTER TO THE HUNGARIAN MATHEMATICIAN FARKAS BOLYAI, 1808

1799

ALEXANDER VON HUMBOLDT

German naturalist Alexander von Humboldt set out on a series of journeys to South America during which he made pioneering advances in physical and plant geography and meteorology. Among his achievements, he uncovered how air and water move to create bands of climate and tracked the ocean current named after him, which flows north from Antarctica along the west coast of South America.

▷ **A chart of plant distribution with altitude** drawn by Alexander von Humboldt

1798

MASS OF EARTH

In what is now known as the Cavendish Experiment, Henry Cavendish used a variation of Coulomb's torsion balance to determine the density of Earth through experiment. His result was within one per cent of the figure generally accepted today, and his work paved the way for later calculations of Earth's mass and specific gravity.

△ **Gravitational experiment** model

1799

1797 French chemist Louis Nicolas Vauquelin detects the new element chromium in a red lead ore from Siberia

△ **Thomas Malthus**

1798

POPULATION GROWTH

English clergyman and economist Thomas Malthus published the first edition of his *Essay on the Principle of Population*, a pioneering work on demography. Malthus argued that the world's population was rising at a faster rate than food production and other resources, and that this would inevitably increase the chances of famine.

△ **Benjamin Thompson** at a Munich cannon factory

1798

HEAT TRANSFER

The American-born physicist Benjamin Thompson (later ennobled as Count Rumford) made significant advances in the study of heat generation in his *Experimental Enquiry Concerning the Source of the Heat which is Excited by Friction*. He studied the frictional heat generated when cannon barrels were being bored, and this led him to argue against the contemporary theory that heat was a fluid called "caloric" that flowed from hotter to colder bodies.

1800
THE BATTERY

Italian scientist Alessandro Volta realized that it was the dissimilar metals (iron and brass) used in Galvani's experiments with frogs (*see p.102*) that produced electricity when touched together, not animal electricity. He subsequently built a "pile" of alternating copper and iron discs, separated by brine-soaked cardboard, creating the first electric battery.

◁ **Voltaic** pile

1800
INVISIBLE LIGHT

In May 1800, William Herschel found that thermometers placed just beyond the red end of the light spectrum showed an increase in temperature. He had discovered an invisible form of "light", which he called Calorific rays. Today, this light is known as infrared radiation.

▽ **Herschel's** experimental apparatus

1800

1800 French physician Marie François Xavier Bichat classifies 21 types of body tissue, each with a distinct function

1800 British chemist William Nicholson electrolyses water using the newly invented battery

BATTERIES AND ELECTROCHEMISTRY

Batteries are portable sources of energy that store energy in chemical form to be converted into electrical energy. A battery has a positive electrode (cathode) at one terminal and a negative electrode (anode) at the other, separated by a conductive electrolyte. When a circuit connects the terminals, free electrons flow through it from anode to cathode: this is an electric current. There are many different types of battery, the most common being the alkaline battery, which contains alkaline electrolytes.

How an alkaline battery discharges
In a battery, chemical reactions free electrons from metal atoms. These electrons flow between the electrodes during discharging, producing an electric current that can power a device.

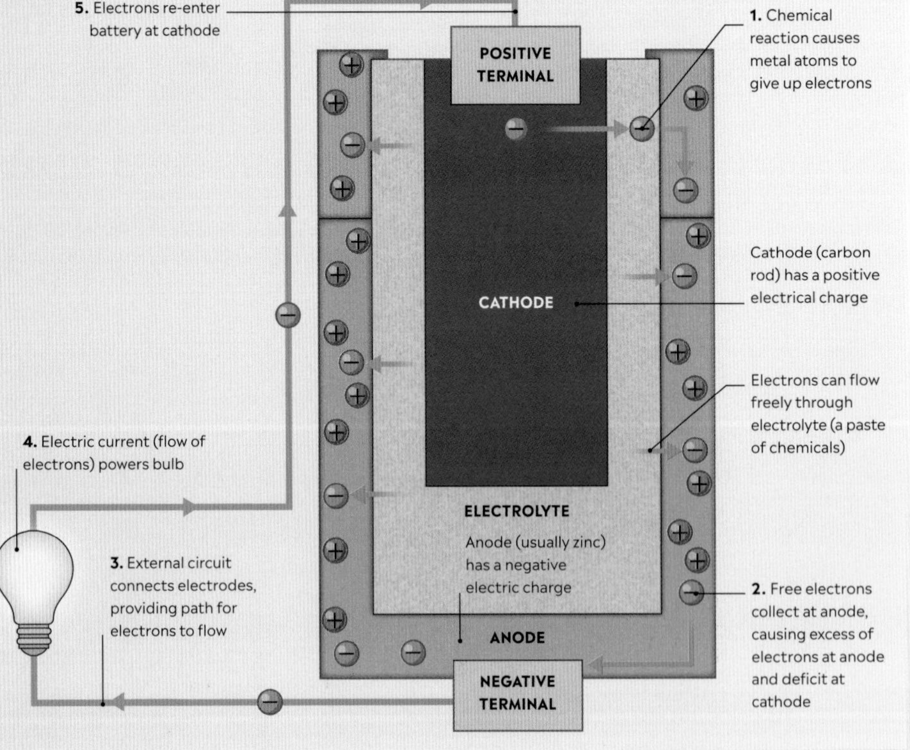
5. Electrons re-enter battery at cathode
POSITIVE TERMINAL
1. Chemical reaction causes metal atoms to give up electrons
Cathode (carbon rod) has a positive electrical charge
CATHODE
Electrons can flow freely through electrolyte (a paste of chemicals)
4. Electric current (flow of electrons) powers bulb
ELECTROLYTE
Anode (usually zinc) has a negative electric charge
3. External circuit connects electrodes, providing path for electrons to flow
ANODE
NEGATIVE TERMINAL
2. Free electrons collect at anode, causing excess of electrons at anode and deficit at cathode

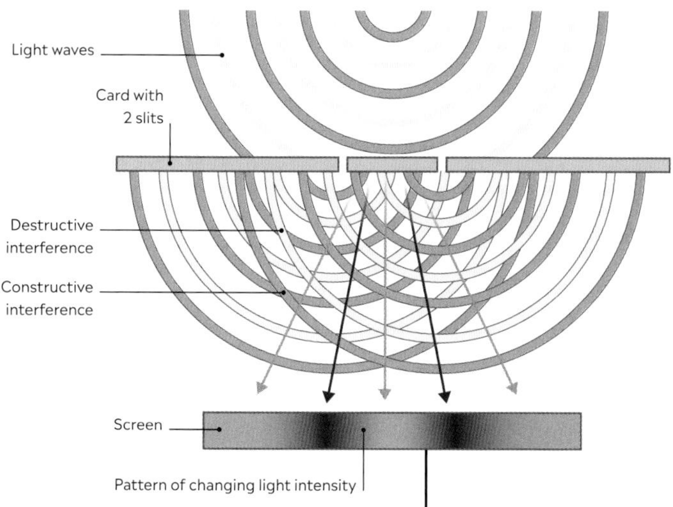

1801

LIGHT WAVES AND INTERFERENCE

Since the publication of Isaac Newton's *Opticks* in 1704, most scientists thought of light as a stream of particles. By duplicating phenomena associated with light using ripples in a tank of water, British scientist Thomas Young provided evidence that light behaves as a wave.

Diffraction and interference of waves
Waves spread out (diffract) when they move past an edge or through a gap. When two waves combine, they "interfere": two peaks add up to form a larger wave, while a peak and a trough cancel one another out. Young showed that interference in light waves creates a series of bright and dark fringes.

1801 German scientist Johann Wilhelm Ritter discovers invisible radiation, now called ultraviolet light, beyond the blue end of the spectrum

1802 Swedish chemist Anders Gustaf Ekeberg discovers the acid-resistant metal tantalum; isolating the pure element proves difficult

1802

1802 French chemist Joseph-Louis Gay-Lussac works out that, at constant pressure, a volume of gas increases in proportion to its temperature

1801

CERES AND THE ASTEROIDS

Italian astronomer Giuseppe Piazzi discovered Ceres (now classed a dwarf planet), the first and largest member of the asteroid belt. Three more asteroids were discovered before 1808, and more than a million asteroids are now known – most just a few kilometres across.

△ Ceres

1745–1827

ALESSANDRO VOLTA

While working as a professor of physics in his city of birth (Como, Italy), Italian physicist and chemist Alessandro Volta discovered methane gas and carried out many experiments with static electricity. He spent the rest of his career at the University of Pavia, Italy.

▽ **Shells from** Lamarck's *Histoire Naturelle*

1801

INVERTEBRATE CLASSIFICATION

French biologist Jean-Baptiste Lamarck studied medicine and botany in Paris before being appointed Professor of the natural history of insects and worms at the Musée National d'Histoire Naturelle. Tackling this new subject, he coined the word "invertebrate" and developed a system for classifying this huge array of animals that had hitherto been little studied. He published his work in the *Histoire Naturelle des Animaux sans Vertèbres* (*Natural History of Animals without Vertebrae*).

1803

△ **Gay-Lussac and Biot** prepare to ascend

1804
HIGH-ALTITUDE STUDIES

French scientists Joseph-Louis Gay-Lussac and Jean-Baptiste Biot became the first to carry out scientific research from a balloon. They ascended to 4,000 m (13,000 ft), studying the Earth's magnetic field at high altitudes and collecting samples to determine the composition of the atmosphere. In a solo flight later that year, Gay-Lussac reached more than 7,000 m (23,000 ft), a record that was unbroken for more than 50 years.

c. 1805 German pharmacist Friedrich Setürner isolates morphine (an active alkaloid) from opium, laying the foundations of alkaloid chemistry

1803 English meteorologist Luke Howard develops a classification of clouds, naming cumulus, stratus, and cirrus for the first time

1804 Swiss chemist Nicolas-Théodore de Saussure is the first to establish the basic principles of photosynthesis in the growth of green plants

1803
METEORITES

On 26 April, a spectacular shower of stones fell from a clear blue sky onto the town of L'Aigle, Normandy. The eminent French physicist Jean-Baptiste Biot was sent to investigate the 37 kg (82 lb) of fallen stones. His detailed report showed that the stones were quite different from those of the local rock formations, and he correctly concluded that meteorites come from outer space rather than from a volcanic eruption.

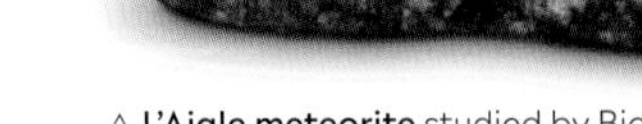

△ **L'Aigle meteorite** studied by Biot

▽ **Dalton's** table of atomic weights

1803
DALTON'S ATOMIC THEORY

English scientist John Dalton proposed that the atoms of different elements varied in size and mass. His list of the relative weights of atoms and his atomic theory convinced many scientists that atoms were real.

ATOMIC THEORY

John Dalton's atomic theory attempted to describe all matter in terms of atoms and their properties. The theory proposes that all matter is made from tiny atoms ("solid, massy, hard, impenetrable, movable particle[s]") that cannot be divided, created, or destroyed. Each element consists of atoms with different properties, such as mass and size, while atoms of the same element have identical properties. According to this theory, chemical compounds are formed by combining atoms of different elements, and chemical reactions are changes in the combinations of atoms.

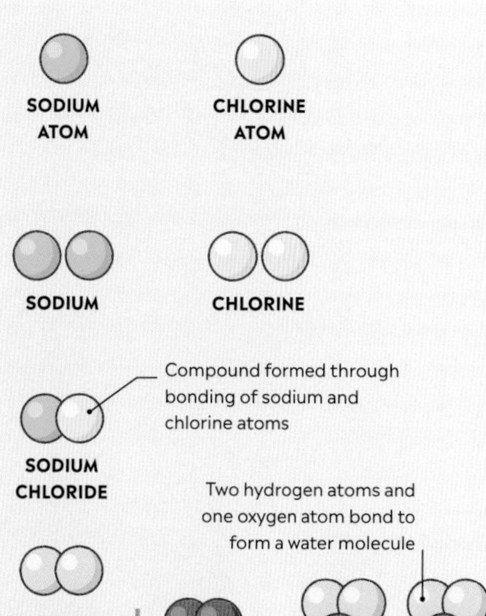

SODIUM ATOM
CHLORINE ATOM
SODIUM
CHLORINE
Compound formed through bonding of sodium and chlorine atoms
SODIUM CHLORIDE
Two hydrogen atoms and one oxygen atom bond to form a water molecule
HYDROGEN
OXYGEN
WATER

Postulate 1
All matter is made from extremely small particles called atoms, which cannot be divided further. This was later proved false.

Postulate 2
All atoms of a given element have identical properties. Atoms of one element differ in properties from atoms of another element.

Postulate 3
Atoms of different elements combine to form chemical compounds. Atoms cannot be divided so always combine in whole-number ratios.

Postulate 4
A chemical reaction is a rearranging of atoms to form a new compound. The atoms themselves are not changed, created, or destroyed.

1807 Swedish chemist Jöns Jacob Berzelius notes that organic (carbon-containing) and inorganic compounds represent distinct branches of chemistry

1808 John Dalton publishes *A New System of Chemical Philosophy*, proposing that atoms of the same element are identical in size and mass

1808

1807

METALS ISOLATED

Using a battery to test the idea that electricity could decompose compounds into their chemical components (a process called electrolysis), English chemist Sir Humphry Davy successfully isolated potassium from molten potash, and sodium from soda. He subsequently isolated barium, strontium, calcium, and magnesium.

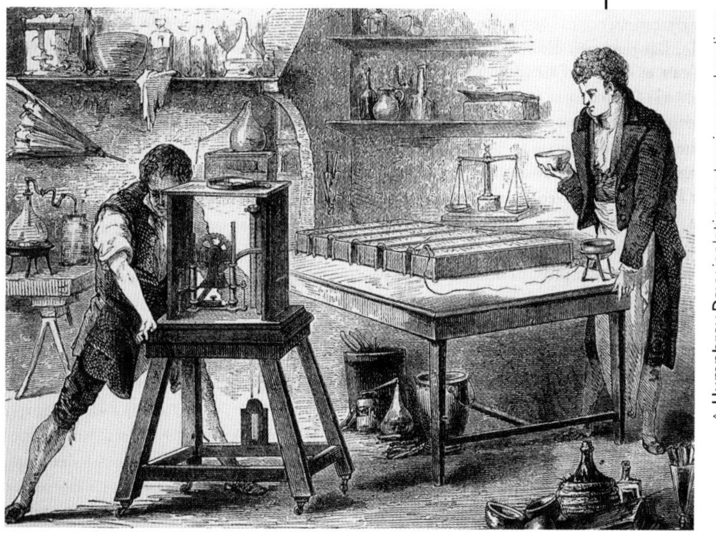

△ **Humphrey Davy** isolating potassium and sodium

Malus' Law experiment
In polarized light, waves are restricted to one plane only. By passing unpolarized light through two polarizing filters at right angles to each other, the light is blocked.

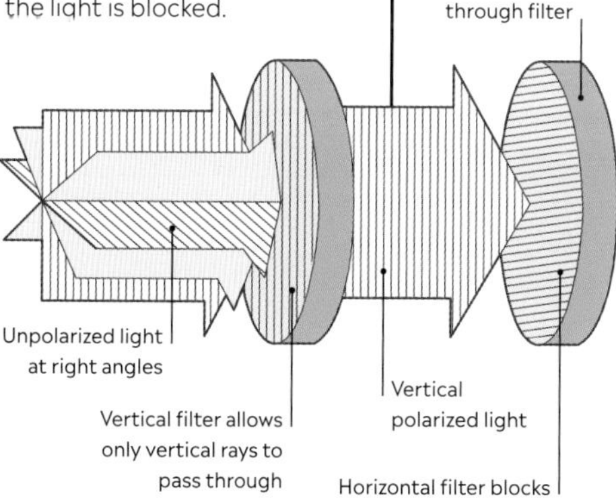

Nothing passes through filter
Unpolarized light at right angles
Vertical filter allows only vertical rays to pass through
Vertical polarized light
Horizontal filter blocks vertical rays

1808

POLARIZED LIGHT

While experimenting with calcite crystals, which act as polarizing filters, French mathematician Etienne Louis Malus established Malus' Law, which relates the amount of light passing through two such filters to the angle between them.

1744–1829
JEAN-BAPTISTE LAMARCK

Born in France, Lamarck studied medicine before turning to natural history, publishing important studies on the flora of France and invertebrate animals. He is best known for his theory of the inheritance of acquired characteristics.

◁ **Giraffe** grazing on a tree

1809
LAMARCKISM

In *Philosophie Zoologique* (*Zoological Philosophy*), Jean-Baptiste Lamarck contributed to the study of evolution by outlining his idea that characteristics acquired in an animal's life could be directly inherited. For example, the offspring of a giraffe that stretched its neck to feed from tall trees would be better adapted to their habitat through inheriting longer necks, though quite how this would work was not clear.

1809

1811 French chemist Bernard Courtois isolates iodine after treating seaweed ash with sulphuric acid

1811 William Herschel proposes that stars originate from the collapse of glowing clouds of gas

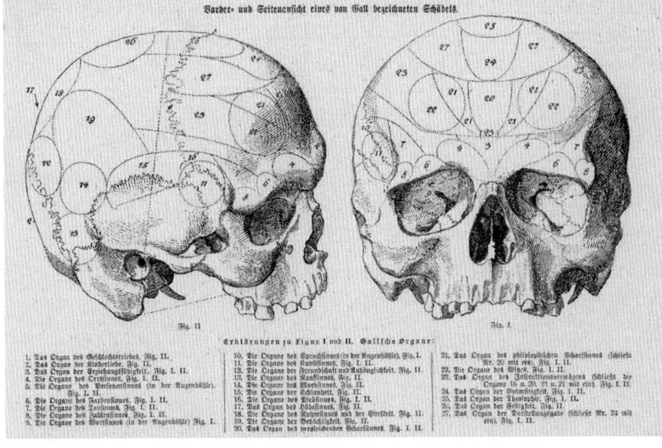

▽ **Phrenology** study by Gall

1810
THE BRAIN AND PHRENOLOGY

German-born physician Franz Joseph Gall published his major work on phrenology, the term he coined to describe his theory that human attributes and abilities could be reflected in physical aspects of the head. The Church regarded Gall's ideas as heretical, but they became popular across Europe and America, partly through lectures he gave at universities including Harvard and Yale.

△ **Amedeo** Avogadro

1811
AVOGADRO'S LAW

In an essay published in this year, Italian scientist Amedeo Avagadro hypothesized that equal volumes of any two gases at the same temperature and pressure contain the same number of particles (atoms or molecules). Despite being largely ignored until the 1860s, Avogadro's Law is now well established as a cornerstone of modern atomic and molecular theory.

1813
CHEMICAL ALPHABET

To aid clarity and brevity in recording chemical information, the Swedish chemist Jöns Jacob Berzelius proposed a system of using one or two letters of each element's Latin name as its chemical symbol. He also suggested adding the relative number of atoms for each element in a molecule. So, for example, ammonia, a compound of nitrogen and hydrogen, is known by the chemical formula NH_3.

▷ **Berzelius'** laboratory

> "A tidy laboratory means a lazy chemist."
>
> *JÖNS JACOB BERZELIUS, LETTER TO NILS SEFSTRÖM, 1812*

1813

1812 Starch is broken down to sugar (glucose) molecules by the Russian chemist Gottlieb Kirchhoff, who uses sulphuric acid as a catalyst

1812 German astronomer Heinrich Olbers suggests comet tails are formed by material from a solid nucleus driven away from the sun

1813 German physician Johann Frank continues writing his nine-volume work on public health

1812
THE PTERODACTYL

The Italian scientist Cosimo Collini had identified the first pterodactyl in 1784, but believed it to be a sea creature. In 1812, debate raged between German anatomist Samuel Thomas von Sömmerring, who argued that it was intermediate between bats and birds, and Georges Cuvier, who correctly described the pterodactyl as a flying reptile.

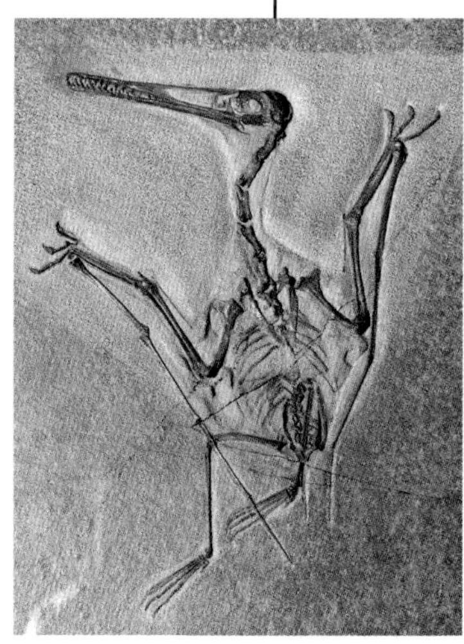

▷ **Pterodactyl** fossil

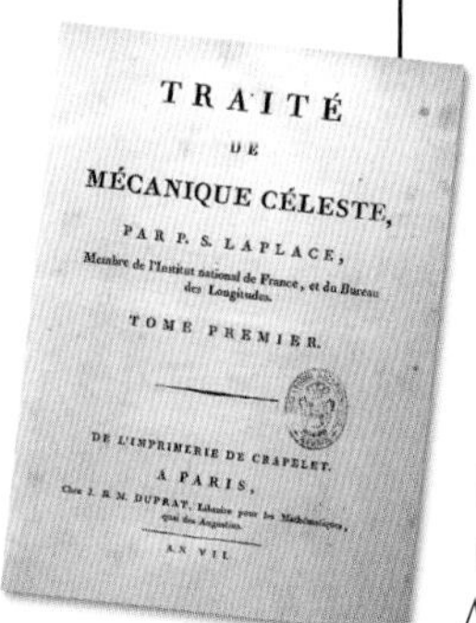

TRAITÉ
DE
MÉCANIQUE CÉLESTE,
PAR P. S. LAPLACE,
TOME PREMIER.
DE L'IMPRIMERIE DE CRAPELET.
A PARIS,

◁ ***Traité de Méchanique Céleste*** front cover

1812
THE MECHANISTIC UNIVERSE

In his *Treatise of Celestial Mechanics*, Pierre-Simon Laplace explored probability, considering the likelihood of any event occurring in terms of the events that preceded it. Expanding on his ideas, he established the doctrine of determinism – the idea that everything that happens is determined by the interaction of particles of matter. His work challenged both religion and concepts of free will.

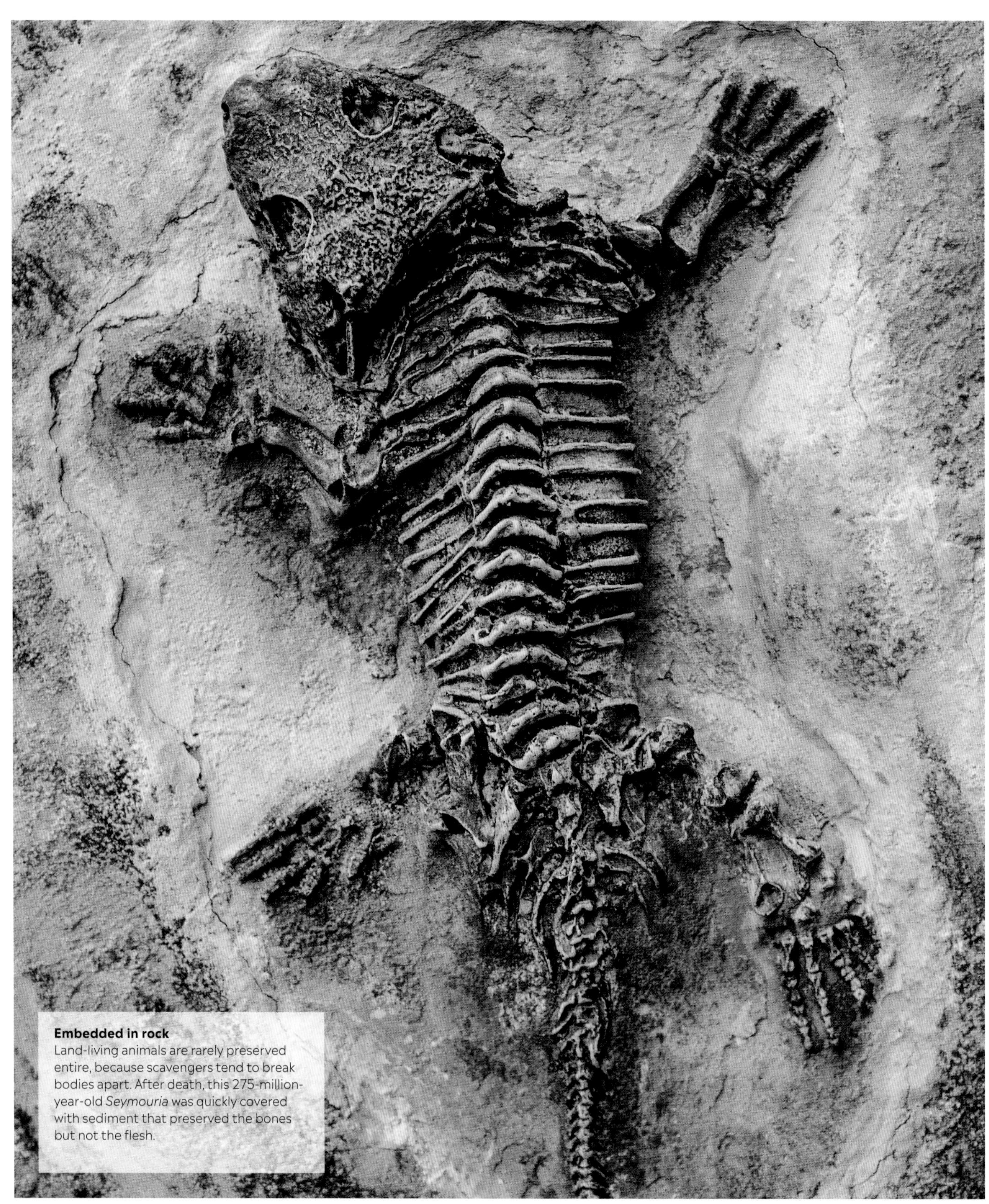

Embedded in rock
Land-living animals are rarely preserved entire, because scavengers tend to break bodies apart. After death, this 275-million-year-old *Seymouria* was quickly covered with sediment that preserved the bones but not the flesh.

FOSSILS AND FOSSILIZATION

The history of life is locked up in Earth's 3.8-billion-year record of fossils buried in sedimentary layers. These fossils chart the path from the earliest marine microbes, through the first shelled animals, to the bones and teeth of vertebrate animals that lived in the seas and on land, alongside the plant seeds, pollen, and wood that reveal how the planet was greened.

However, fossilization is highly selective, with soft-bodied organisms rarely preserved. Preservation usually requires that organic remains be covered and buried after death by sediment, although amber resin, ice, tar, and lava can also preserve fossils. Scientists distinguish between body fossils - hard body parts, such as shells and bones - and trace fossils, which are marks such as footprints and burrows left in the sediment by organisms. There are also chemical fossils that remain after death and decay, such as oil and gas and ancient DNA.

Scientists study the various ways in which life forms are fossilized in a bid to fill the gaps in our knowledge of evolution. They have discovered rare environments that preserve soft bodies and tissues that are not normally fossilized, including jellyfish, feathers, and hair.

Without fossils, we would know nothing about extinct plants and animals such as the dinosaurs. However, fossils can only tell us a certain amount. The continued study of living organisms allows scientists to make informed suggestions about how these creatures might have looked, relate body form to behaviour, and work out evolutionary relationships using genetic information.

Preservation of shape
After death, the calcium carbonate (calcite) skeleton of a modern coral, *Trachyphyllia*, was covered with seabed sediment. As a result, it has not been deformed by compaction.

FOSSIL FORMATION

Most fossils are the remains of marine organisms preserved in seabed sediments. However, there are also sediment traps on land and in coastal areas, such as lakes, swamps, and deltas, where the remains of land-living plants and animals are preserved as fossils. Here, dinosaur remains are buried in swampy sediments.

Prehistoric animal living near soft, marshy ground

Creature dies and is buried in muddy sediment, which protects it from scavengers

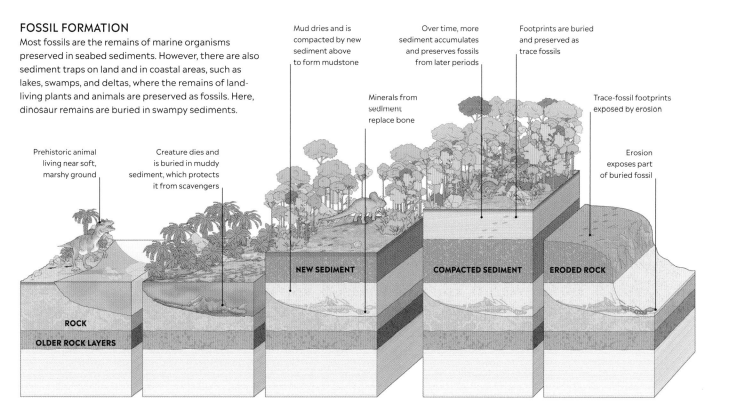

Mud dries and is compacted by new sediment above to form mudstone

Minerals from sediment replace bone

Over time, more sediment accumulates and preserves fossils from later periods

Footprints are buried and preserved as trace fossils

Trace-fossil footprints exposed by erosion

Erosion exposes part of buried fossil

ROCK

OLDER ROCK LAYERS

NEW SEDIMENT

COMPACTED SEDIMENT

ERODED ROCK

1814

FRAUNHOFER LINES

From 1814, German physicist Joseph von Fraunhofer used a spectroscope (which he invented) to record the existence of more than 500 dark lines in sunlight. Scientists have since catalogued many more of these Fraunhofer lines, each of which results from the absorption of light of specific frequency by atoms of particular elements present in the Sun's atmosphere.

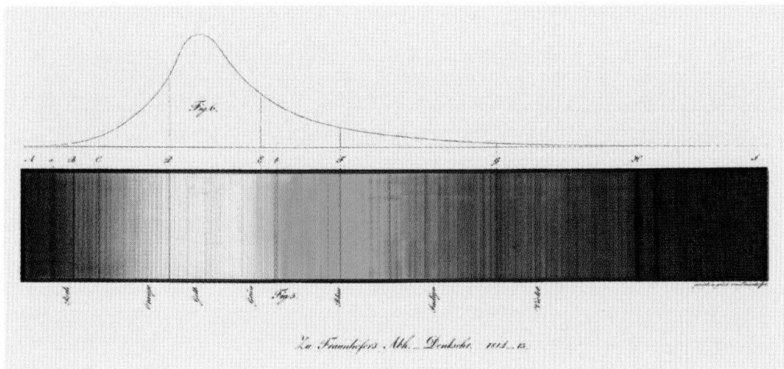

△ **Fraunhofer** solar spectrum

1816

THE STETHOSCOPE

French physician René Laennec created a trumpet-like device for listening to sounds made by internal organs, such as the heart, blood vessels, lungs, and intestines, in a process known as auscultation. He gave the name "stethoscope" to his invention, which evolved over the years to have a flexible tube and two earpieces.

▷ **Wooden stethoscope** for single ear

1814

1814 English engineer George Stephenson's steam locomotives outperform horses pulling coal trucks, laying the foundations of the railways

1815 Heinrich Olbers discovers the comet that will be named after him - 13P/Olbers

1815 English chemist William Prout states that the atomic weights of all elements are whole-number multiples of hydrogen's

1815

ROCK STRATA AND FOSSILS

English geologist William Smith was the first to recognize that fossils can be used to identify different layers of sedimentary strata. This allowed him to make the first large-scale and detailed geological map of the distribution and succession of different rock types across England, Wales, and part of Scotland.

▷ **William Smith's** geological map

> "Organized fossils are to the naturalist like coins to the antiquarian."

WILLIAM SMITH, STRATIGRAPHICAL SYSTEM OF ORGANIZED FOSSILS, *1817*

▷ **Early photograph** by Joseph Nicéphore Niépce

1816

EARLY PHOTOGRAPHY

French inventor Joseph Nicéphore Niépce was the first to use light-sensitive chemicals to preserve images of real scenes produced by lenses or pinholes. He made his first permanent photographs in the mid-1820s. Each one required several hours' exposure time. French inventor Joseph Louis Daguerre later sped up the time taken for the process.

1818 Jöns Jacob Berzelius publishes an accurate table of atomic weights after doing more than 2,000 analyses in ten years

1819 French physicists Dulong and Petit show that the heavier a substance's atoms and molecules, the more heat it takes to raise its temperature

1819

TRANSVERSE WAVE

Direction of travel

Vibration

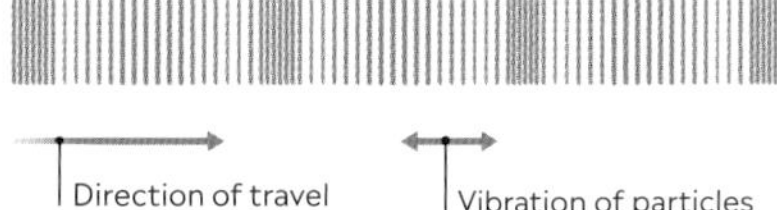

1818

NATURE OF LIGHT WAVES

In 1816, French physicist Augustin-Jean Fresnel had used complex mathematics to settle a debate, proving that light is a wave motion, rather than a stream of particles. Two years later, he established that light waves are transverse by studying the polarization of light, in which transverse waves are restricted to one plane of vibration.

Light and sound and waves

Sound waves are longitudinal: the variations in air pressure that constitute a wave are in the direction of the waves' travel. Light waves are transverse: the vibrations of the waves are at right angles to the direction of the waves' travel.

1819–20

ANTIMALARIAL TREATMENT

Quinine, the active antimalarial ingredient present in the bark of the cinchona, or fever tree, was isolated by French chemists Pierre-Joseph Pelletier and Joseph-Bienaimé Caventou. Cinchona is native to South America. Extracts from its bark were used by native peoples for centuries to treat ailments, including malaria.

△ **Cinchona** (*Cinchona officinalis*)

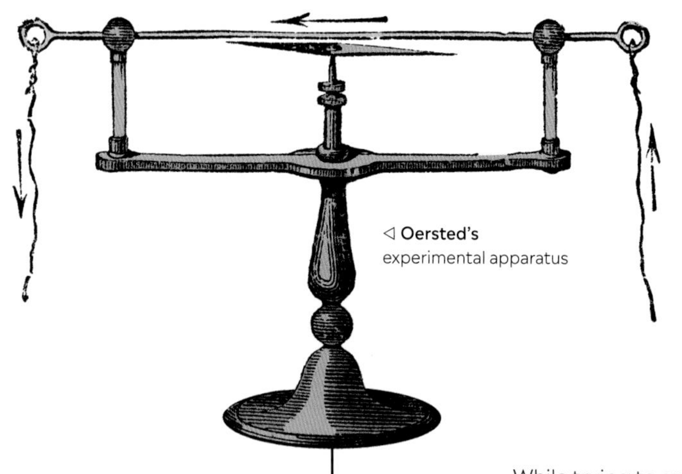
◁ **Oersted's** experimental apparatus

1820
OERSTED'S NEEDLE

For two years, Danish physicist and chemist Hans Christian Oersted experimented with batteries, wires, and magnetic compass needles in his search for a connection between electricity and magnetism. In 1820, he published his results, showing that whenever an electric current passes along a wire near to a magnetic needle, the needle is deflected. This opened the door to a new field of study: electromagnetism.

1820
AMINO ACIDS

While trying to extract sugar from animal products by boiling gelatin in sulphuric acid, French chemist Henri Braconnot isolated glycine, the simplest stable amino acid. His discovery began the work of identifying and understanding the composition of proteins and how they are made of sequences of amino acids.

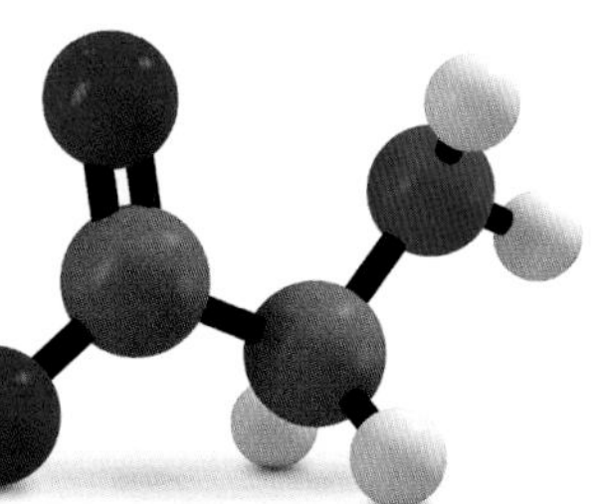
▽ **Glycine molecule** model

1820

1820 French mathematician Augustin-Louis Cauchy lays the foundations for mathematical analysis, using calculus to analyse physical systems

1821 Baltic-German physicist Thomas Seebeck discovers the Seebeck effect, the basis of the thermocouple

ELECTROMAGNETISM

Electricity and magnetism are deeply intertwined. A changing magnetic field produces an electric field (the basis of electric power generation), and a changing electric field produces a magnetic field. In fact, electricity and magnetism are different elements of a single phenomenon: electromagnetism. Electric and magnetic fields travel together through space as electromagnetic waves.

Electric current and magnetic fields
An electric current produces a local magnetic field. Electromagnets are devices that act as magnets when a current is switched on.

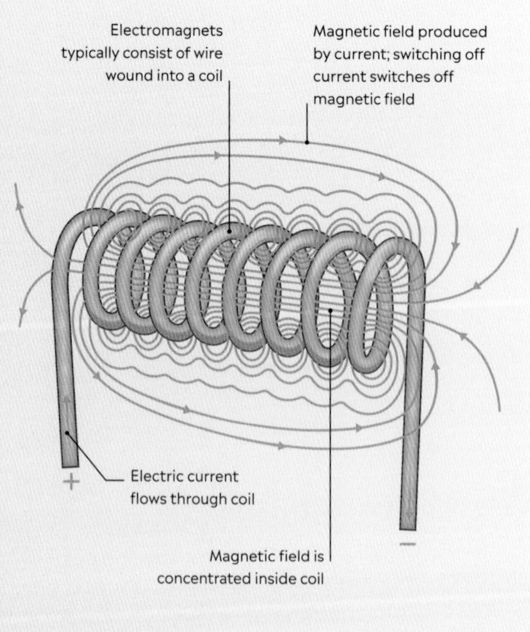
Electromagnets typically consist of wire wound into a coil
Magnetic field produced by current; switching off current switches off magnetic field
+
Electric current flows through coil
Magnetic field is concentrated inside coil
–

1791–1867
MICHAEL FARADAY

A pioneering English physicist and chemist, Faraday made contributions to the study of electricity and magnetism. His interest in science was sparked by attending lectures at London's Royal Institution, where he went on to work for over forty years.

1822
DATING ROCK STRATA

French geologist and naturalist Alexandre Brongniart wrote a broad-ranging study of the trilobites - invertebrates that prevailed in the seas for 270 million years from the early Cambrian period. Brongniart classified species from Europe and the Americas, attempting to group them by relative age. He introduced the idea of geological dating by the identification of distinctive fossils found in each rock stratum.

▷ **Trilobite** fossil

1822 French mathematician Joseph Fourier publishes his work on heat flow, which becomes important in many scientific fields

1823 German chemist J.W. Döbereiner discovers that platinum acts as a catalyst, speeding up certain reactions of hydrogen

1823

1821
FIELDS OF FORCE

In 1821, Michael Faraday began his decades-long series of experiments on the invisible forces of electricity and magnetism, and how the two are related to each other. He visualized the space around a magnet or a current-carrying wire as filled with lines of force, and in 1852, chose the term "field" to describe it.

◁ **Electromagnet** built by Faraday

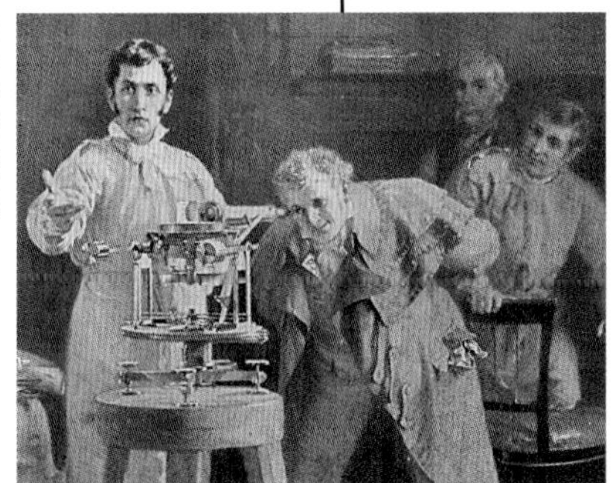

▽ **Fraunhofer** demonstrates his spectrometer

1823
SPECTRA OF STARS

When observing some of the brightest stars in the sky, Joseph von Fraunhofer found that the dark absorption lines in their spectra varied considerably from those seen in sunlight (*see p.130*), and from one another. Fraunhofer concluded that the lines arose from properties of the stars themselves rather than the passage of their light through Earth's atmosphere.

> "Nothing is too wonderful to be true, if it be consistent with the laws of nature."
>
> *MICHAEL FARADAY, LABORATORY JOURNAL, 1849*

1824
DINOSAUR DISCOVERIES

The fossil remains of two new kinds of giant lizard-like animals were discovered and named by English naturalists. William Buckland attributed a jawbone from Jurassic strata near Oxford to a creature called *Megalosaurus*, and Gideon Mantell named the *Iguanodon* from remains found in Cretaceous strata in Sussex. English naturalist Richard Owen would later give the general name "dinosaurs" – meaning terrible reptiles – to these extinct animals.

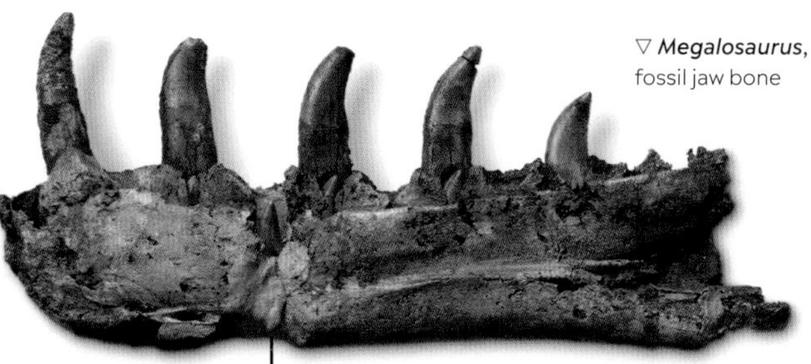

▽ ***Megalosaurus***, fossil jaw bone

△ **First steam rail** passenger service

1825
STEAM RAIL SERVICE

The first passenger railway in the world opened, between Stockton and Darlington in the UK. The engine, Locomotion No. 1, was driven on its first official outing by its maker, and "Father of Railways", George Stephenson. It was capable of pulling 450 passengers at 24 km/h per hour (15mph) and remained in service for three years before its boiler exploded.

1824

1824 Silicon is isolated by Jöns Jacob Berzelius when he reacts potassium fluorosilicate with potassium metal

1826 Heinrich Olbers asks why the night sky is dark if the Universe is infinite and unchanging and stars are evenly scattered

1824 Joseph Gay-Lussac notes the existence of compounds (later named isomers) that have exactly the same number of atoms but different properties

1825 French physician François-Joseph-Victor Broussais advocates treating disease by bloodletting with medicinal leeches

1826
AMERICAN ORNITHOLOGY

Self-trained American artist and ornithologist John James Audubon set off for Britain with his collection of detailed paintings of birds. There, he worked with Scottish ornithologist William MacGillivray to add descriptions of the life histories of the bird species and found a publisher for his work. The result, *Birds of America*, contained 435 life-sized watercolour paintings, printed from hand-engraved plates, accompanied by notes about the 489 species depicted. Today there are only 120 complete copies of this remarkable work.

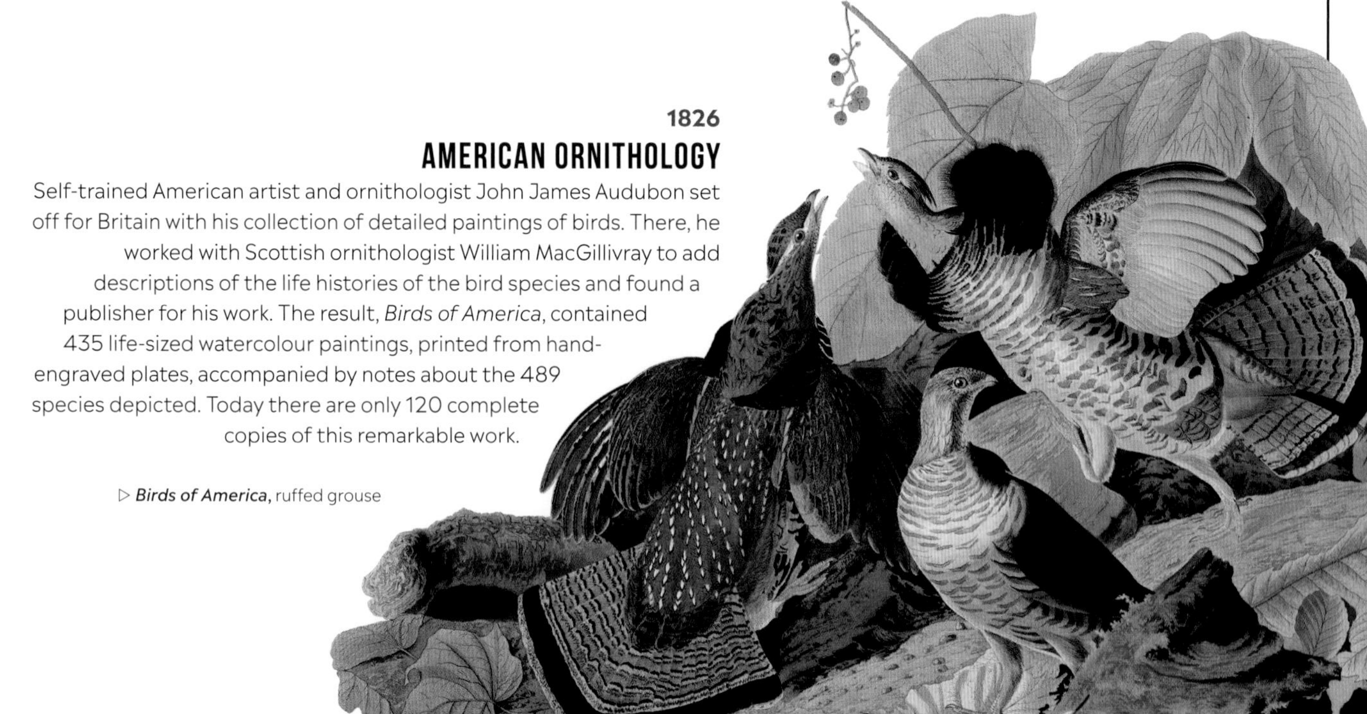

▷ ***Birds of America***, ruffed grouse

ORGANIC COMPOUNDS

Molecules that contain carbon bonded with other elements (mostly hydrogen, oxygen, and nitrogen) are called organic compounds. Carbon-carbon bonds are stable and can create a huge variety of molecules that may have long chains, rings, or functional groups. These variations give them a wide range of properties, as in gases such as methane (CH_4), biological substances essential for life (DNA, proteins, and carbohydrates), and manufactured long-chain molecules (plastics).

Functional groups and families

Carbon can bond with up to four other atoms or molecules to form functional groups that give the larger molecule specific properties. A series of organic compounds that have the same functional groups and similar chemical formulae is called a family.

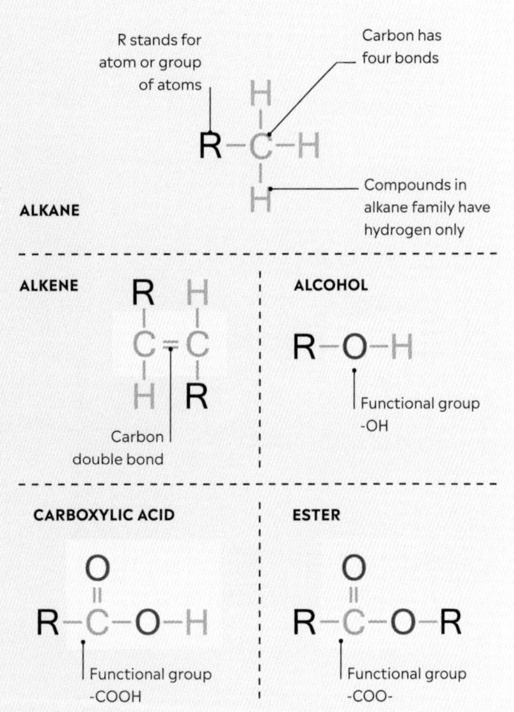

1828
ORGANIC SYNTHESIS

German chemist Friedrich Wöhler became the first person to synthetically create an organic compound when he heated the inorganic compound ammonium cyanate and produced urea.

△ Friedrich Wöhler

1829

1827 German physicist Georg Ohm determines the relationship between the voltage of a battery and the electric current in a circuit

1827 Scottish botanist Robert Brown observes the random motion of pollen grains in water, now known as Brownian motion

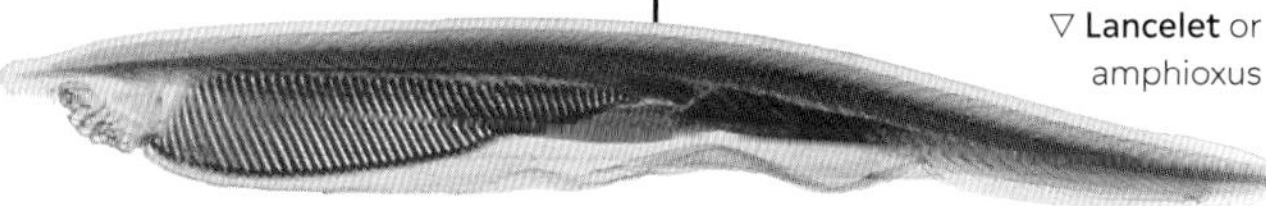
▽ **Lancelet** or amphioxus

1828
THE NOTOCHORD

First described by German scientist Karl Ernst von Baer during his studies of embryonic development, the notochord is a rod-like structure found during development in chordate animals. In most adult vertebrate chordates (including humans), it is replaced by a spinal column of cartilage or bone. In most invertebrate chordates (such as lancelets), the notochord is retained into the adult stage.

1785–1851
JOHN JAMES AUDUBON

John James Audubon was born in Saint Domingue (now Haiti). After a childhood in France, he was sent to America aged 18. He developed a passion for drawing birds, building an extraordinary collection of life-sized paintings of America's birds.

Audubon painted birds from the enormous whooping crane to the tiny ruby-throated hummingbird

1830

THE PRINCIPLES OF GEOLOGY

In *Principles of Geology*, Scottish geologist Charles Lyell claimed that Earth and its climate change very slowly over long periods of time and that Earth must be hundreds of millions of years old. He argued that present-day processes, such as volcanic eruptions, are subject to the same natural laws as they were in the past.

▷ ***Principles of Geology*** illustration

1791–1875

CHARLES LYELL

Trained as a lawyer, Lyell became one of the most influential geologists of the 19th century. His *Principles of Geology* greatly influenced the thinking of Charles Darwin and laid the foundations of a modern scientific approach to the study of Earth.

1830

1830 English physician Thomas Southwood Smith highlights the connection between poverty and epidemic diseases such as cholera

1830 Scottish physiologist Charles Bell is knighted for his research into nerves, especially those associated with touch and movement

△ **Joseph Lister** and his microscope

1830

LISTER MICROSCOPE

English amateur scientist Joseph Jackson Lister made a breakthrough in microscopy, creating the first achromatic light microscope and improving the quality of lenses. His microscope overcame the problem of distortion in the image caused by chromatic aberration (when different wavelengths of light are diffracted at different angles) and became a reliable tool for detailed medical research.

1830

ROCKS UNDER THE MICROSCOPE

Scottish naturalist William Nicol discovered how to cut sections of rock, mount them on glass, and grind them down so that light could pass through the mineral grains. His slides of a fossil tree were the first to show its cellular structure. Nicol also developed a method of polarizing light with a calcite prism. This allowed thin sections of rock to be viewed under a microscope and their mineral composition determined by the way light behaved as it passed through different minerals.

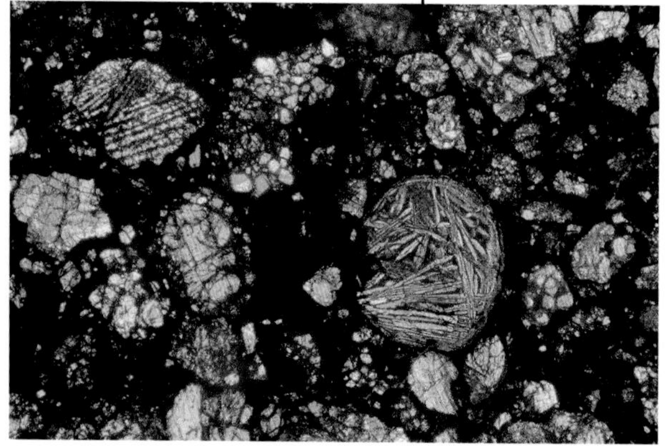

△ **Polarized photomicrograph** of a thin section of meteorite

1831

VOYAGE OF THE BEAGLE

Charles Darwin's voyage as naturalist aboard the HMS *Beagle* began. It proved to be one of the most important events in the history of biology. Darwin made copious notes about many aspects of nature and geology as the ship travelled on its five-year journey, mainly around South America. He published the first version of his observations in 1839, but in later editions he included some of his ideas about evolution, inspired especially by the diversity of finch species on the Galápagos Islands.

◁ **HMS *Beagle*** on its scientific expedition

1831 Chloroform is discovered through the distillation of concentrated ethyl alcohol and chloride of lime

1831

1831 Scottish botanist Robert Brown coins the term nucleus for a subcellular structure

△ **Faraday's** disc

1831

MOTORS AND GENERATORS

During the 1830s, scientists across the world produced the first practical electric motors and generators. In the previous decade, scientists including Michael Faraday and Hungarian physicist Ányos Jedlik had built simple but impractical electric motors. It was Michael Faraday who invented the first generator, in 1831, following his discovery of electromagnetic induction – the production of an electric voltage in a conductor such as a metal when a changing magnetic field passes through it.

△ **Hurricane** viewed from space

1831

CYCLONIC STORMS

After observing the pattern of trees felled by the Long Island hurricane of 1821, American amateur meteorologist William Redfield concluded that such storms were gigantic whirlwind bodies of air. He was proved right, and the winds in such storms are now understood to increase in speed towards the centre and circulate with greater speed than that of the storm itself as it passes over land or water.

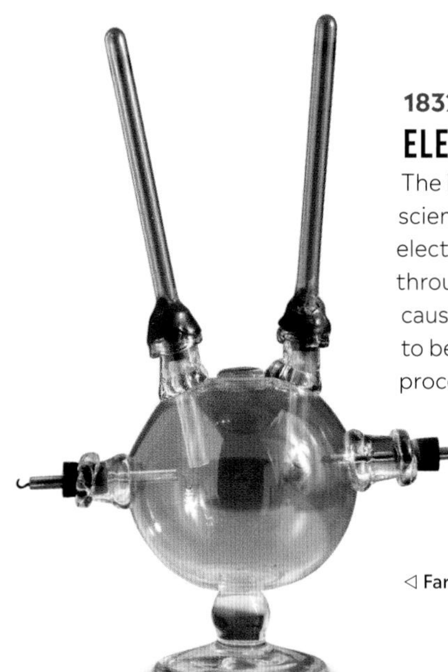

1832
ELECTROLYSIS

The invention of the battery in 1800 enabled scientists to investigate the effects of passing electric current from one electrode to another through solutions of chemical compounds. Current causes some compounds to separate, and elements to be deposited on one of the electrodes - a process called electrolysis. Michael Faraday discovered the connection between the amount of electric charge passing through the solution and the mass of the elements deposited in the process.

◁ **Faraday's** electrolysis globe

1791-1871
CHARLES BABBAGE

A mathematician, engineer, inventor, and more, Charles Babbage co-founded the Royal Astronomical Society, taught mathematics at Cambridge University, and was an expert in cryptography. He even invented a device to save cows from oncoming trains.

1832

1833 German astronomer Friedrich Bessel compiles a catalogue with highly accurate positions for 50,000 stars

1834 Russian physicist Emil Lenz's law relates the direction of an induced electric current to the direction of the magnetic field that produced it

1833 American physician William Beaumont directly observes digestive processes in a patient with an unhealed abdominal wound

△ **Anselme Payen**

1833
ENZYMES

The first enzyme (proteins that work as biological catalysts, speeding up chemical reactions) was discovered by French chemist Anselme Payen. Discovering a chemical from malt extract that helped convert starch into sugar, he named the substance diastase (now known as amylase). The enzyme pepsin was found by German physiologist Theodor Schwann three years later.

1834
EXPLORING THE SOUTHERN SKIES

From his observatory at the Cape of Good Hope in South Africa, English astronomer John Herschel began a detailed five-year survey of the skies visible from the southern hemisphere. Herschel's catalogue, eventually published in 1847, includes stars and other objects in the Large and Small Magellanic Clouds, now known to be satellite galaxies of the Milky Way.

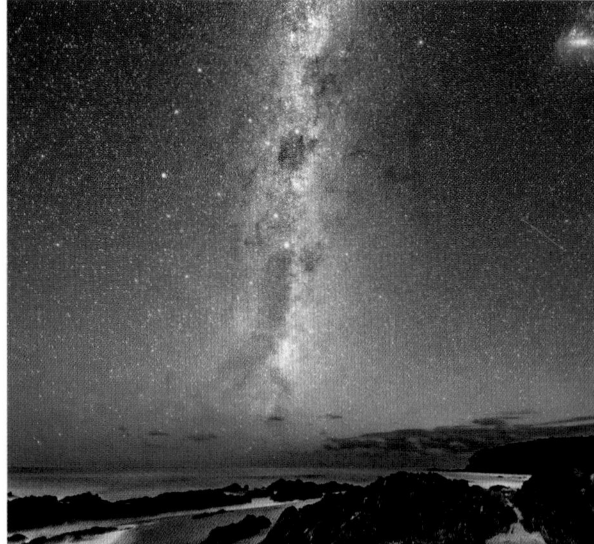

△ **The southern Milky Way** and Magellanic Clouds (right)

1834

THE ANALYTICAL ENGINE

During the 1820s, English polymath Charles Babbage built prototypes of a "difference engine" that could calculate the values of complex mathematical functions. In 1834, he began a more ambitious project, his "analytical engine". The machine was to be a general-purpose mechanical computer that could be programmed via punched cards. It was to have an arithmetical unit, a control unit, a memory, and even a printer. Only small parts of the machine were built in Babbage's lifetime.

◁ **Part of Babbage's** analytical engine

1835 Charles Darwin reaches the Galápagos Islands where he makes observations that will later underpin his theory of natural selection

1836 English chemist John Daniell develops a more reliable electrochemical cell; it becomes standard for batteries used in telegraphy

1836

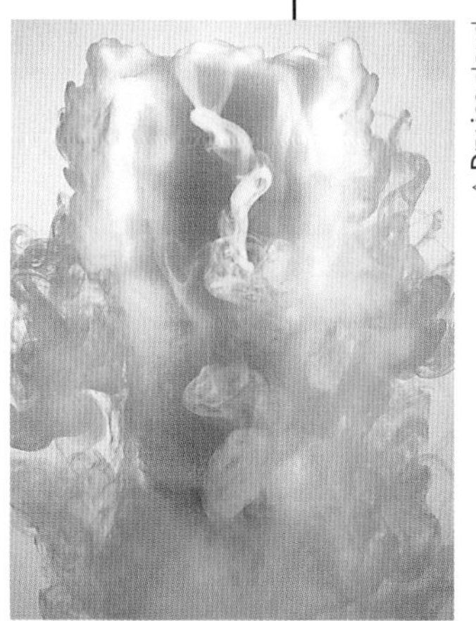

△ **Dry ice** cloud

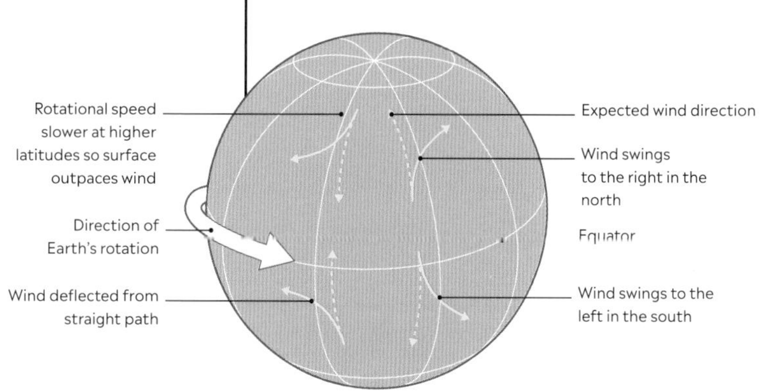
Rotational speed slower at higher latitudes so surface outpaces wind
Direction of Earth's rotation
Wind deflected from straight path
Expected wind direction
Wind swings to the right in the north
Equator
Wind swings to the left in the south

The force at work
The Coriolis effect influences global wind and weather patterns. Earth's rotational speed is faster at the equator than at the poles, and this causes air to curve as it travels across Earth's surface. In the northern hemisphere, wind is deflected to the right; in the southern hemisphere, wind is deflected to the left.

1835

DRY ICE

French inventor Adrien-Jean-Pierre Thilorier made a device that could liquefy carbon dioxide by putting the gas under pressure. When released, the pressurized gas formed a snowy white solid (dry ice) – solid carbon dioxide.

1835

CORIOLIS FORCE

In his essay *"On the equations of relative motion of systems of bodies"*, the French mathematician Gaspard-Gustave de Coriolis investigated the mathematics of energy transfer by rotating bodies, such as waterwheels. He showed that an inertial force acts to the left with clockwise rotation and to the right with anticlockwise rotation.

◁ **Morse-Vail** telegraph key

1837

TELEGRAPHY

The 1830s saw rapid developments in telegraphy, the ability to send messages across long distances using electrical signals. In 1837, English inventors William Fothergill Cooke and Charles Wheatstone patented and built a system that ran alongside railway tracks. However, it was the cheaper, simpler system built by Americans Samuel Morse and Alfred Vail that underpinned the network of telecommunications that soon criss-crossed the world.

1838

MEASURING DISTANCES TO THE STARS

Using an instrument called a heliometer, Friedrich Bessel accurately measured the distance to a star (star 61 Cygni) for the first time. The discovery was based on measurements of the star's parallax – its apparent shift in direction throughout the year when seen from opposite sides of Earth's orbit. Parallax measurements were soon applied to other stars.

▷ **Heliometer** telescope

1837

1837 French botanist Henri Dutrochet demonstrates that chlorophyll, the green pigment found in most plants, helps them assimilate carbon dioxide

1837 Swiss-born geologist Louis Agassiz shows that an ice age covered Europe and North America in snow and ice

1837

YEAST AND FERMENTATION

Examining yeast microscopically, German physiologist Theodor Schwann showed that it has a cellular structure resembling the cells of plant tissues. He established that yeast is a living organism and that it is the action of yeast cells that is responsible for fermentation, disproving a widely-held view at the time that oxygen caused the process.

△ **Yeast cells**

1839

ANTARCTICA EXPLORED

American naval officer Charles Wilkes began a four-year-long expedition, with seven armed ships and 350 men, to explore and survey the Pacific Ocean and surrounding lands. Sailing south from Australia in December 1839, the men first spotted the land of Antarctica on 16 January 1840. They went on to survey 2,400 km (1,500 miles) of the Antarctic coast but ice prevented them from landing.

1839
THE FUEL CELL

In 1839, British scientist William Grove published a description of a new device for generating electricity: the fuel cell. Grove's device, which he called a gaseous voltaic battery, consisted of cells, each made of a pair of platinum-coated electrodes in contact with hydrogen and oxygen gas. Although the device worked well, it would be nearly 100 years before the first practical, commercial fuel cell became available.

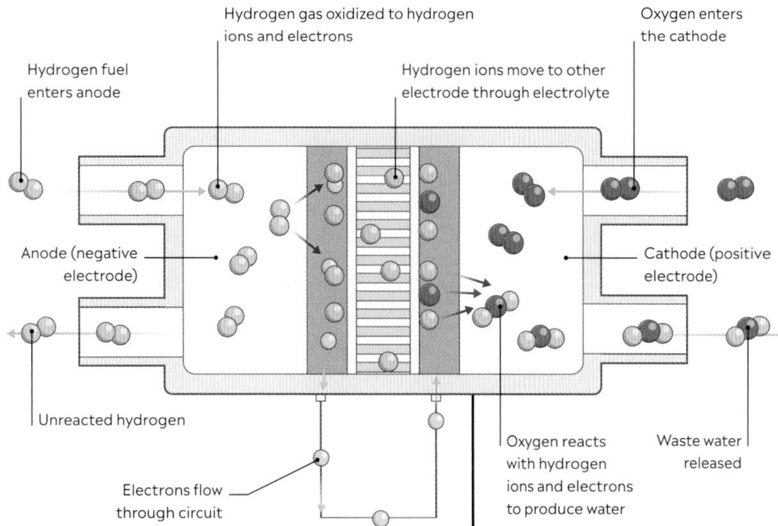
Hydrogen gas oxidized to hydrogen ions and electrons
Oxygen enters the cathode
Hydrogen fuel enters anode
Hydrogen ions move to other electrode through electrolyte
Anode (negative electrode)
Cathode (positive electrode)
Unreacted hydrogen
Oxygen reacts with hydrogen ions and electrons to produce water
Waste water released
Electrons flow through circuit

How fuels cells work

A fuel cell makes use of the energy released by the chemical reaction between a fuel (such as hydrogen or methane) and oxygen – just as burning the fuel would. In a fuel cell, the energy is released as electricity; in a flame, it becomes heat and light.

1838 The word protein is first used to describe important large molecules, such as those in egg white, by Jöns Jacob Berzelius

1839 Vulcanization is discovered when American inventor Charles Goodyear combines natural rubber and sulphur, creating a stronger material

1839

1839 French physicist Louis Daguerre speeds up photograph processing times by using silver iodide plates

△ **Wilkes'** expedition

1839
CELL SCIENCE

German physiologist Theodor Schwann laid many of the foundations of cell biology in his masterwork, the title of which was translated as *Microscopic investigations on the similarity of structure and growth of animals and plants.* His main finding was that all living things are composed of cells and cell products.

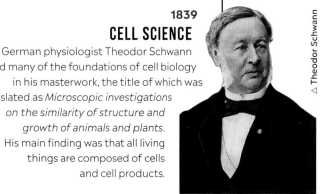

△ Theodor Schwann

"The elementary parts of all tissues are formed of cells."

THEODOR SCHWANN, MICROSCOPICAL RESEARCHES, *1839*

1840
THERMOCHEMISTRY

Swiss-Russian chemist Germain Henri Hess investigated the change in energy during chemical reactions. He found that the amount of heat generated, or absorbed, when reacting chemicals to form new products is constant, regardless of the route taken to create them. His studies of the heat of reaction formed the basis of thermochemistry.

◁ **Thermometer** from c.1840

1842
CONSERVATION OF ENERGY

While investigating the limitations of water power, German physicist and chemist Julius von Mayer uncovered one of the most fundamental tenets of physics: the law of the conservation of energy. This states that, in a closed system, energy cannot be created or destroyed, but only converted from one form to another.

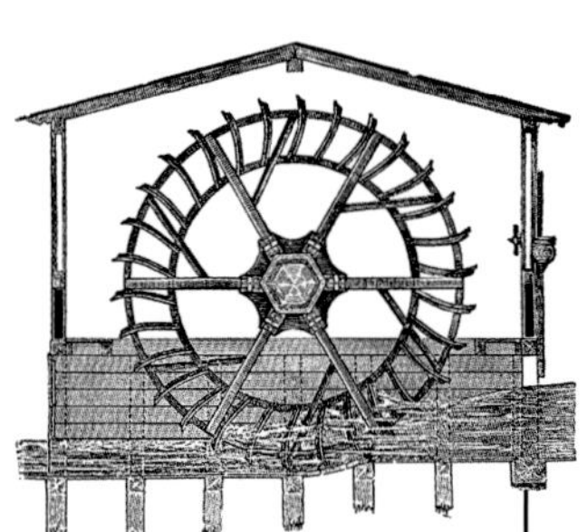

△ **A waterwheel,** Mayer's inspiration

1840

1840 German-Swiss chemist Christian Schönbein isolates and names ozone, a pungent allotrope of oxygen

1841 American scientist John William Draper photographs the Moon, pioneering astrophotography as a new technique for astronomical discoveries

1840 German physician Friedrich Henle speculates that infectious diseases are transmitted by living organisms, many of them microscopic

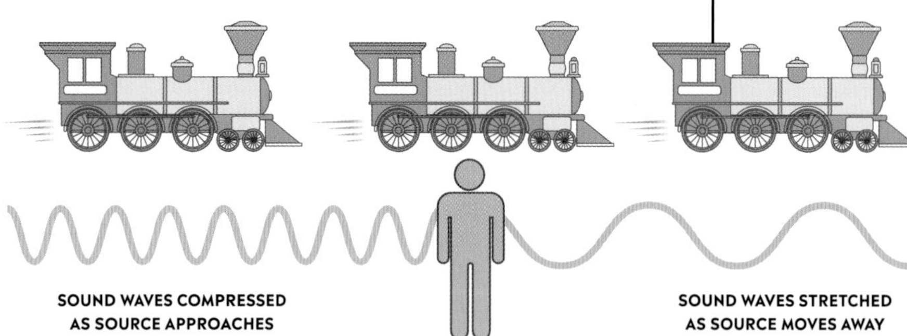

Changing sound frequencies
As a train approaches, the sound waves from its engine become bunched up. The waves' peaks and crests arrive at a higher frequency, which we hear as a higher pitch. As the train recedes, the pitch drops.

1842
DOPPLER EFFECT

In a scientific paper, Austrian physicist Christian Doppler suggested that the colours of some stars are due to their movement towards and away from us. Specifically, the frequency of light from a star moving towards Earth would appear shifted towards the shorter (blue) end of the spectrum. His mathematical analysis also applied to sound waves. The effect has since become known as the Doppler effect.

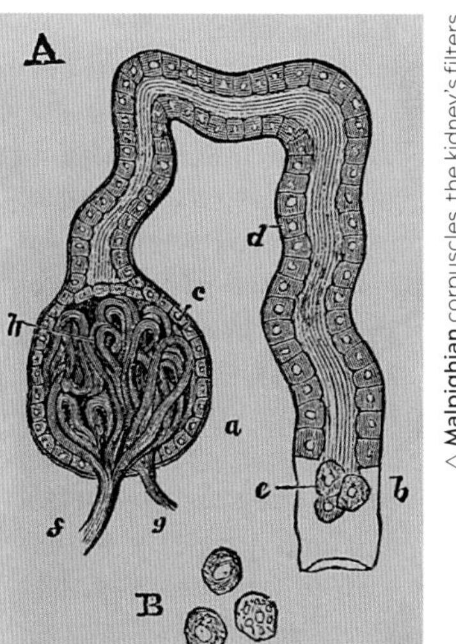

△ **Malpighian** corpuscles, the kidney's filters

1842
KIDNEY FUNCTION

It was widely thought that the kidney acted simply as a gland, secreting urine, until German physiologist Carl Ludwig demonstrated that the excretion of urine involves a two-stage process – one of filtration and another of reabsorption. He was the first scientist to correctly describe the way in which the kidney achieves this filtration process.

1843
SUNSPOTS

Based on 17 years of continuous observation of dark spots on the Sun, German astronomer Samuel Heinrich Schwabe announced the existence of a recurring pattern in their numbers. This sunspot cycle is now known to cause both the number of sunspots and the locations where they occur to vary over a period of around 11 years.

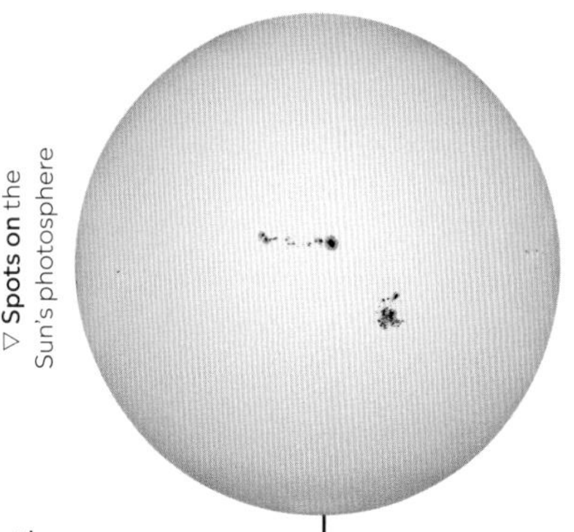

▽ **Spots on the** Sun's photosphere

1843

1842 English palaeontologist Richard Owen first calls a group of extinct land-living reptiles Dinosauria

1843 English mathematician John Couch Adams attempts to calculate the position of a new unknown planet based on its influence on the orbit of Uranus

1843
MECHANICAL EQUIVALENT OF HEAT

In the 1840s, scientists began to understand how energy transfers between different forms. Working out the equivalence of one form to another was a key part of this development. In 1843, English physicist James Joule carried out the ground-breaking experiment that allowed him to work out the numerical equivalence between mechanical work, such as that done by a steam engine, and heat. His experiment involved measuring the small temperature rise of water when paddle wheels were turned by falling weights.

> "My object has been, first to discover correct principles and then to suggest their practical development."
>
> *JAMES JOULE,* ANNALS OF ELECTRICITY, *1840*

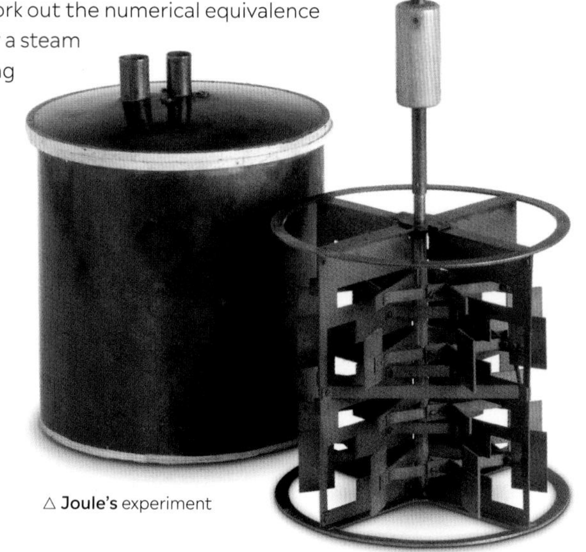

△ **Joule's** experiment

TYPES OF ENERGY

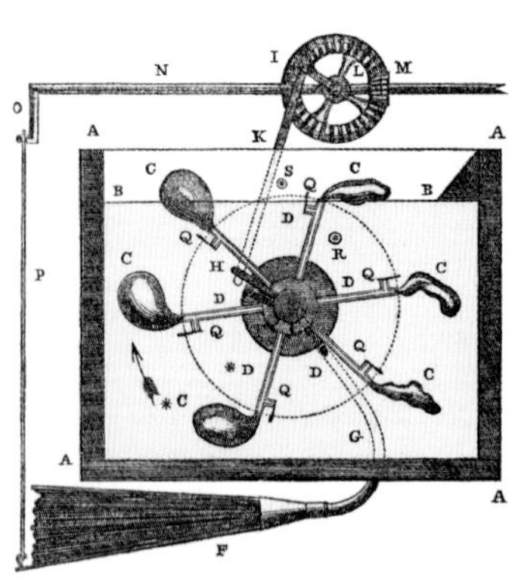

Perpetual motion machines
This fantasy device is one that can work forever without any energy input. Such a machine would violate the law of conservation of energy.

Energy can be understood as the capacity to change things (do work), such as moving or heating an object. Every phenomenon – whether natural or artificial – involves work.

Energy exists in many forms. The best-known types are kinetic, thermal, and nuclear, but there are several others. When work is done, energy is converted from one form to another. When an energy company generates electricity, it is converting energy from various forms (such as the kinetic energy of moving water, or the chemical energy found in fossil fuels) into electrical energy that is more usable for its customers' needs.

The many forms of energy can be divided into two broad families: potential energy and kinetic energy. Potential energy is the energy an object has due to the relative positions of different parts of a system – for example, a spring has elastic potential energy when it is stretched, while a tennis ball has gravitational potential energy when it is held above the ground. Kinetic energy is associated with movement, such as in thermal energy, which is the energy of moving particles in a substance.

One of the most important rules in physics is that energy is conserved, meaning that the total amount of energy in the Universe remains the same. It is not possible to create or destroy energy; it can only be changed from one form to another. Another important rule is the concept of entropy, which describes how, over the course of time, energy becomes more spread out and less useful for carrying out work.

ENERGY FORMS

A simple task such as pushing a load up a slope and tipping it off the edge involves changes between many forms of energy. Chemical energy is converted into kinetic energy, then the load's gravitational potential energy changes to kinetic energy as it falls.

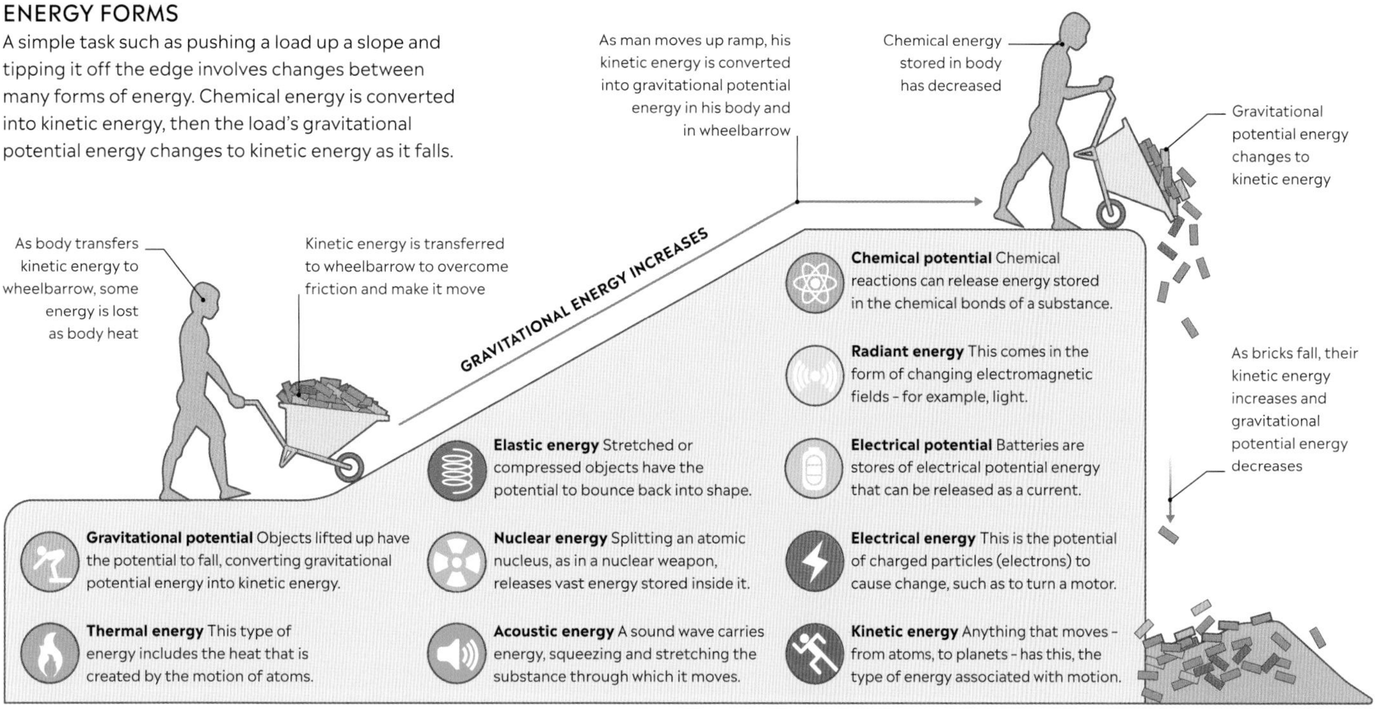

As body transfers kinetic energy to wheelbarrow, some energy is lost as body heat
Kinetic energy is transferred to wheelbarrow to overcome friction and make it move
GRAVITATIONAL ENERGY INCREASES
As man moves up ramp, his kinetic energy is converted into gravitational potential energy in his body and in wheelbarrow
Chemical energy stored in body has decreased
Gravitational potential energy changes to kinetic energy
As bricks fall, their kinetic energy increases and gravitational potential energy decreases

Gravitational potential Objects lifted up have the potential to fall, converting gravitational potential energy into kinetic energy.

Thermal energy This type of energy includes the heat that is created by the motion of atoms.

Elastic energy Stretched or compressed objects have the potential to bounce back into shape.

Nuclear energy Splitting an atomic nucleus, as in a nuclear weapon, releases vast energy stored inside it.

Acoustic energy A sound wave carries energy, squeezing and stretching the substance through which it moves.

Chemical potential Chemical reactions can release energy stored in the chemical bonds of a substance.

Radiant energy This comes in the form of changing electromagnetic fields - for example, light.

Electrical potential Batteries are stores of electrical potential energy that can be released as a current.

Electrical energy This is the potential of charged particles (electrons) to cause change, such as to turn a motor.

Kinetic energy Anything that moves - from atoms, to planets - has this, the type of energy associated with motion.

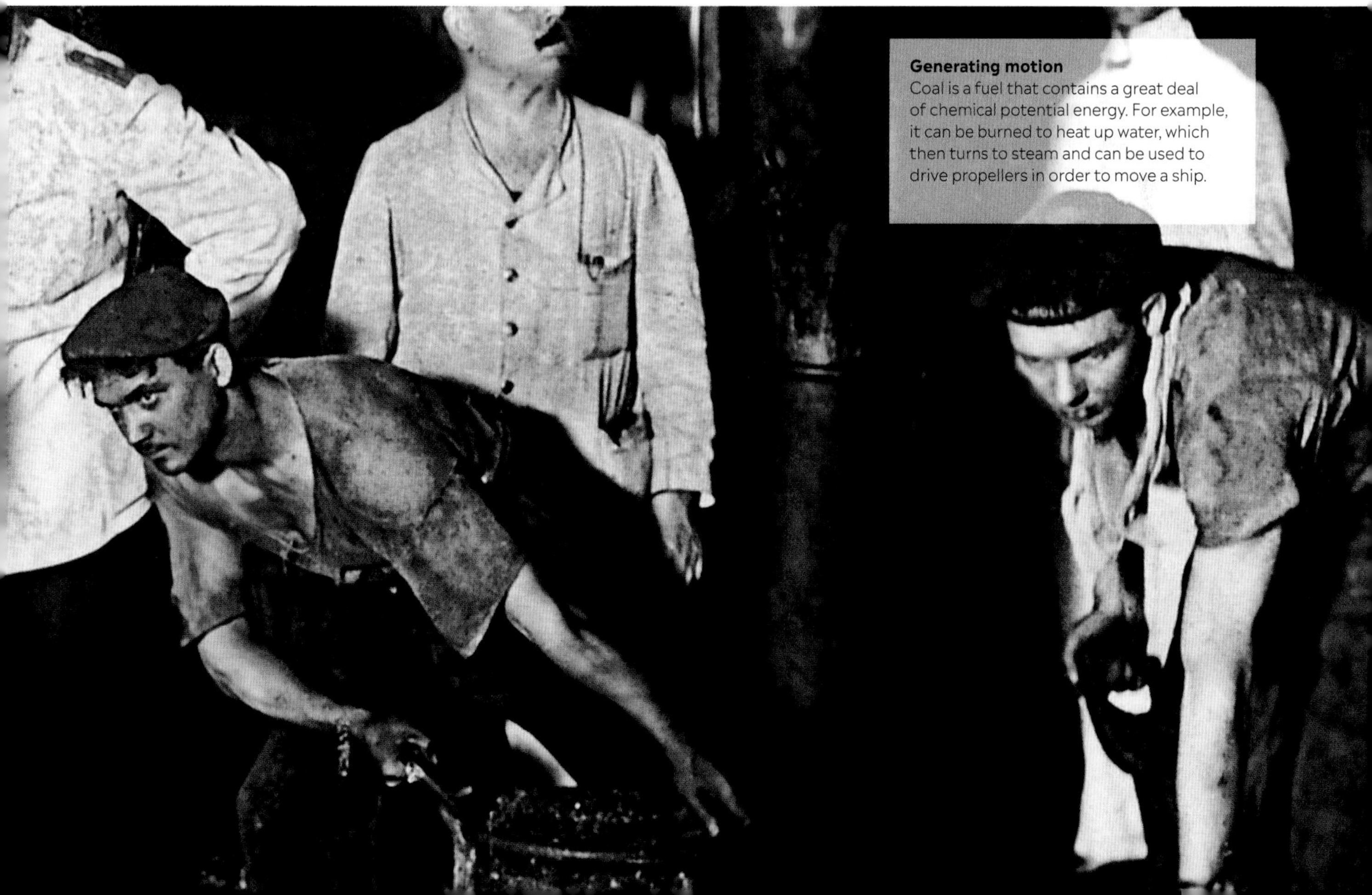

Generating motion
Coal is a fuel that contains a great deal of chemical potential energy. For example, it can be burned to heat up water, which then turns to steam and can be used to drive propellers in order to move a ship.

1845
SPIRAL NEBULAS

Irish astronomer William Parsons, Lord Rosse, used his newly completed giant telescope (nicknamed the Leviathan of Parsonstown) to show that the fuzzy-looking nebula known as Messier 51 appeared to be a spiral containing countless faint stars. The confirmation of other "spiral nebulae" led to a debate about their nature and speculation that they might be solar systems seen during formation or independent star systems beyond the main Milky Way.

▷ **Rosse's 1.8 m (6 ft)** reflecting telescope

1844

1844 Samuel Morse sends the first long-distance telegraph message, from Washington to Baltimore, USA

1844 Friedrich Bessel finds that the star Sirius is influenced by an unseen heavy companion (later found to be a white dwarf)

1845 Michael Faraday suggests light is a form of electromagnetism in *Thoughts on Ray Vibrations*

1846 American dentist William Morton first demonstrates the use of inhaled ether (diethyl ether) as an anaesthetic

"What hath God wrought."

THE FIRST TELEGRAPH MESSAGE

1846
PROTOPLASM

German botanist Hugo von Mohl spent much of his career making careful studies of the physiology and anatomy of plant cells. Most notably, he showed that the nucleus of the cell lies within a colloidal, active substance that he named protoplasm. He described this as a living mass within each active cell and as the site of stored energy driving cell activity.

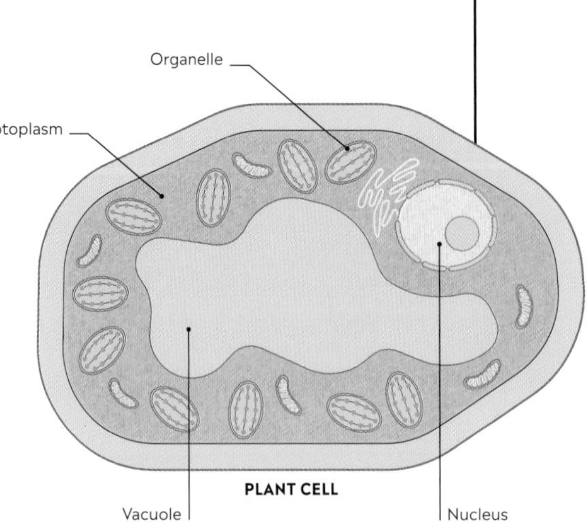

PLANT CELL

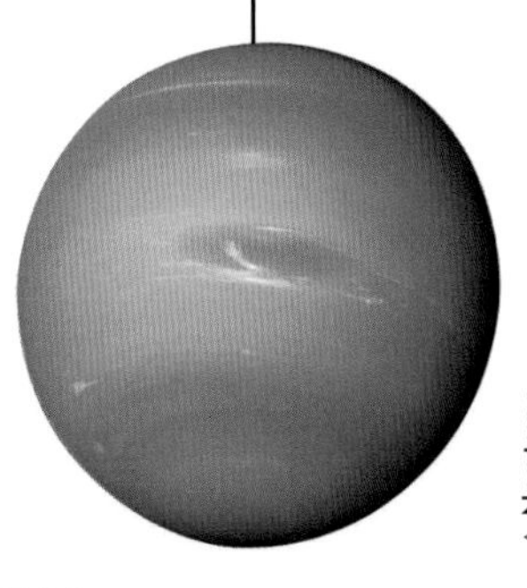

△ **Neptune**

1846
NEPTUNE SIGHTED

Following a prediction for the position of a planet beyond Uranus made by French mathematician Urbain Le Verrier, German astronomer Johann Galle of Berlin Observatory made the first observations of the Solar System's outermost major planet: Neptune. Astronomer William Lassell found the planet's largest moon, Triton, 17 days later.

1847
BERGMANN'S RULE
A student of comparative anatomy, German biologist Carl Bergmann observed that within related animals (for example, penguins), species of larger body size tend to be found in colder environments, while those with smaller body size are normally found in warmer habitats.

◁ **Emperor** penguin, to scale

◁ **Galápagos** penguin, to scale

1848
ISOMERS
French chemist Louis Pasteur discovered and separated the two types of tartaric acid crystal. The crystals had the same chemical formula, but when dissolved in solution, they rotated light shone through them in opposite directions. This was due to the slightly altered shape of each solution's molecules, which were mirror images of one other, or isomers.

△ **Tartaric acid** crystals

1849

1846 Italian chemist Ascanio Sobrero creates the explosive nitroglycerine by adding glycerol to a mix of nitric and sulphuric acids

1847 John Herschel publishes his survey of the southern sky. It includes chapters on nebulae, the satellites of Saturn, and sunspots

1849
FIZEAU AND THE SPEED OF LIGHT
In the first serious attempt to measure the speed of light, French physicist Hippolyte Fizeau shone a beam of light through a rapidly rotating toothed wheel. The light bounced off a mirror 8 km (5 miles) away and was sent back through the wheel. At certain speeds of rotation, the wheel's teeth blocked the return light, and this enabled Fizeau to calculate the light's speed. His value was accurate to within 5 per cent.

▷ **Fizeau's apparatus**

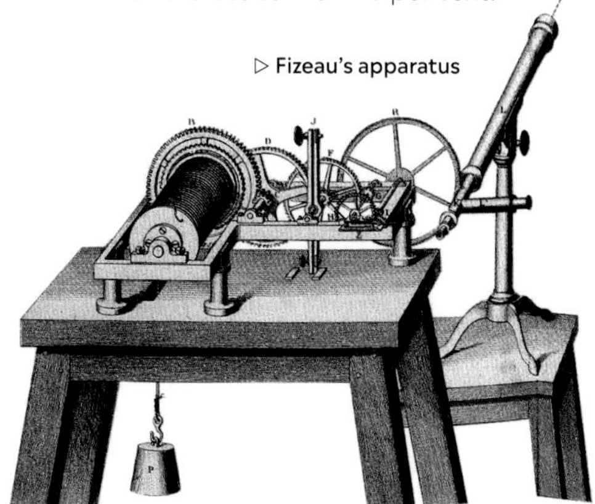

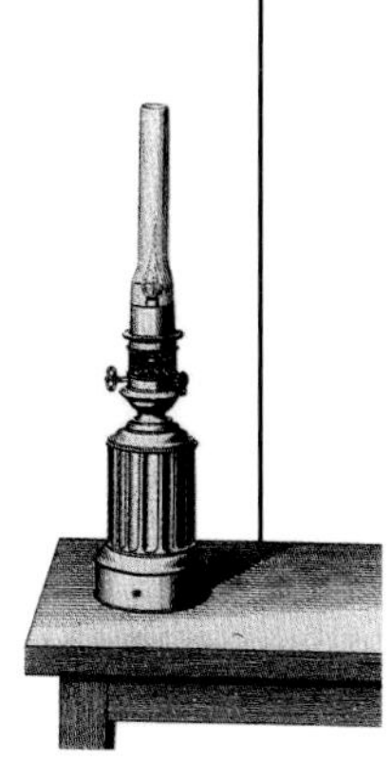

1851
EARTH'S ROTATION

An ingenious experiment by French physicist Léon Foucault directly demonstrated Earth's rotation for the first time. A pendulum with a long oscillation period was suspended from a single point so that as it slowly swung back and forth, Earth rotated beneath it and the axis of the pendulum's sweep slowly changed orientation relative to the ground below.

▷ **Foucault's** pendulum, Paris

1852
VALENCY

English chemist Edward Frankland was one of the first to investigate organometallic compounds, a combination of metal ions with organic molecules. His observations led him to propose a theory of valence, which stated that each type of atom has a specific number of ways it can combine with other atoms. This insight was an important step in understanding bonding and structure in chemistry.

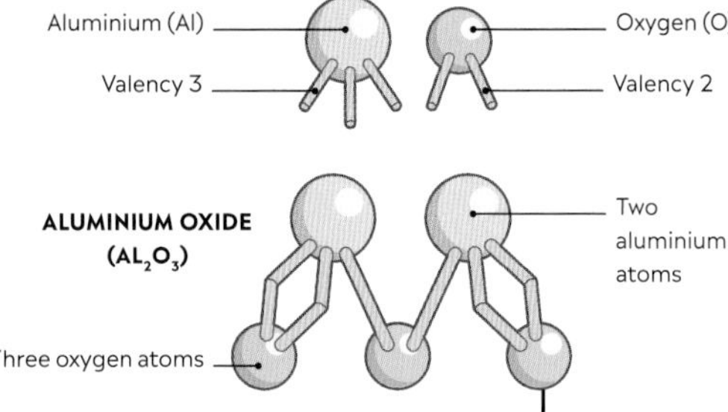

Number of bonds
Atoms with a valency of 3, such as aluminium, make three bonds; oxygen has a valency of 2, so makes two bonds.

1850

1850 The London Epidemiological Society is founded to promote the study of epidemic diseases

1850 Scottish chemist Thomas Graham helps to lay the foundations of colloid chemistry

1852 Léon Foucault uses a rapidly spinning gyroscope to demonstrate, and observe the effects of, Earth's rotation

▷ **Broad Street** water pump, London

1850
THE DRIVING POWER OF HEAT

In a paper he read to the Berlin Academy, German physicist and mathematician Rudolf Clausius explored the fact that heat never flows from a cold to a hot object unless "work" is done, that is, unless energy is put into the system. This challenged the prevailing "caloric" theory, in which heat was considered a substance. Clausius's statement was the first formulation of what is now called the second law of thermodynamics.

▷ **Rudolph Clausius**

1854
EPIDEMIOLOGY

A cholera outbreak in London sent English physician John Snow, one of the founders of the study of epidemiology, looking for its origins. He traced one source of the outbreak to a public water pump in Broad Street and also found that the incidence of the disease was higher where polluted water was drawn from the River Thames than from cleaner sources upstream, showing that cholera was waterborne.

△ **Cartoon** depicting death supplying cholera

Foucault's pendulum was suspended from the dome of the Panthéon in Paris

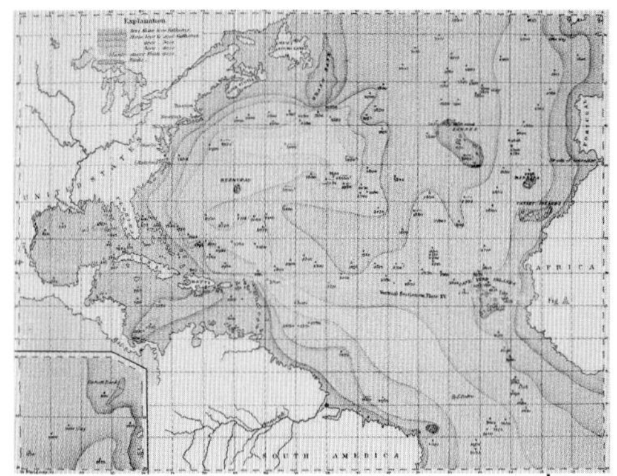

1855

MODERN OCEANOGRAPHY

Using data from ships' logbooks on ocean winds, currents, and depths, US naval officer Matthew Maury published *Physical Geography of the Sea and Its Meteorology*. The book's maps in particular proved invaluable to navigators.

◁ **Maury's map** of the depths of the North Atlantic

1831–1879

JAMES CLERK MAXWELL

A Scottish mathematician and physicist, Maxwell made major contributions to the field of electromagnetism (particularly in demonstrating that light is a form of electromagnetic radiation), thermodynamics, and colour vision.

1854 German physicist Hermann von Helmholtz wrongly proposes that the Sun produces energy by gravitational contraction

1855 Scottish scientist James Clerk Maxwell publishes a groundbreaking mathematical description of Michael Faraday's experiments with electricity and magnetism

1855

1855

VACUUM TUBES

German physicist and glassblower Johann Geissler developed a hand-cranked pump that could create a high vacuum and used it to evacuate all air from glass tubes of his own design. Scientists could then fill the tubes with rarefied gases and pass electric current through the gases. Geissler tubes became an important tool in physics and led to the discovery of the electron, the development of neon lighting, and the creation of thermionic valves, which sparked the beginning of modern electronics.

△ **Geissler** tubes

NATURAL SELECTION

In organisms that reproduce by sexual reproduction, there is natural variation between individuals. Those that are better suited to their environment are more likely to survive and reproduce, passing on their genes and leading to long-term changes in the population. The probability that an organism will survive and reproduce is a measure of its reproductive fitness, which has led to the phrase "survival of the fittest".

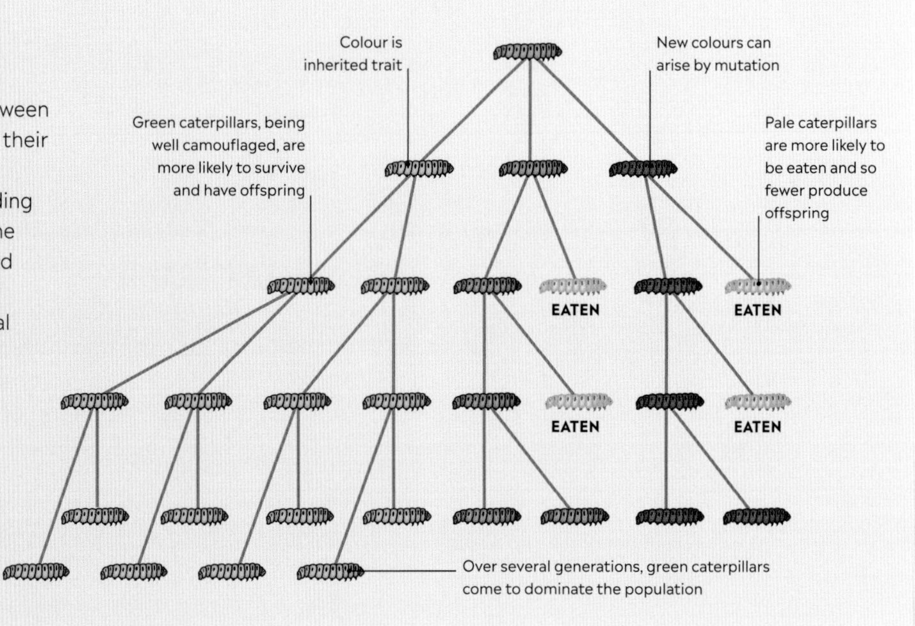

Colour is inherited trait
New colours can arise by mutation
Green caterpillars, being well camouflaged, are more likely to survive and have offspring
Pale caterpillars are more likely to be eaten and so fewer produce offspring
EATEN
EATEN
EATEN
EATEN
Over several generations, green caterpillars come to dominate the population

Survival of the fittest
In this example of natural selection, a caterpillar's chance of survival depends on how well its colour camouflages it from predators. Caterpillars that are not well camouflaged are selected out of the population.

1856

1856 The first skeleton recognized as belonging to a predecessor of humans is discovered in the Neander Valley, Germany

1856 Cheap steel becomes available after English inventor Henry Bessemer develops a process that blows oxygen into molten iron

1857 French physiologist Claude Bernard announces he has isolated glycogen from liver tissue

1857 German chemist Friedrich August Kekulé suggests that carbon can bond with up to four other atoms, including other carbon atoms

1856

PASTEURIZATION

While investigating fermentation, French chemist Louis Pasteur discovered that microbes are responsible for rendering beverages such as beer, wine, and milk unpalatable. He also found that heating the liquids to 60–100 °C (140–212 °F) destroys most of the bacteria and moulds that cause the problem. This way of preserving food became known as pasteurization.

△ **Louis Pasteur** in his laboratory

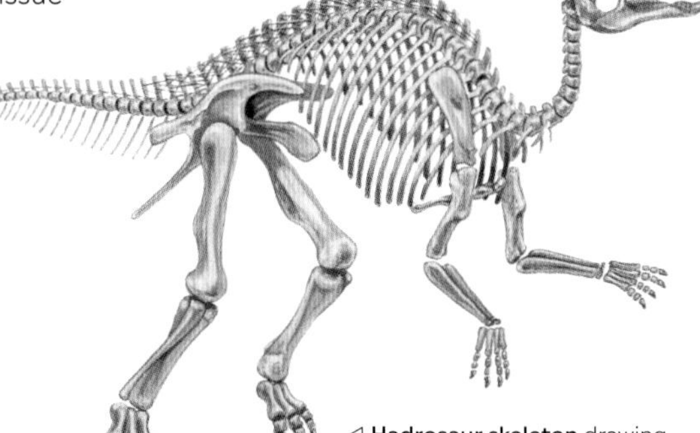

◁ **Hadrosaur skeleton** drawing

1858

HADROSAURUS

Hadrosaurus foulkii was the first mostly complete dinosaur skeleton found in North America after bones were retrieved from Haddonfield, New Jersey. The generic name means "heavy lizard" and the species was named in honour of its discoverer, American lawyer and amateur geologist William Parker Foulke. This giant herbivore lived in the middle Cretaceous Period, from about 100 million years ago.

1809–1882
CHARLES DARWIN
Born in Shrewsbury, England, Darwin studied medicine and theology but was most passionate about natural history. He gathered large collections of fossils, animals, and plants, and his theory of evolution by natural selection made him hugely influential.

"I am convinced that Natural Selection has been the main but not exclusive means of modification."

CHARLES DARWIN, THE ORIGIN OF SPECIES, *1859*

1859
THE ORIGIN OF SPECIES

This year saw the first printing of the most influential book on evolutionary biology, Charles Darwin's *On the Origin of Species by Means of Natural Selection, or the Preservation of Favoured Races in the Struggle for Life*. It sold out quickly, attracting attention and controversy, and eventually ran to six editions.

▽ ***The Origin of Species***, first edition

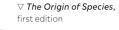

1858 English surgeon Henry Gray publishes *Gray's Anatomy*; it becomes the standard reference on human anatomy

1858 Italian chemist Stanislao Cannizzaro publishes convincing evidence supporting Avogadro's hypothesis

1859

1859
SOLAR FLARES

While observing a large group of sunspots, English amateur astronomer Richard Carrington spotted a brilliant burst of light, identified as the first recorded solar flare. Just hours later, dramatic displays of aurorae (northern and southern lights) were seen around the world, while electrical systems such as telegraph wires went haywire. Scottish physicist Balfour Stewart demonstrated that the disruption was caused by a cloud of solar particles striking the Earth and its atmosphere.

▷ **Flares forming** above the Sun's surface

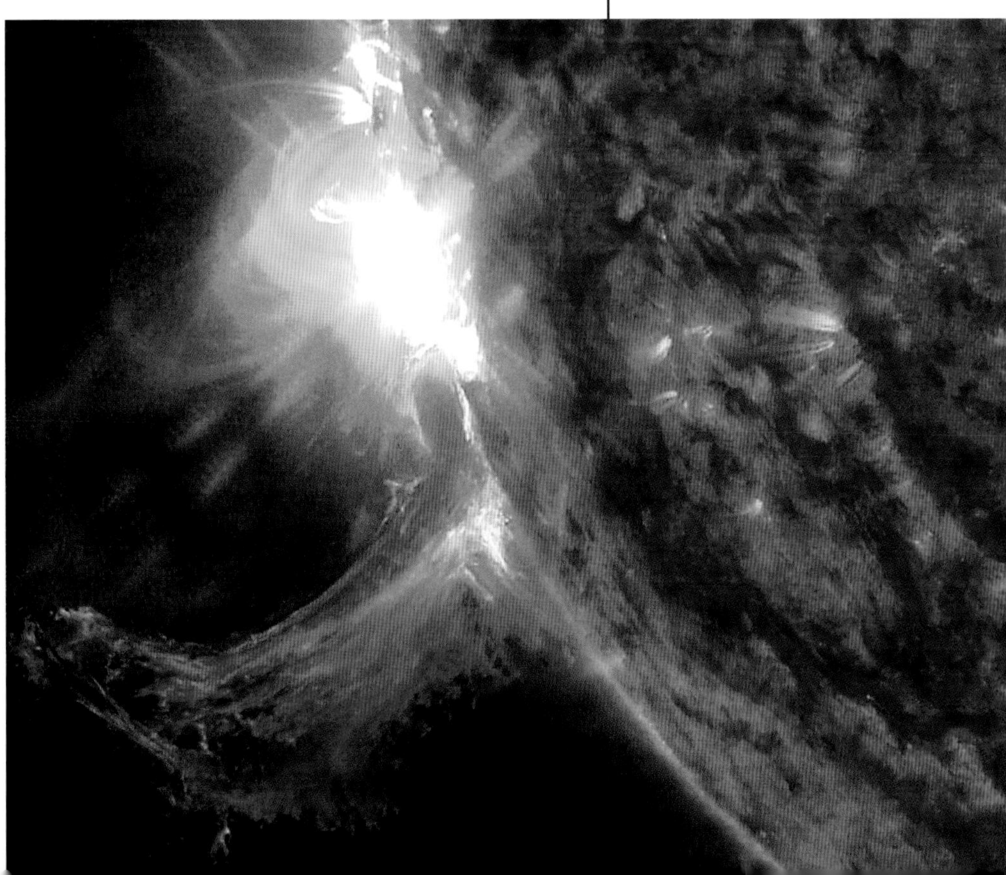

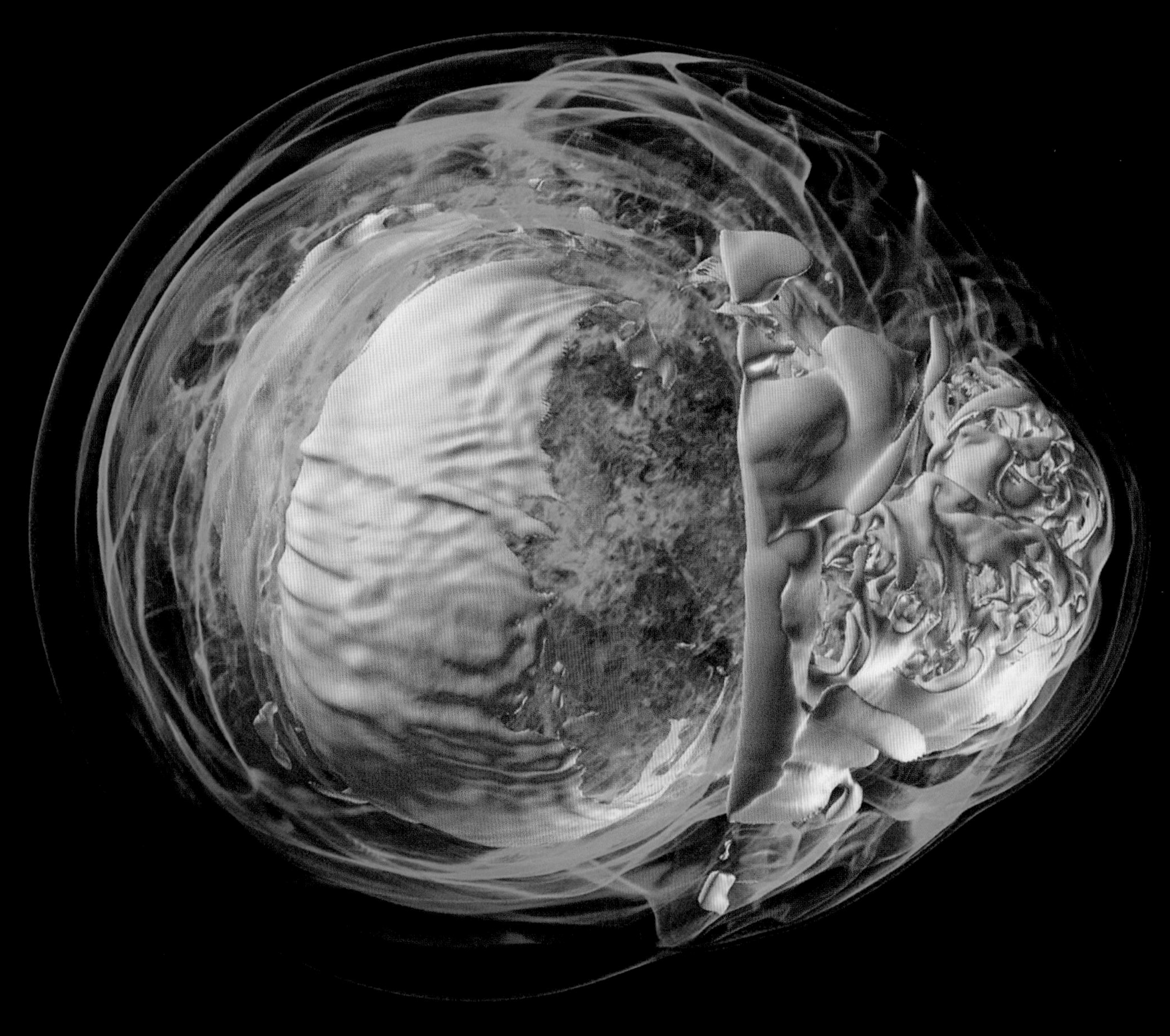

Entropy
Once a star has been destroyed in a supernova (shown here in a computer simulation), its energy dissipates into space, and the original star cannot put itself back together.

THE LAWS OF THERMODYNAMICS

Thermodynamics is the study of the physics of heat, temperature, work, and energy. The fundamental laws of thermodynamics describe various properties of thermodynamic systems - for example, the temperature of gas inside a balloon - and how they change under different circumstances.

Traditionally, there are three laws of thermodynamics, although another law was later added. This zeroth law states that if two systems are in thermal equilibrium - meaning there is no heat flowing between them - with a third system, they are in thermal equilibrium with each other.

The first law is an expression of conservation of energy: the rule is that energy cannot be created or destroyed, so there is a fixed amount of energy in the Universe. The law states, in simple terms, that the energy gained (or lost) by a system is equal to the energy lost (or gained) by its surroundings.

The second and third laws deal with the concept of entropy. This is a measure of the disorderliness of a system. An ordered system (one with low entropy) can do more work, such as driving a piston. The second law states that the entropy of a system never decreases. In other words, systems do not spontaneously become more ordered. The third law states that a system's entropy approaches a constant value as the temperature moves towards absolute zero. For a perfect crystal at absolute zero, this value would be zero, because the system would be in perfect order.

Cold atom laboratory
Absolute zero (0°K or -273.15°C) is impossible to reach, although scientists have cooled systems to trillionths of a degree above it.

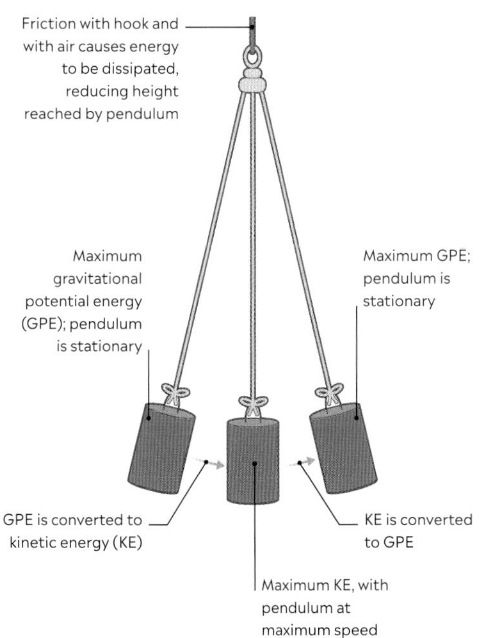
Friction with hook and with air causes energy to be dissipated, reducing height reached by pendulum
Maximum gravitational potential energy (GPE); pendulum is stationary
Maximum GPE; pendulum is stationary
GPE is converted to kinetic energy (KE)
KE is converted to GPE
Maximum KE, with pendulum at maximum speed

THE FIRST LAW

Energy cannot be created or destroyed, but it can change forms, such as from gravitational potential to kinetic. The change in the system's energy is equal to the difference between heat added to the system from its surroundings (energy in) and work done by the system on its surroundings (energy out).

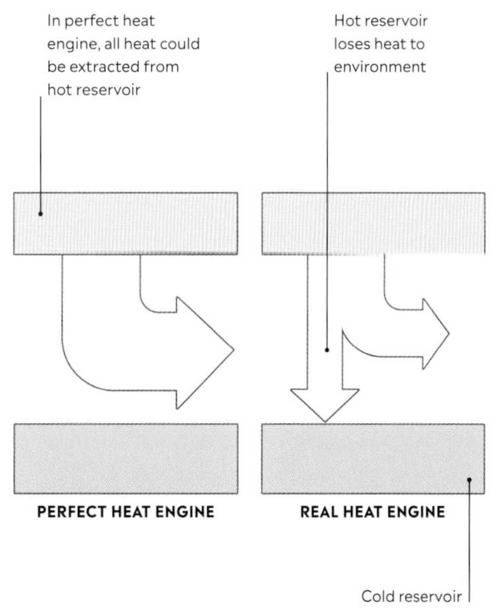
In perfect heat engine, all heat could be extracted from hot reservoir
Hot reservoir loses heat to environment
PERFECT HEAT ENGINE
REAL HEAT ENGINE
Cold reservoir

THE SECOND LAW

The second law states, in essence, that heat always moves spontaneously from warmer to cooler objects and not the other way round. A heat engine - a machine that uses heat to do work in a cyclical process - cannot use all its heat to do work, because some is lost to its cool surroundings.

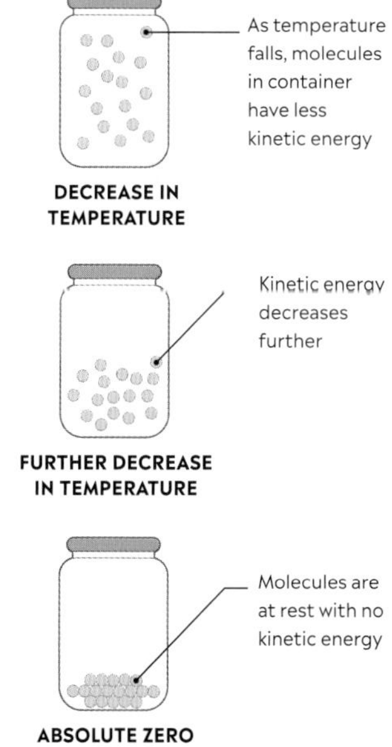
As temperature falls, molecules in container have less kinetic energy
DECREASE IN TEMPERATURE
Kinetic energy decreases further
FURTHER DECREASE IN TEMPERATURE
Molecules are at rest with no kinetic energy
ABSOLUTE ZERO

THE THIRD LAW

According to the third law, the cooler a system gets, the more orderly it is - and the lower its entropy. This reaches a minimum when the temperature hits a theoretical absolute zero.

SPECTROSCOPY

Atoms and molecules absorb and emit radiation, and this radiation - from light, microwaves, and radio frequencies, among others - creates a unique spectrum of wavelengths. Spectroscopy - the study of these spectra - allows for the identification of atoms and the determination of the structures of complex molecules.

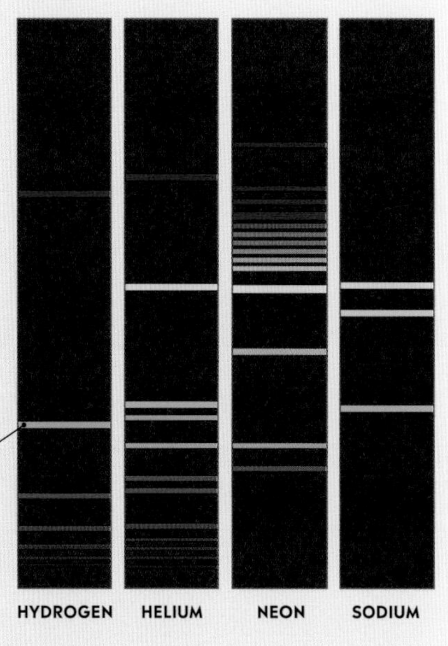

Each line indicates a wavelength of light

Emission spectra
It is possible to identify pure elements by the unique pattern of wavelengths of light that they emit.

1860

COLOUR ANALYSIS

German chemist Robert Bunsen and physicist Gustav Kirchhoff discovered two previously unknown elements: rubidium and caesium. They did so by using a spectroscope to study the spectra of coloured light given off when samples were heated using a gas burner invented by Robert Bunsen. Their approach led to the discovery of many other elements.

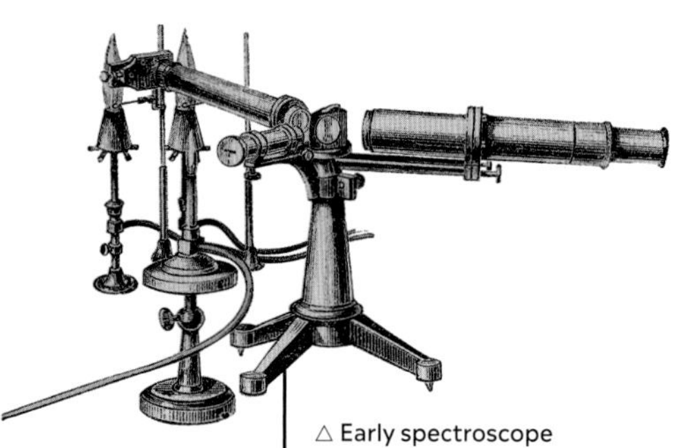

△ Early spectroscope

1860

1860 Mathematical tools known as Maxwell-Boltzmann statistics are developed that allow the analysis of the behaviour of gas molecules

1860 French chemist Pierre-Eugène-Marcellin Berthelot uses organic synthesis to make compounds from nature, something many thought impossible

1860 Belgian engineer Jean J. Lenoir develops the first internal combustion engine

△ Gustav Kirchoff

1860

BLACK BODY RADIATION

German physicist Gustav Kirchhoff introduced the idea of an hypothetical idealized substance that would be a perfect emitter and absorber of radiation: a black body. His attempts to characterize the emission of black bodies were later taken up by Max Planck and eventually led to the introduction of the ideas of quantum physics.

1861

ARCHAEOPTERYX

Long believed to be the earliest known bird, *Archaeopteryx* (literally, old wing) was first described from a single feather found in a limestone quarry near Solnhofen, Germany in 1861. *Archaeopteryx* lived about 150 million years ago and had features of both reptiles and birds, suggesting an evolutionary link.

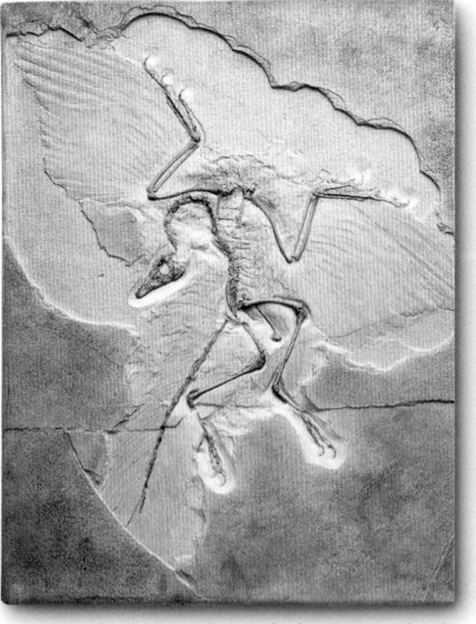

△ ***Archaeopteryx*** fossil

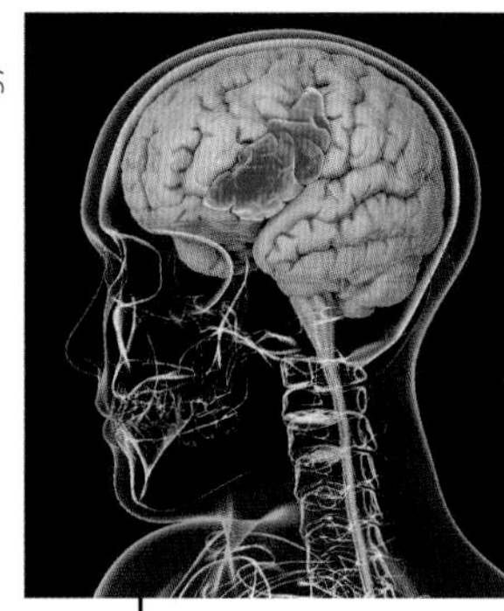
▽ **Inferior** frontal gyrus

1861

BROCA'S AREA

Studying the brain of a patient, Louis Victor Leborgne, who had lost much of his ability to speak, French physician Pierre Paul Broca discovered a lesion on the man's left frontal lobe. Broca correctly surmised that the affected area, which is in the inferior frontal gyrus, was linked to speech.

1862

CHLOROPLASTS

German botanist and physiologist Julius von Sachs found that chlorophyll grains (chloroplasts) are the site of the formation of starch from inorganic chemicals, that this occurs under the influence of light, and that starch is necessary for plant growth. Sachs is widely regarded as the pioneer of studies on photosynthesis.

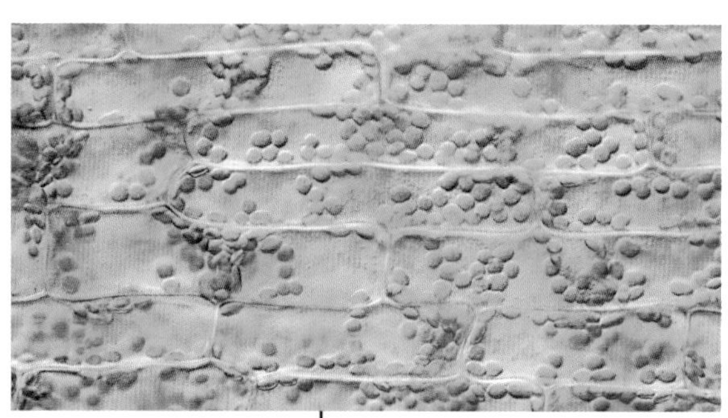
△ **Chloroplasts** in pond weed cells

"I am on the verge of mysteries and the veil is getting thinner and thinner."

LOUIS PASTEUR, LETTER, 1851

1862

1861 German biologist Max Schultze describes cells as consisting of protoplasm that contains a nucleus; this opens up the study of cell biology

▽ **Haemoglobin** molecule model

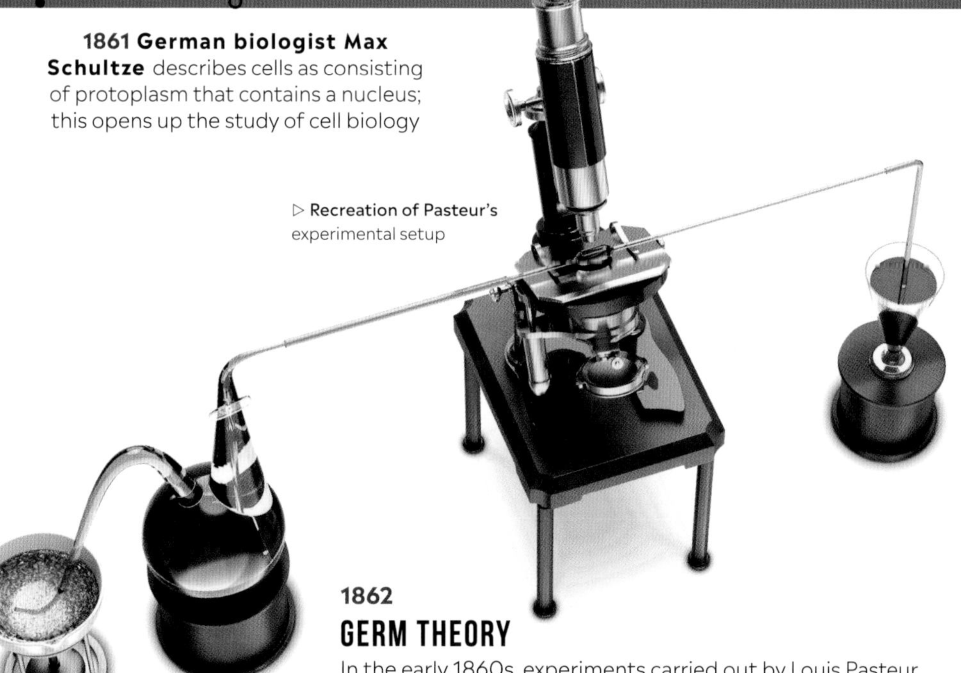
▷ **Recreation of Pasteur's** experimental setup

1862

GERM THEORY

In the early 1860s, experiments carried out by Louis Pasteur helped to confirm that many diseases are caused by germs entering the body from outside. He established, for example, the existence of tiny organisms in the blood of a patient suffering with a fever. This germ theory of disease replaced the earlier, rather vague miasma theory, which proposed that diseases were caused by "bad air".

1862

HAEMOGLOBIN

German physiologist and chemist Ernst Felix Hoppe-Seyler was one of the founding fathers of biochemistry and molecular biology. Among other investigations, he studied and named the red blood pigment haemoglobin and demonstrated that it binds oxygen in red blood cells to form oxyhaemoglobin.

The greenhouse effect
Sunlight passing through the atmosphere warms Earth's surface, with some heat reflected back to space. However, heat is also retained in the atmosphere by greenhouse gases.

SUN
THE ATMOSPHERE
EARTH
Some heat escapes back into space
Some heat is trapped in the atmosphere by greenhouse gases
Some heat is reflected back from Earth's surface
Sunlight heats Earth's surface

1863
THE GREENHOUSE EFFECT

The Irish physicist John Tyndall and American scientist Eunice Foote independently discovered the role of gases such as carbon dioxide and water vapour in absorbing heat. They were able to explain the discovery that Earth's surface temperature was higher than expected by identifying that these atmospheric gases absorb incoming heat from sunlight and prevent that heat from leaving the atmosphere. This "greenhouse effect" means that heat tends to accumulate at the surface of the planet. Tyndall also suggested that changes in the amount of these gases in the atmosphere could alter climate.

1863

1863 English scientists William Huggins and William Allen Miller observe the spectra of stars, using them to identify elements in their atmospheres

1863 English chemist John Newlands identifies a repeating pattern in the properties of the elements ordered by atomic weight and suggests it follows a "law of octaves"

1863
ANTICYCLONES AND WEATHER MAPS

English polymath Francis Galton was the first to collect, map, and interpret weather data for a particular region and date. In doing so, he discovered atmospheric zones of high pressure called anticyclones and established meteorology, the modern scientific study of weather. His first weather map for northwest Europe was for 31 March 1875 and was published in *The Times* newspaper the following day. It showed changes in the state of the sea and sky, air pressure, and temperature, along with wind direction and strength.

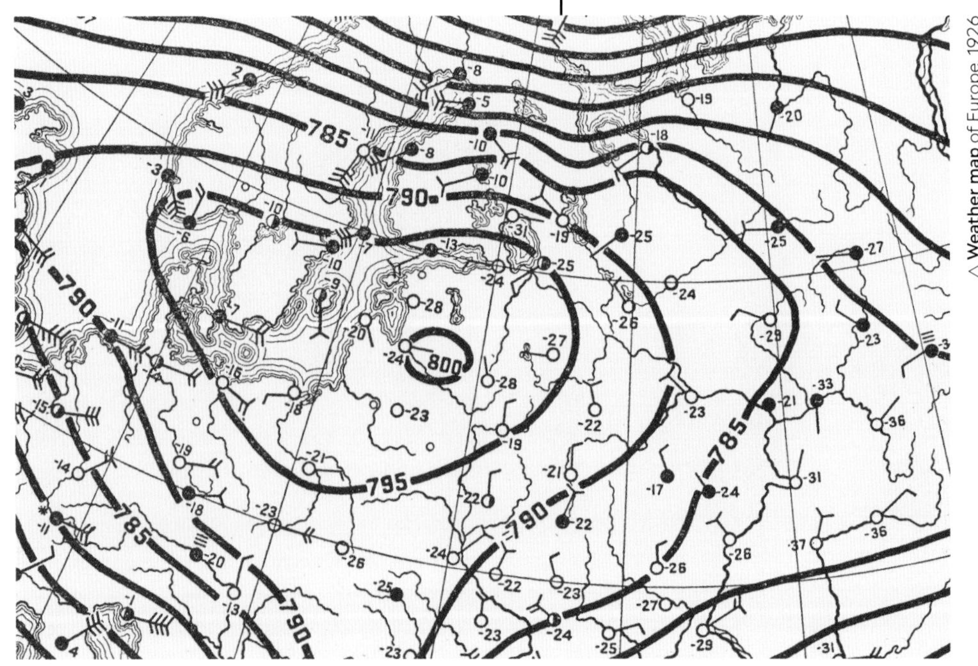

△ **Weather map** of Europe, 1926

1825–95

THOMAS HENRY HUXLEY

British biologist Huxley was born in London and studied medicine before becoming a naval surgeon. He became a strong supporter of Charles Darwin's theory of evolution after meeting him in 1856.

"The method of scientific investigation is nothing but the expression of the necessary mode of working of the human mind."

THOMAS HUXLEY, ON OUR KNOWLEDGE OF THE CAUSES OF THE PHENOMENA OF ORGANIC NATURE, *1863*

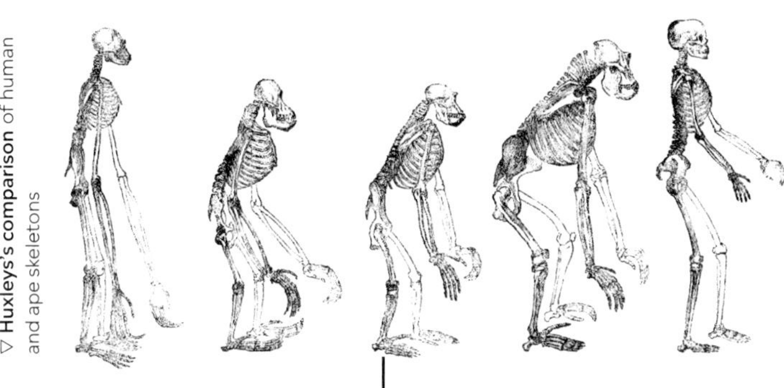

▽ **Huxleys's comparison** of human and ape skeletons

1863

HUMAN EVOLUTION

English biologist Thomas Huxley applied Darwin's theory to human evolution, publishing his ideas in *Evidence as to Man's Place in Nature*, a popular work that dealt mainly with the ancestry of the human species. His enthusiastic support for Darwin had earlier earned him the nickname "Darwin's Bulldog".

1864

1864 English biologist Herbert Spencer, influenced by Darwin's work, coins the famous phrase "survival of the fittest"

1864 US astronomer Hubert Anson Newton identifies a 33-year cycle in the intensity of the Leonid meteor storm

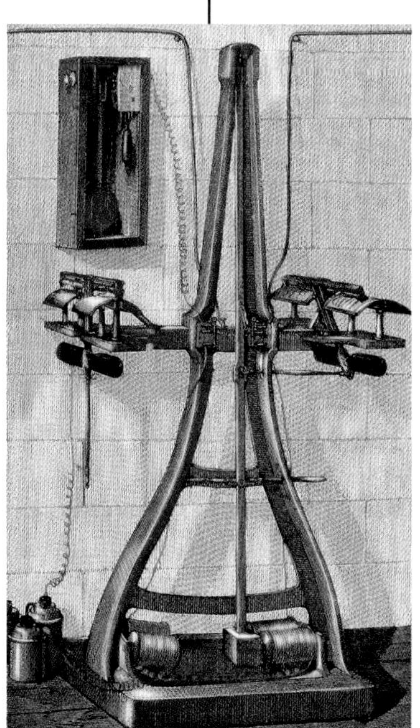

◁ **A pantelegraph**

1863

THE PANTELEGRAPH

Italian physicist Giovanni Caselli's invention, the pantelegraph, which transmitted handwritten messages over the telegraph network, was first made available to the public in France in 1863. A sender used electrically insulating ink to mark a sheet of paper. A long pendulum passed an electrode to and fro across the paper, and the resulting electrical signals passed along telegraph cables to a second machine, with a pendulum synchronized to the first. At the receiving end was a sheet of paper soaked in a solution that darkened when electrical current flowed through it, and so the message was reproduced.

▽ **A mammoth** carved on mammoth ivory

1864

MAN AND MAMMOTH

The discovery of an ancient engraving at La Madeleine rock shelter in the Dordogne region of France shed new light on the lives of early humans. Led by French palaeontologist Édouard Lartet, the Anglo-French expedition team uncovered a piece of mammoth ivory carved with a depiction of a mammoth. The artefact showed that humans had once lived alongside extinct species.

ELECTROMAGNETIC RADIATION

Wavelengths and vision
The human eye can see only a small section of the electromagnetic spectrum: visible light (*left*). Other animals detect different parts - for example, bees see in UV (*right*).

Along with matter, electromagnetic radiation is one of the main components of the Universe. It consists of waves that move through space at the speed of light. The waves themselves are electric and magnetic fields oscillating in perfect synchronicity. Changing electric fields create magnetic fields, and changing magnetic fields create electric fields, so they sustain each other. The interactions between electric and magnetic fields - not separate phenomena but two aspects of a fundamental force called the electromagnetic force - are described by equations formulated by physicist James Clerk Maxwell.

There are many sources of electromagnetic waves, both natural and artificial, from which they emanate into space like ripples across the surface of a pond. For example, outer space is filled with radio waves from objects like pulsars (dense, rotating stars) but also from communication satellites. All forms of ordinary matter emit a certain amount of electromagnetic radiation.

The different types of electromagnetic radiation form an extremely wide range known as the electromagnetic spectrum. Although all electromagnetic waves have the same speed, they have different frequencies and wavelengths. Radio waves have the lowest frequency and longest wavelength; gamma rays have the highest frequency and shortest wavelength. The various parts of the spectrum have distinct properties - for example, visible light can be detected by the human eye, while the highest-frequency (most energetic) waves can damage living cells, making them powerful tools in medical treatments.

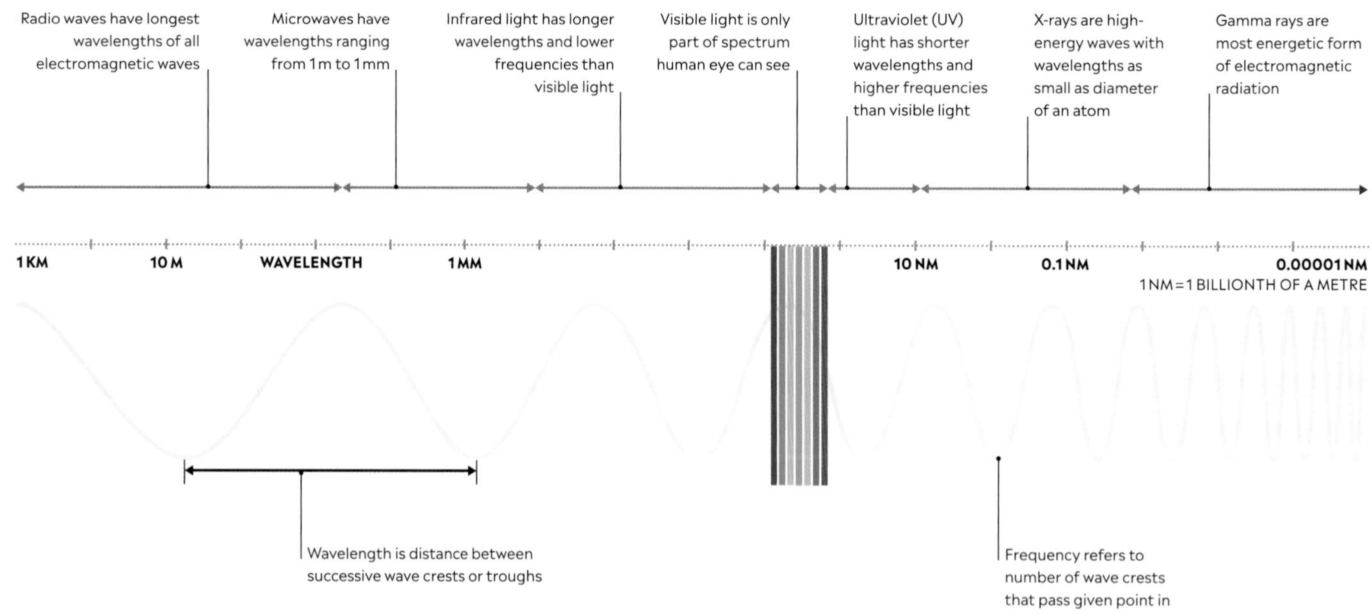
Radio waves have longest wavelengths of all electromagnetic waves
Microwaves have wavelengths ranging from 1 m to 1 mm
Infrared light has longer wavelengths and lower frequencies than visible light
Visible light is only part of spectrum human eye can see
Ultraviolet (UV) light has shorter wavelengths and higher frequencies than visible light
X-rays are high-energy waves with wavelengths as small as diameter of an atom
Gamma rays are most energetic form of electromagnetic radiation
1 KM
10 M
WAVELENGTH
1 MM
10 NM
0.1 NM
0.00001 NM
1 NM = 1 BILLIONTH OF A METRE
Wavelength is distance between successive wave crests or troughs
Frequency refers to number of wave crests that pass given point in 1 second

THE ELECTROMAGNETIC SPECTRUM

The electromagnetic spectrum is often represented as a horizontal band with radiations of different wavelengths and frequencies spaced out along it. The spectrum is usually divided into seven regions - from lowest frequency and longest wavelength, to highest frequency and shortest wavelength. Wavelengths can be longer than the radius of Earth or smaller than an atomic nucleus.

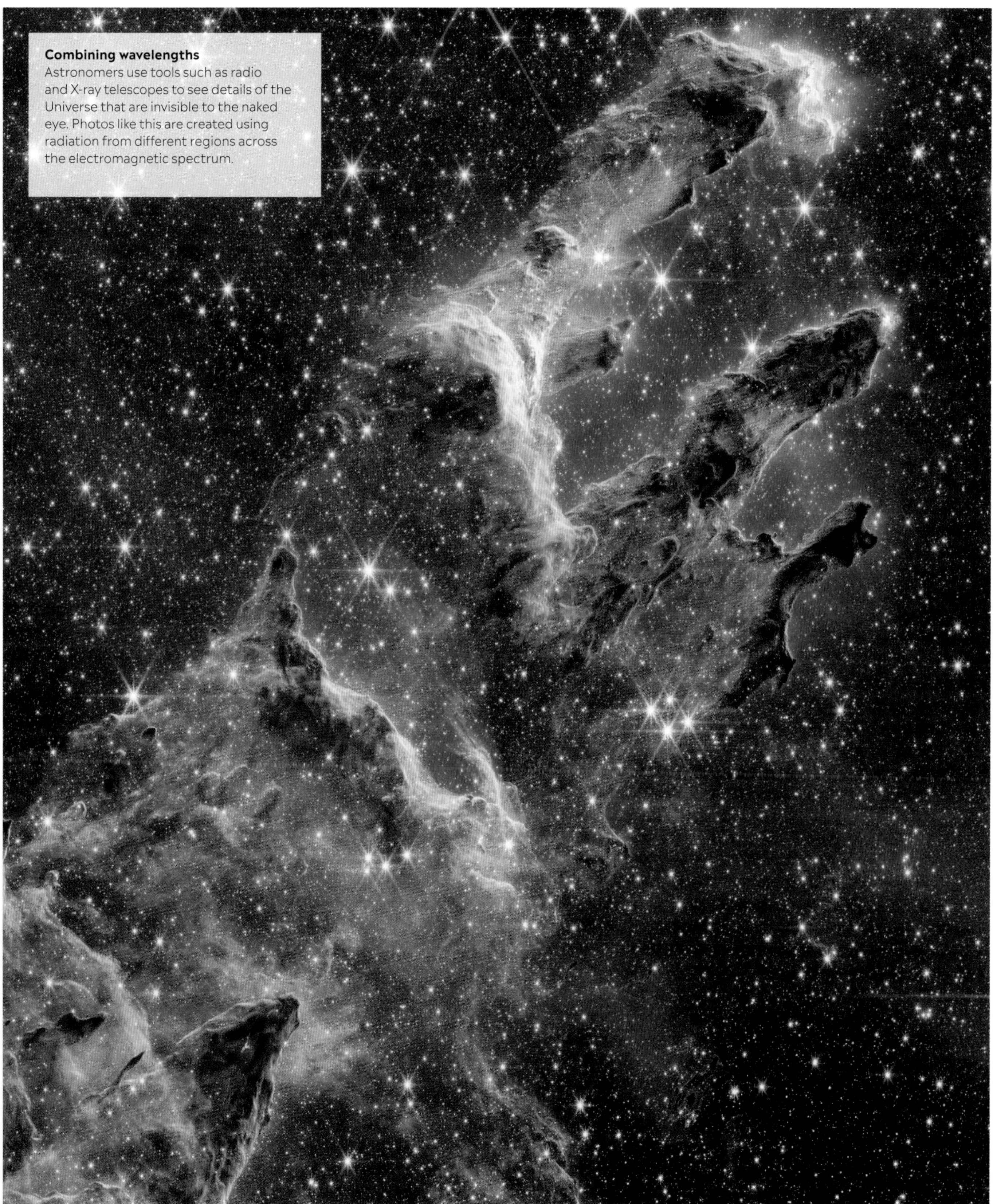

Combining wavelengths
Astronomers use tools such as radio and X-ray telescopes to see details of the Universe that are invisible to the naked eye. Photos like this are created using radiation from different regions across the electromagnetic spectrum.

1865

STRUCTURE OF BENZENE

Benzene was discovered in 1825, but scientists remained puzzled about the structure of this important chemical until 1865. Then, German chemist August Kekulé, having dreamed about a snake eating its own tail, realized that a ring of six carbon atoms, each with a hydrogen atom attached, solved the problem of the benzene molecule's structure.

△ **The motif of the snake** eating its tail dates back to Ancient Egypt.

◁ **Benzene's** carbon atoms form a ring in which each is bonded to a single hydrogen atom

carbon atom

hydrogen atom

1822–84

GREGOR JOHANN MENDEL

Austrian biologist and monk Gregor Mendel is known as the father of genetics for his pioneering work on the inheritance of traits such as flower colour in pea plants, which he grew in the garden of the monastery in Brno (now in the Czech Republic).

1865

1865 Gregor Mendel presents his work on crossbreeding varieties of peas with different flower colour, establishing the basic principles of genetic inheritance

1865 German physicist Rudolf Clausius coins the term "entropy" (disorder), which becomes a key concept in thermodynamics

1865 Joseph Lister uses carbolic acid as an antiseptic, greatly increasing surgical survival rates

▽ **Paper Möbius strip**

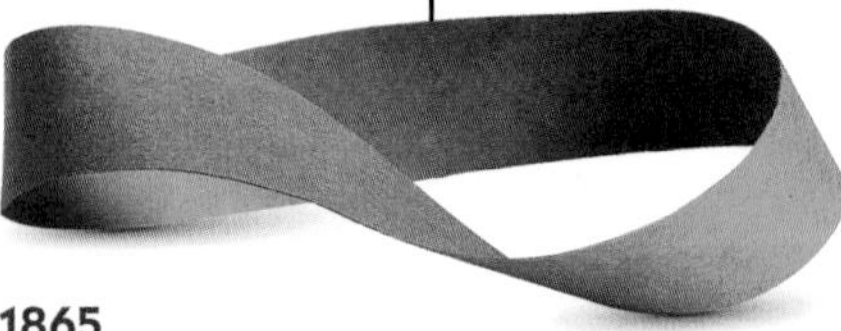

1865

MÖBIUS STRIP

German mathematician Rudolf Möbius was a key figure in the development of topology, the study of the geometrical properties of shapes during transformations (such as stretching) that underpins many aspects of physics. In 1865, Möbius wrote an influential paper about the Möbius strip, a one-sided, one-edged object that he and fellow mathematician Johann Benedict Listing had studied in 1858.

1865

MAXWELL'S EQUATIONS

In an 1865 paper, Scottish physicist and mathematician James Clerk Maxwell combined his equations describing the behaviour of, and interactions between, electric and magnetic fields into a single formula. This turned out to be a wave equation: the speed of the wave the formula describes is the speed of light. Maxwell proved that light is an electromagnetic wave and, at the same time, predicted the existence of a whole range of other electromagnetic waves that travel at the speed of light.

▷ **James Clerk Maxwell**

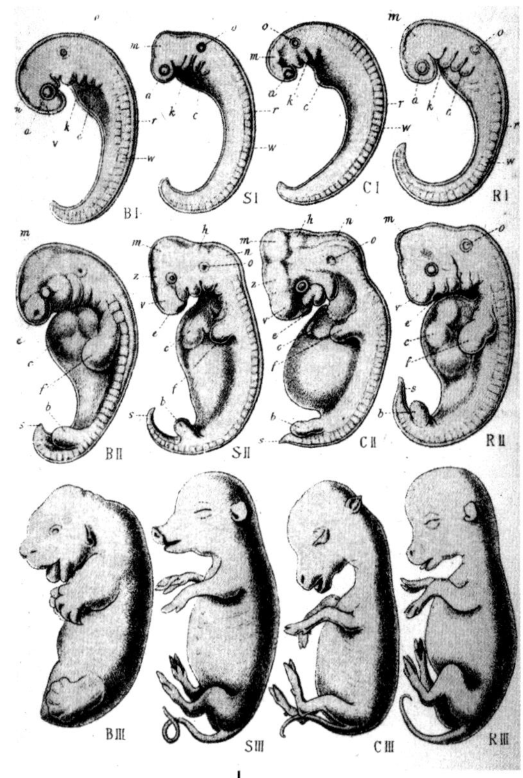

◁ **Haeckels's** drawings of embryo development

1866
THE BIOGENETIC LAW

German zoologist Ernst Haeckel's evolutionary theory, known as the biogenetic law, suggested that the stages of a developing animal embryo resembled stages of the adult bodies of other animals, and each embryonic developmental stage represented the adult form of an evolutionary ancestor. Thus, evolutionary relationships between animals could be discerned by their embryonic stages. Haeckel's theory was later shown to be inadequate to acount for enbyonic development.

Protists are organisms that do not fit into animal, plant, bacterial, or fungal groups

1866 Ernst Haeckel coins the term Protista for a third kingdom of organisms to be added to the then established kingdoms of Animalia and Plantae

1866

HEREDITY

Also known as inheritance, heredity is the passing of characteristics from parents to offspring. Many characteristics are determined by genes. During sexual reproduction, offspring inherit a random selection of each parent's genes. Some genes are dominant, some recessive, and some work in combination. Pea plants have been used to study inheritance because the pods and their seeds have distinct, measurable features, such as height, seed shape, and colour.

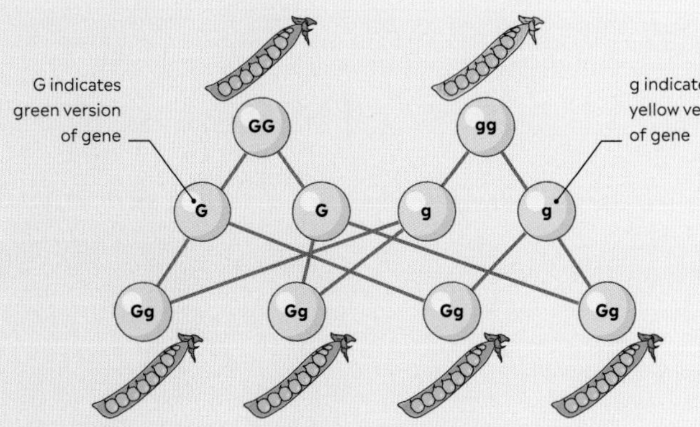

First cross
When green pods are crossed with yellow pods, the offspring are all green. This is because the greenness is determined by a dominant gene, or hereditary unit.

Second cross
When the green offspring are bred, one in four of the next generation is yellow, as a result of the greenness gene not being present in those plants.

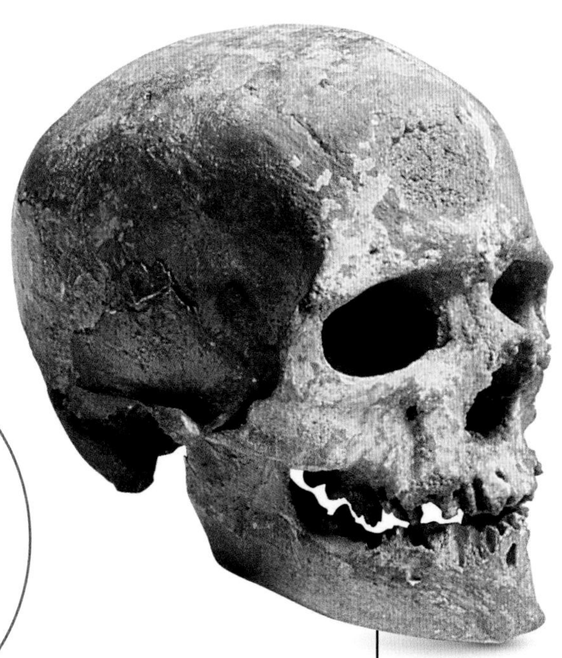

1868

CRO-MAGNON MAN

The first specimen of Cro-Magnon Man, some 30,000 years old, was discovered this year by French palaeontologist Louis Lartet. It is one of the first early examples regarded as belonging to our own species, *Homo sapiens*, and is named after the site of discovery – a rock shelter at Cro-Magnon near Les Eyzies in France. The cave contained the skeletons of four humans, along with a few ornaments. The skull has features unique to modern humans, being tall and rounded, and lacking obvious brow ridges.

◁ ***Homo sapiens*** skull

1847–1931

THOMAS EDISON

Raised in the American Midwest, Edison was a prolific inventor and successful businessmen, famed for his invention of the phonograph, development of electric lighting, and as a pioneer of movie cameras.

1867

1867 Italian astronomer Angelo Secchi publishes a star catalogue introducing a scheme for classifying stars according to features in their spectra

1867 Swedish chemist Alfred Nobel patents dynamite, a safer way to use the explosive nitroglycerine

1868 French astronomer Jules Janssen discovers the element helium by spotting an unexplained spectral line in the Sun's atmosphere during an eclipse

1868 Prolific American inventor Thomas Edison patents a machine to record votes; it is the first of more than a thousand patents in his name

"My dynamite will sooner lead to peace than a thousand world conventions."

ATTRIBUTED TO ALFRED NOBEL

1868

DEEP SEA LIFE

Fascinated by the biology of the deep oceans, Scottish naturalist Charles Wyville Thomson embarked on a series of expeditions aboard the Royal Navy ships, *Lightning*, *Porcupine*, and, most famously, *Challenger* (*see p.154*). During his research, he took hundreds of soundings, dredgings, and temperature readings from across the oceans, and he showed that animal life existed down to depths of 1,200 m (4,000 ft). He also collected around 4,500 invertebrate species, many new but some previously thought to be extinct.

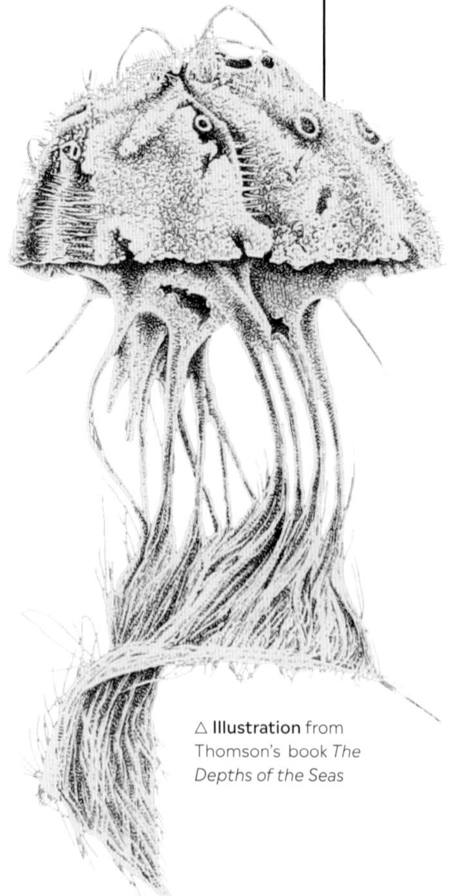

△ **Illustration** from Thomson's book *The Depths of the Seas*

1869
ZOOGEOGRAPHY

In *The Malay Archipelago*, Welsh naturalist Alfred Russel Wallace published his discovery, made during fieldwork in the area, of a divide between the animals of western Indonesia, mainly of Asian origin, and those further east, which had affinities with Australasian species. This divide is now referred to as the Wallace Line, and Wallace is widely regarded as the father of zoogeography.

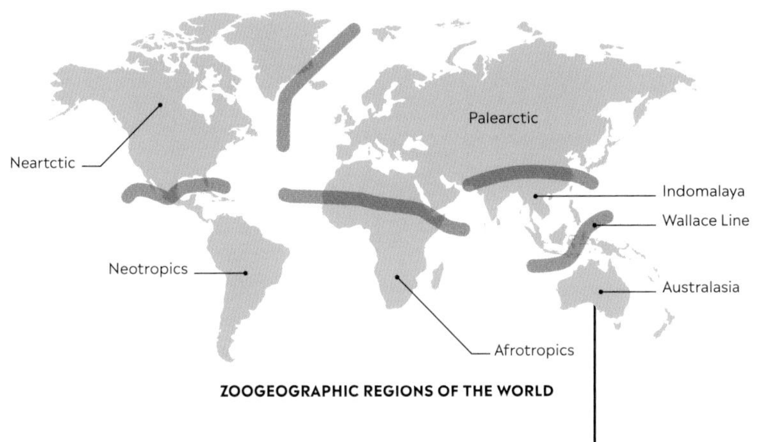

ZOOGEOGRAPHIC REGIONS OF THE WORLD

1869
DNA ISOLATED

While investigating the composition of cells, Swiss physician Friedrich Miescher isolated a new molecule from the nucleus of white blood cells. He named it "nuclein" and showed that it consisted of hydrogen, oxygen, nitrogen, and phosphorus. In effect, he was the first to isolate DNA (deoxyribonucleic acid). Although the roles of nucleic acids were not known at the time, Miescher believed that his isolates were somehow involved in heredity.

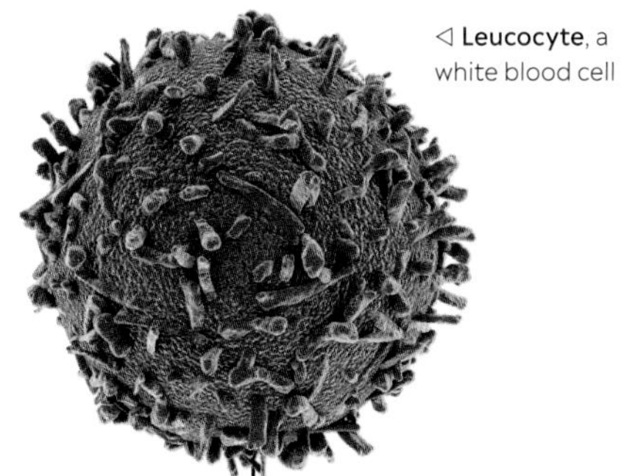

◁ **Leucocyte**, a white blood cell

1869

1868 William Huggins measures a Doppler shift in the light from Sirius, caused by its motion away from Earth

1869 Russian chemist Dmitri Mendeleev publishes his first table of elements sorted by atomic weight and shows there are periodic, or repeating, properties

1869
PANCREATIC CELLS

German physiologist Paul Langerhans studied the anatomy of the pancreas in a range of organisms. While working for his doctorate, he discovered nine distinct cell types in the pancreas and postulated that certain small polygonal cells might have a special function. More than 20 years later, French histologist Gustave-Édouard Laguesse named these cells the Islets of Langerhans; they are now known to produce hormones including insulin.

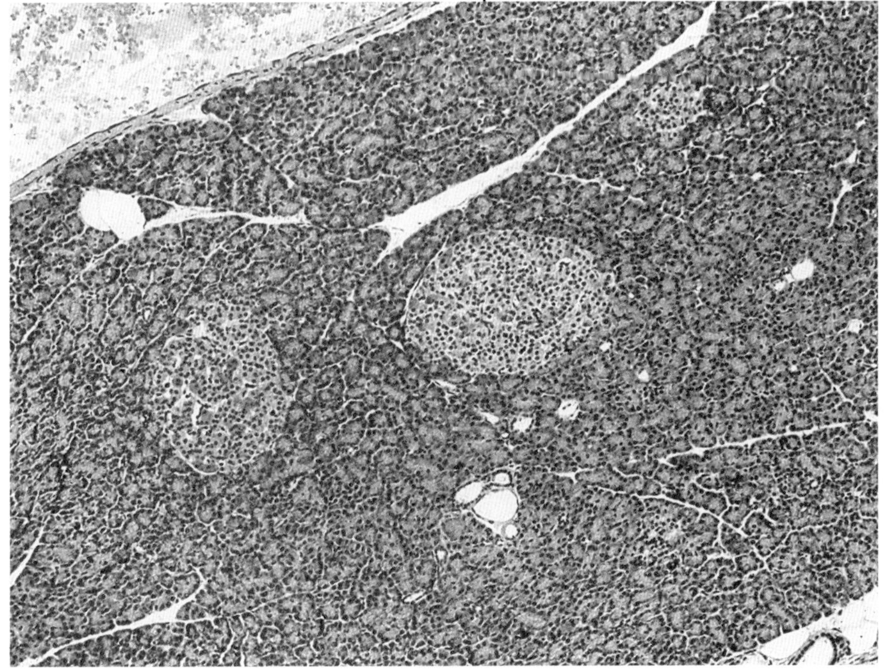

▷ **Islets of Langerhans,** light micrograph

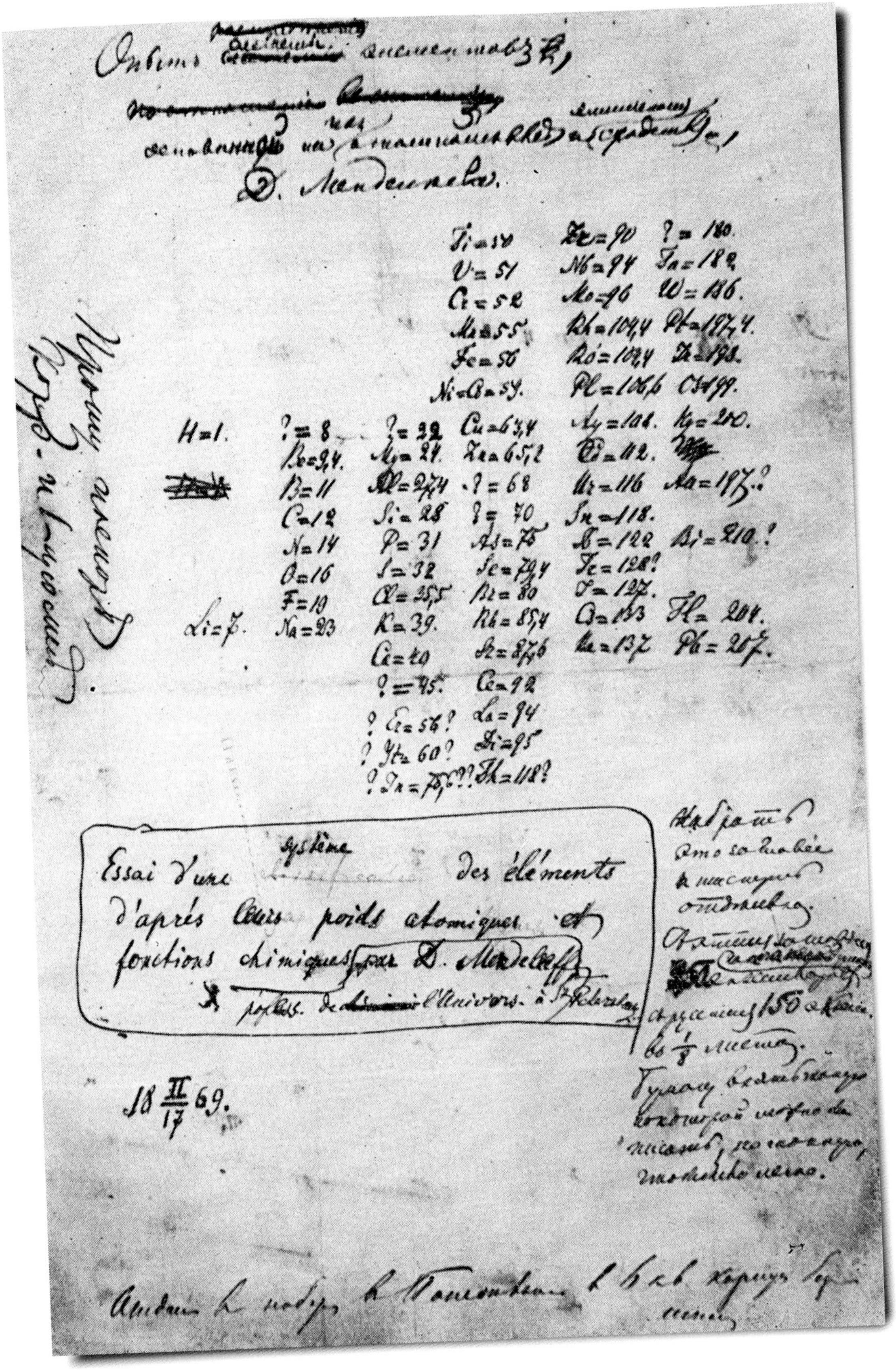

Mendeleev's notes
Russian chemist Dmitri Mendeleev's first description of a periodic table of elements included question marks to represent elements that had not yet been discovered but that he predicted should exist.

THE PERIODIC TABLE

The periodic table is an arrangement of the chemical elements that have been either discovered or created in laboratories. By arranging elements with similar properties into columns (groups), chemists realized that there were periodic (recurring) patterns of chemical and physical properties. Characteristics such as the state of the elements - as well as melting point, density, and hardness - all show similar properties.

The order of the table was formed without understanding the scientific explanation, and it was not until subatomic particles (protons, electrons, and neutrons) were discovered that the reason for the periodic table's order and structure became clear. The unique atomic number of each element is the number of protons - the positively charged particles - in the nucleus of the atom. There are also the same number of negatively charged electrons circling the nucleus in regions known as shells.

The pattern of the periodic table is explained by the way electrons fill these shells around the nucleus. All the elements in a row, or period, have the same number of electron shells. The shells may contain either 2, 6, 10, or 14 electrons.

The properties and chemical behaviour of the elements are determined both by their size and the number and distribution of electrons in the shells around the nucleus. For example, the noble gases of group 8 have a complete outer electron shell and are chemically unreactive because they do not need to share, or do not easily lose, electrons.

Missing elements
Mendeleev correctly predicted missing elements and their properties by seeing gaps in his periodic table. He called them eka-silicon, eka-aluminium, and eka-boron. They were discovered later and named germanium (shown here), gallium, and scandium respectively.

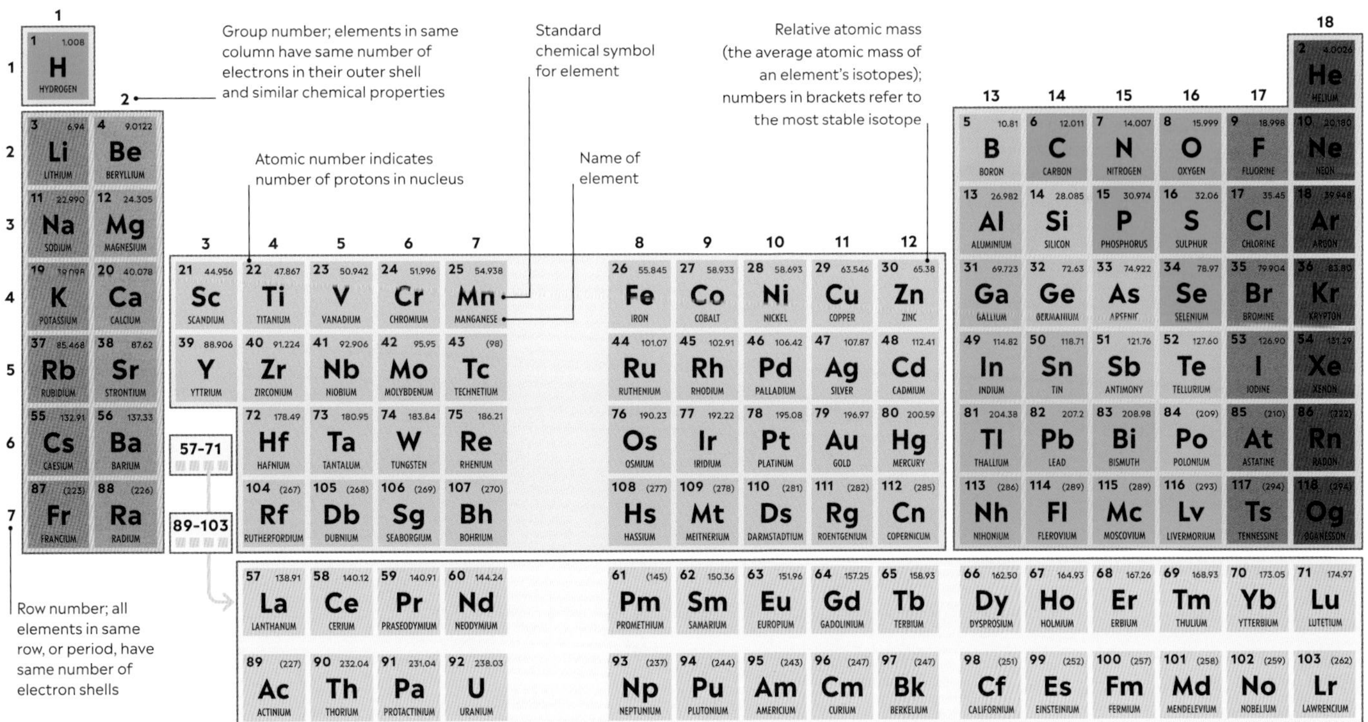

THE ELEMENTS

Chemistry is built on the understanding of the elements and of their unique physical and chemical properties. Elements rarely exist in their pure form in nature and have gradually been discovered and isolated. The elements are arranged in order of increasing atomic number, from 1 (hydrogen) to 118 (oganesson).

KEY
- Hydrogen

Reactive metals
- Alkali metals
- Alkaline earth metals

Transition elements
- Transition metals

Mainly nonmetals
- Metalloids
- Other metals
- Carbon and other nonmetals
- Halogens
- Noble gases

Rare earth metals
- Also called lanthanoids and actinoids, these are reactive metals - some are rare or synthetic

△ **A colourful male** bird of paradise with his mate

1871
SEXUAL SELECTION

In *The Descent of Man, and Selection in Relation to Sex*, English biologist Charles Darwin presented his sexual selection theory, proposing a role for the sexes in driving aspects of evolution. For example, in some species, females choose to mate with the most "attractive" males, which can lead to elaborate breeding plumage, such as that seen on the male bird of paradise.

1826–1911
GEORGE STONEY

Irish physicist Stoney taught at and was later secretary of Queen's University, Dublin. His work in molecular physics included identifying the electron as the basic unit of electrical charge.

1870

1870 German scientists Gustav Theodor Fritsch and Eduard Hitzig provide experimental evidence that the brain's cortex is involved with the control of movement

1871 Russian chemist Dmitri Mendeleev leaves gaps in his periodic table for undiscovered elements he predicts exist

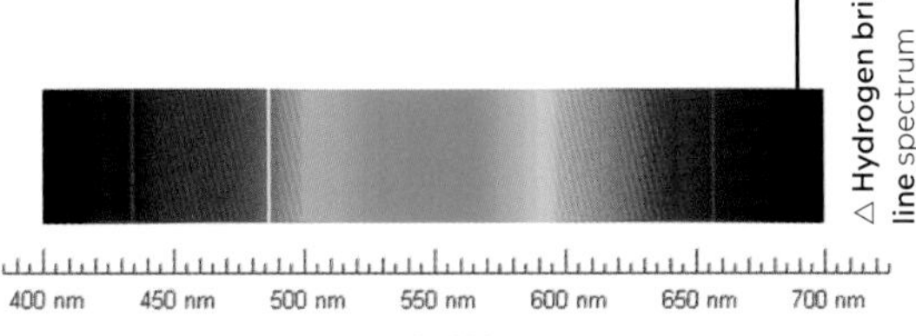

△ **Hydrogen bright** line spectrum

Early 1870s
HYDROGEN SPECTRUM

When in an excited state (its electrons temporarily occupying an energy state greater than their ground state), each element emits a characteristic spectrum of electromagnetic radiation – a set of bright lines of specific wavelength. Focusing on hydrogen, Irish physicist George Stoney was the first to notice the patterns in its spectrum. The relationships between the lines' wavelengths helped to found the science of quantum physics.

"The measurements of the *Challenger* expedition set the stage for all branches of oceanography."

DR JAKE GEBBIE, WOODS HOLE OCEANOGRAPHIC INSTITUTION, USA, 2020

1872
HMS CHALLENGER

Over four years and 127,000km (79,000 miles), the British Royal Navy's HMS *Challenger* began the modern mapping of the oceans. During the round the world expedition, naturalists Charles Wyville Thomson and John Murray systematically measured sea depths and discovered the shape of the Atlantic ocean basin.

△ HMS *Challenger*

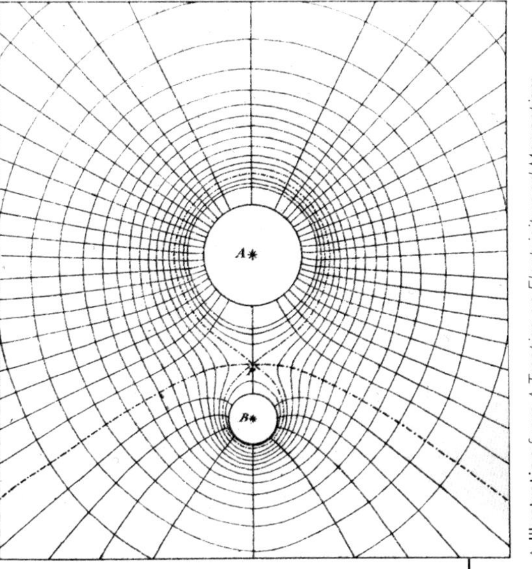

△ **Illustration** from *A Treatise on Electricity and Magnetism*

1873

STUDIES ON THE STATES OF MATTER

In his influential thesis *On the Continuity of Gas and Liquid States*, Dutch physicist Johannes van der Waals published an equation that described the behaviour of gases, vapours, and liquids. The equation took account of the attractive forces between molecules at a time when the very existence of molecules was still in doubt among most physicists. For his insight, Van der Waals was awarded the 1910 Nobel Prize in Physics.

◁ **Johannes van der Waals** depicted on a commemorative medal

1873

LAWS OF ELECTROMAGNETISM

James Clerk Maxwell published *A Treatise on Electricity and Magnetism*. It was a comprehensive and extremely influential treatise on all aspects of electromagnetism.

1873

1872 American astronomer Henry Draper captures the first detailed photograph of a stellar spectrum

1873

NERVE CELL STUDIES

While working with nerve tissue at a hospital in Abbiategrasso, the Italian biologist Camillo Golgi developed the staining technique that would be named after him. By making entire neurones clearly visible, the Golgi stain (or black reaction) enabled the detailed study of nerve tissue under the microscope. The technique paved the way for scientists to investigate many aspects of the brain and sensory neurones, and to classify cell organelles.

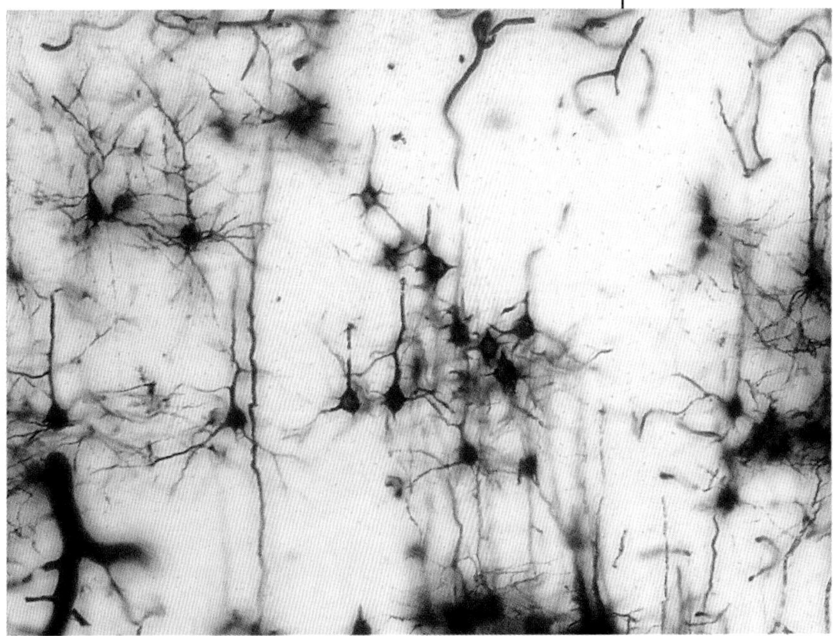

▷ **Golgi stain** of a section of mouse brain

1874
THREE-DIMENSIONAL MOLECULES

Dutch chemist Jacobus van 't Hoff and French chemist Joseph-Achille Le Bel suggested that four atoms bonding to a carbon atom form a tetrahedral shape. This insight explained why two forms of some organic molecules are mirror images of each other.

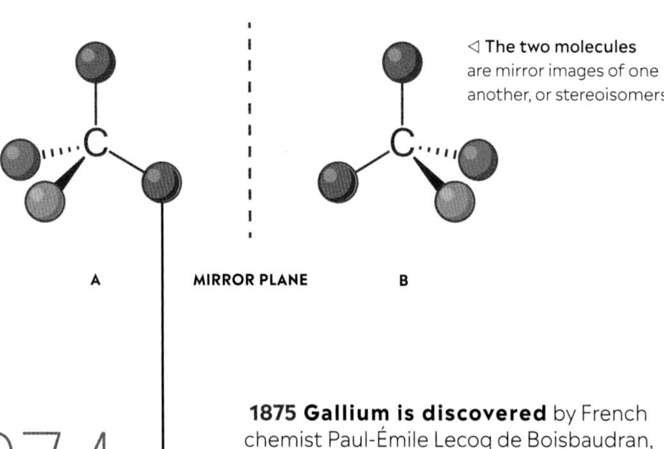

◁ **The two molecules** are mirror images of one another, or stereoisomers

1875
MECHANISM OF CELL DIVISION

German biologist Walther Flemming made important discoveries about cell division, which he named mitosis. Using aniline dyes, he identified chromatin in the thread-like structures within the cell nucleus. He also discovered the centrosome, a structure involved in the process of mitosis. In 1882, he published his findings in his book *Zellsubstanz, Kern und Zelltheilung* (*Cell Substance, Nucleus, and Cell Division*).

▷ **Drawing** of mitosis by Flemming

1874

1875 Gallium is discovered by French chemist Paul-Émile Lecoq de Boisbaudran, who shows it is one of the predicted elements on Mendeleev's periodic table

1875 James Clerk Maxwell produces compelling evidence that atoms have internal structure

1. The DNA within each chromosome is duplicated to form two identical copies, which are joined in the middle by the centromere.

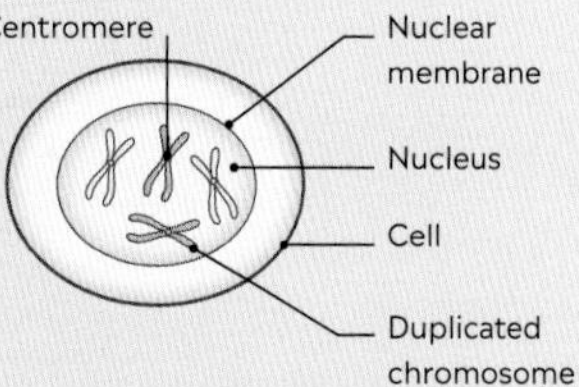

2. The membrane around the nucleus breaks down, and threads form across the cell. The chromosomes line up along the threads.

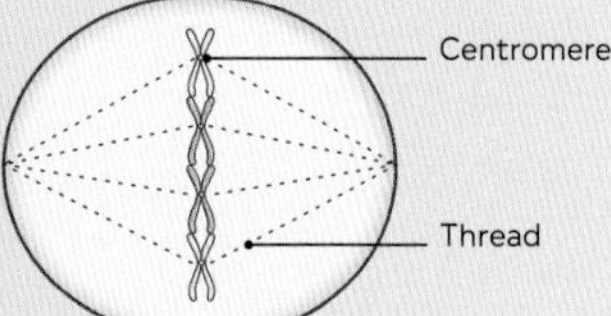

3. The newly duplicated chromosomes are pulled apart by the threads, and the single chromosomes move to opposite sides of the cell.

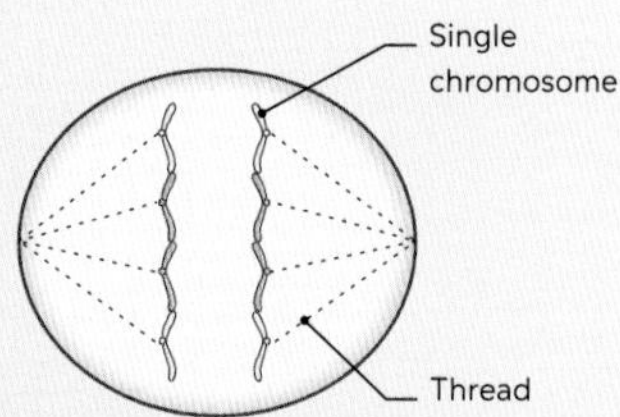

4. A nuclear membrane forms around each set of chromosomes, and the cell begins to divide into two new cells.

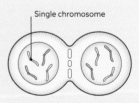

Single chromosome

5. Two new cells form. Each cell has a nucleus that contains an identical set of chromosomes with exactly the same genes.

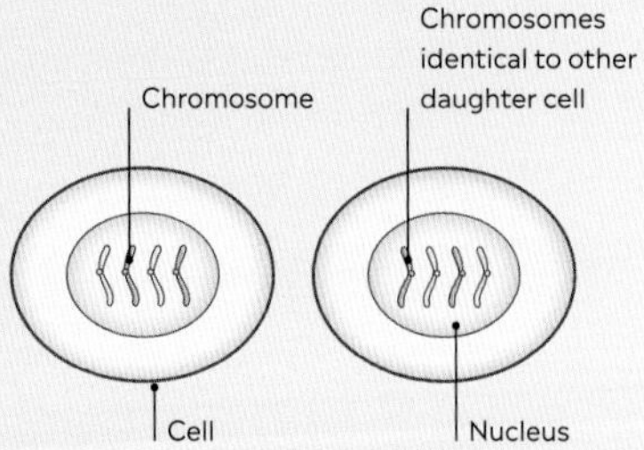

MITOSIS

The type of cell division that produces two identical daughter cells is called mitosis. It occurs during growth and to replace worn out cells. Before a cell divides, its DNA must first be duplicated. Each of the two strands in the original DNA acts as a template for the building of a new strand. This is followed by the division of the cell contents into two daughter cells with identical genomes.

"Mr Watson, come here — I want to see you."

ALEXANDER GRAHAM BELL, FIRST WORDS SPOKEN ON THE TELEPHONE, 1876

▷ **Early telephone** design

1876
FIRST TELEPHONE

Two people invented a viable telephone at about the same time, and applied for patents at the US patent office on the same day in 1876. They were Scottish-born American inventor Alexander Graham Bell and American engineer Elisha Gray. Bell was awarded the patent, and went on to found the American Telephone and Telegraph Company (AT&T).

1876 German physicist Eugen Goldstein coins the term "cathode rays" for electrical discharge in evacuated glass tubes

1876

1876 German engineer Nicolaus Otto develops the first practical four-stroke internal combustion engine, which leads the way to motor vehicles

c. 1875
CROOKES RADIOMETER

English physicist William Crookes noticed that the thin pans of his accurate chemical balance were affected by sunlight falling on them. After investigating this phenomenon, he devised the radiometer, an evacuated glass tube with lightweight vanes on a rotor. The rotor spins when light falls on the vanes. This phenomenon was the source of debate and disagreement among physicists for several years.

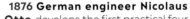

△ **Crookes** radiometer

△ **Josiah Williard Gibbs**

1876
CHEMICAL THERMODYNAMICS

American physicist Josiah Willard Gibbs published important ideas about the forces at work in chemical reactions. He defined the concepts of free energy and chemical potential, which are responsible for helping chemicals react. He also deduced the phase rule, which helps to determine how substances in contact may be changed by varying temperature, pressure, or concentration.

1877

MARTIAN DISCOVERIES

With Mars and Earth close to one another in the night sky, US astronomer Asaph Hall discovered Phobos and Deimos, two Martian satellites. Italian astronomer Giovanni Schiaparelli reported seeing straight lines linking the darker regions of Mars' surface, leading to speculation about Martian-made canals; the lines were later proved to be an optical illusion.

◁ **Schiaparelli's map of** Mars

1877

1877 The term biochemistry is used by German physician Ernst Hoppe-Seyler to describe the study of the reactions and molecules in living systems

1877 Thomas Edison develops the phonograph, a device for recording and playing back sound

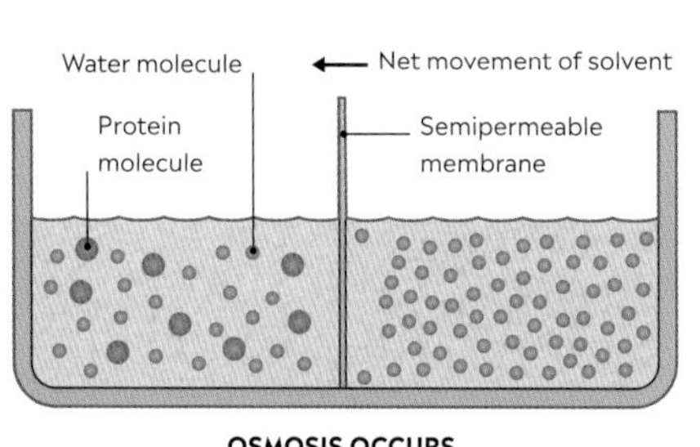

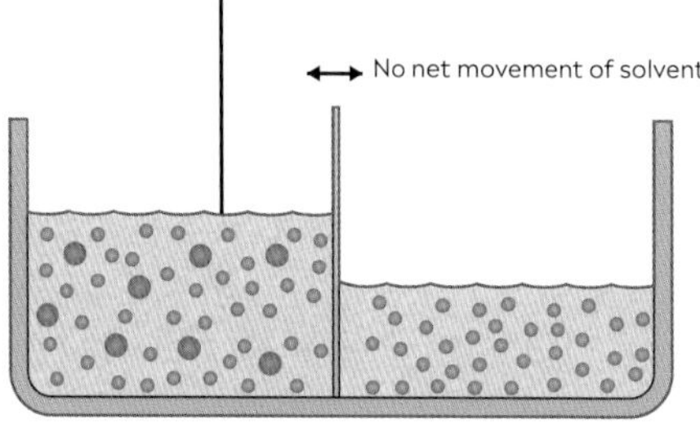

Osmotic pressure measurement
A semipermeable membrane separates a solution of water (the solvent) and protein molecules (solute) from water. Water molecules move through the membrane until the pressure on both sides of the membrane is the same.

1877

OSMOTIC PRESSURE

German plant physiologist Wilhelm Pfeffer pioneered studies into semi-permeable membranes that allow small water molecules in a solution to pass through while holding larger ones, such as proteins, back. Because the solvent (in this case water) can pass through the membrane (a process called osmosis), pressure builds up between the two solutions. Pfeffer used this measure of osmotic pressure to determine the molecular weights of proteins.

1877

THE BONE WAR

The discovery of dinosaurs in the American West fuelled rivalry between palaeontologists at the Academy of Natural Science of Philadelphia and the Peabody Museum of Natural History, Yale, leading to theft and bribery in the hunt for the best fossils.

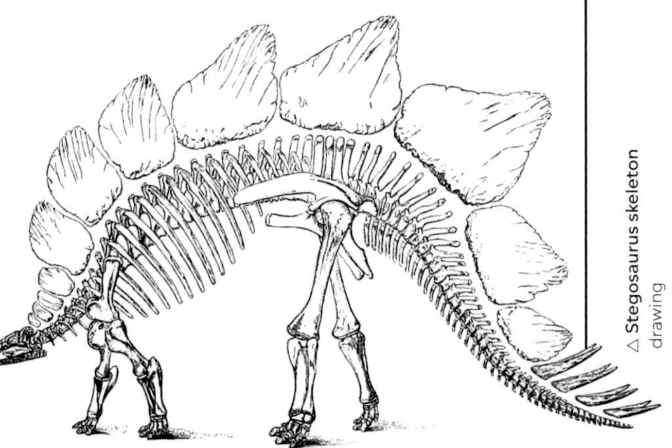

△ **Stegosaurus skeleton** drawing

1878

MULLERIAN MIMICRY

German naturalist Johann Friedrich Theodor (Fritz) Müller proposed that two distinct species may evolve to resemble each other closely. For example, the North American viceroy butterfly and the Central American monarch are unrelated but look very similar. Both are distasteful to predators, and their resemblance is a way of "sharing the cost" of producing a deterrent to predation.

△ **Monarch** butterfly

△ **Viceroy** butterfly

1879

RADIATIVE ENERGY

In 1879, Austrian physicist Josef Stefan derived an important equation that related the total amount of energy radiated by an object to the object's temperature. His work was extended in the 1880s, by Austrian physicist Ludwig Boltzmann, to become the Stefan-Boltzmann law.

▷ **Joseph Stefan**

1879 Louis Pasteur discovers the first attenuated vaccine when a live culture of chicken cholera bacteria that had been exposed to the air immunizes chickens

1879

1878 French physiologist Paul Bert connects the painful cramps (bends) experienced by divers to their descent and ascent in deep water

1879 American physicist Edwin Hall discovers the Hall effect, which is exploited in sensitive magnetic measuring devices

Edison and his associates tested thousands of materials before they chose carbonized cotton thread as the filament of their electric bulb

◁ **Edison's** first light bulb (replica)

1879

ELECTRIC LIGHT

The first electric light bulbs contained a thin filament made of carbonized bamboo or paper, which glowed white hot when electric current passed through it. Thomas Edison in the USA and Joseph Swan in Britain produced the first commercially successful bulbs, in 1879, though several inventors had produced bulbs before, with various degrees of success.

1880
SYNAESTHESIA

English polymath Francis Galton investigated the rare condition now known as synaesthesia ("joined perception"), whereby individuals experience unusual sensations triggered by unrelated phenomena, for example, associating musical sounds with different colours. The phenomenon had been known since early in the 19th century, though the term "synaesthesia" was not coined until around 1892. Many artists, including Russian abstract painter Wassily Kandinsky, are thought to have experienced this neurological phenomenon.

▷ ***The Waterfall,*** Wassily Kandinsky

1880

1880 English geologist John Milne detects earthquakes with his new seismograph

1880 French physicist Emile Amagat carries out experiments with gases under extreme pressures

1880
CATHODE RAYS AS PARTICLES

In the 1870s, physicists began investigating rays that travelled from one electrode to the other inside an evacuated glass tube called a Crookes tube (after British physicist William Crookes). Some believed the rays were electromagnetic radiation, others that they were a stream of particles. In 1880, Crookes found evidence that they were particles and in 1897, British physicist J.J. Thomson discovered the particles, now known as electrons.

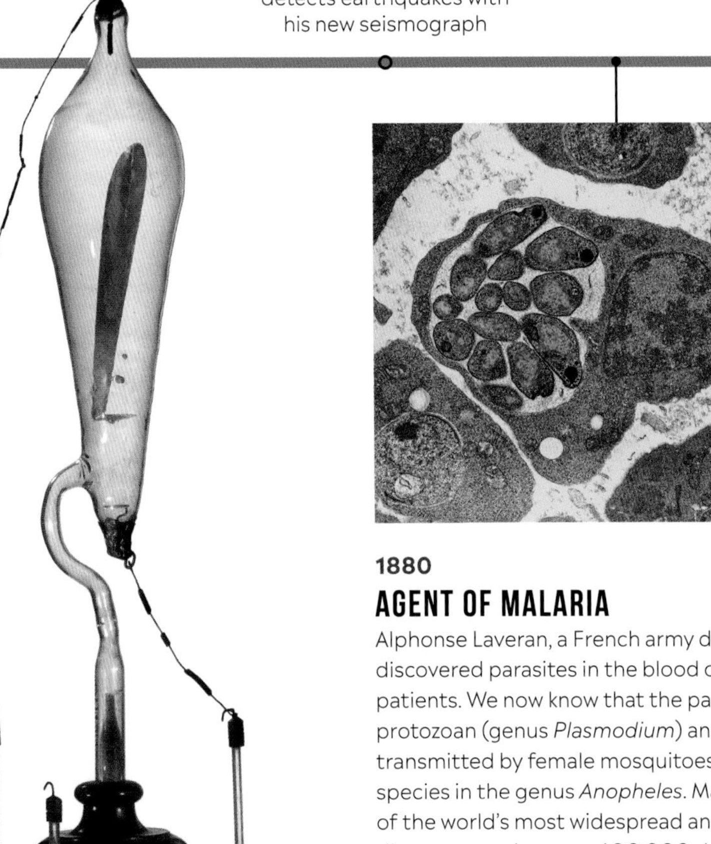

▷ **Crookes** tube

△ **Malaria-infected** blood cell

1880
AGENT OF MALARIA

Alphonse Laveran, a French army doctor, first discovered parasites in the blood of malaria patients. We now know that the parasite is a protozoan (genus *Plasmodium*) and that it is transmitted by female mosquitoes of certain species in the genus *Anopheles*. Malaria is one of the world's most widespread and debilitating diseases, causing over 600,000 deaths annually.

1880

PIEZOELECTRICITY

The brothers Pierre and Jacques Curie discovered that applying mechanical pressure to crystals such as quartz produces an electrical charge on the crystal surfaces. This is known as the piezoelectric effect from the Greek *piezo* ("push"). The effect also works in reverse, with charge generating a mechanical force in the crystal – a phenomenon applied in quartz timepieces.

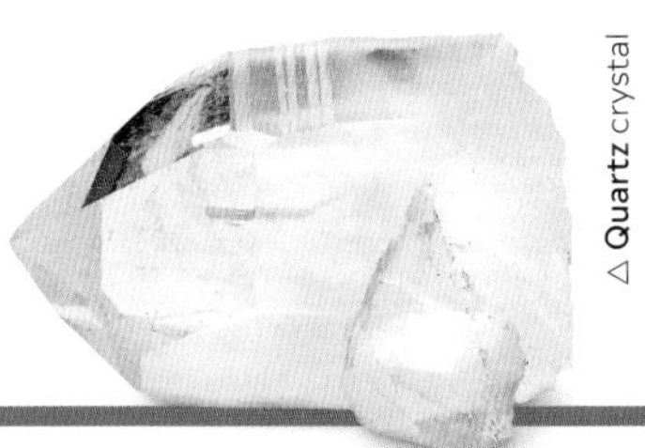

△ **Quartz** crystal

1881

ATTENUATED VACCINES

French microbiologist Louis Pasteur is often regarded as the father of immunology for his pioneering research into vaccines. Following an outbreak of the deadly disease anthrax in sheep in 1879, Pasteur confirmed that a bacterium was responsible. He prepared attenuated (weakened) cultures of the bacterium and, in a controlled experiment, immunized sheep, which then proved resistant to the disease.

◁ **Anthrax** spores

1881 Prussian inventor Hermann Ganswindt proposes a method of launching a vehicle into space

1881 J.J. Thomson works out that the mass of an object changes when electric charge is added, laying one foundation of relativity

1881

1881 German pathologist Karl Joseph Erberth isolates the bacterium that causes typhoid

1822–1911

FRANCIS GALTON

English scientist Francis Galton was an expert explorer, geographer, and statistician interested in meteorology, forensic science, and psychology. Cousin to Charles Darwin, he was also interested in heredity and introduced the term eugenics.

△ An interferometry experiment

1881

THE SPEED OF LIGHT

In 1881, and again in 1887 (with US scientist Edward Morley), American physicist Albert Michelson carried out an interferometry experiment to detect any change in the speed of light as Earth moves in different directions through space in its orbit. There was no detectable change in speed. This confounding result is explained by the fact that the speed of light is absolute, and not relative – one of the foundations of the special theory of relativity.

1882

PALAEOBIOLOGY

Appointed to the Royal Belgian Institute of Natural Sciences, palaeontologist Louis Antoine Marie Joseph Dollo became head of its vertebrate fossil section and excavated and reconstructed fossils, notably of iguanodons. Dollo was one of the first to consider dinosaurs as part of past ecosystems, thereby developing the study of palaeobiology. He later proposed that structures or organs lost in evolution would not reappear in an organism (Dollo's Law).

△ Iguanodon

1882

PHAGOCTYES

While studying the transparent larvae of starfish, Russian scientist Elie Metchnikoff noticed that some specialized cells were moving freely about inside the tissues and consuming microbes. From these observations he proposed that the white cells of human blood have the same function of clearing infection by consuming the microbes responsible for disease; he named such cells phagocytes. Metchnikoff is regarded as the discoverer of phagocytosis, the process by which certain cells protect the body from infection by consuming pathogenic microorganisms.

How phagocytosis works
1 The phagocyte stretches out parts of its cell into pseudopods (false feet) that surround a bacterium. **2** As the pseudopods envelop the bacterium, the cell membrane joins, trapping the target inside a fluid-filled sac called a food vacuole. **3** Digestive enzymes then pour into the food vacuole to kill and break down the bacterium.

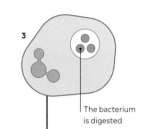

1
Pseudopod
Bacterium
Nucleus
Phagocyte

2
Cell membrane
Food vacuole
The bacterium is engulfed

3
The bacterium is digested

1882

1882 German microbiologist Robert Koch shows that tuberculosis is caused by bacteria and is not an inherited disease

1882 German mathematician Ferdinand von Lindemann proves that pi (π) belongs to a class called transcendental numbers

1883

USING ALTERNATING CURRENT

Serbian engineer Nikola Tesla made a working model of an induction motor, which ran on alternating current (AC) rather than direct current (DC). The motor was part of his vision to use high-voltage AC to supply large areas with electricity. In the USA, his ideas were taken on by Westinghouse Electric; however, the Edison Electric Light Company's bulbs ran on DC. There followed a vicious "war of the currents", which AC eventually won.

△ Induction motor

"Of all the frictional resistances, the one that most retards human movement is ignorance, what Buddha called 'the greatest evil in the world'."

NIKOLA TESLA, THE PROBLEM OF INCREASING HUMAN ENERGY, *1900*

MEIOSIS

Meiosis is cell division that produces gametes (eggs and sperm). A parent cell, with a full complement of genetic information, divides twice to produce four gametes, each of which contains half the original amount of genetic information.

How meiosis works
Matching pairs of chromosomes exchange genetic material randomly. Each resulting sperm or egg cell ends up with a slightly different mixture of genes.

1 Chromosomal DNA is duplicated to form two identical copies, each pair connected by a centromere.

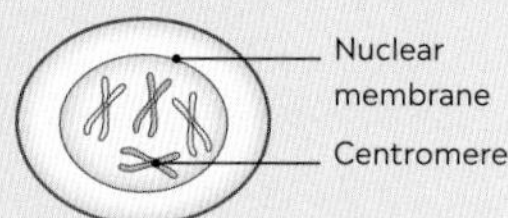

2 The membrane around the nucleus breaks down, and the duplicated chromosomes line up on threads across the cell.

3 The threads pull apart the double chromosomes; the chromosomes move to either side of the cell.

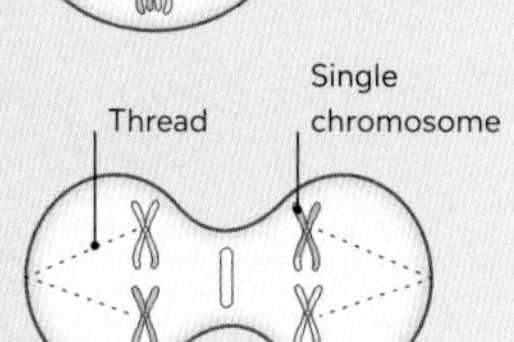
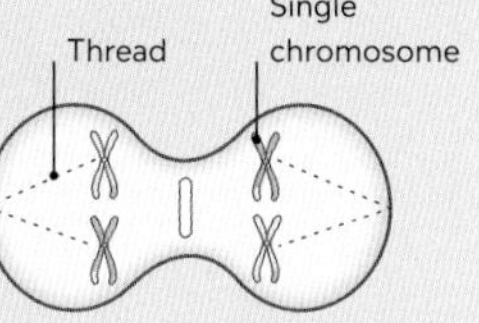

4 The cell divides to make two new cells, each with a full set of duplicated chromosomes from the original cell.

5 As the duplicated chromosomes line up, threads attach to each. They are pulled apart to form single chromosomes.

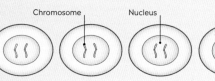

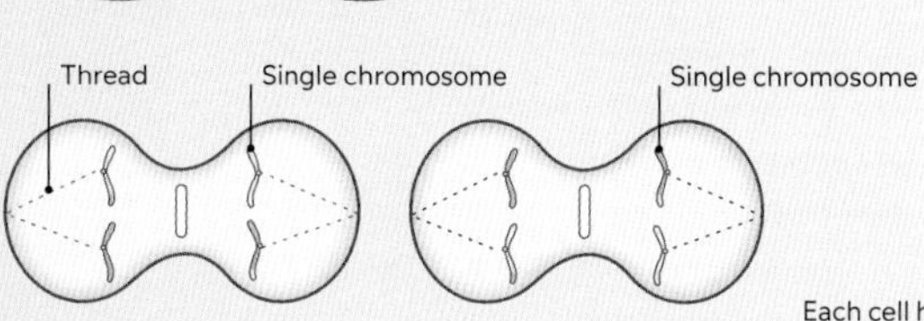

6 Two cells divide to produce four, each with half the genetic material of the original cell.

1883 English metallurgist Robert Hadfield adds manganese and carbon to iron, creating a super-tough steel alloy

1883 Belgian zoologist Edouard van Beneden demonstrates meiosis, the form of cell division involved in the production of sex cells

1883

1883 Francis Galton coins the term eugenics to express the process of improving the heritable qualities of the human race

Beneden discovered meiosis by studying parasitic roundworms

1856–1943
NIKOLA TESLA
Born in the former Austrian Empire, Tesla worked in telecommunications and electrical installation in Europe before emigrating to the USA in 1884. Tesla's great strides in radio, X-ray, and electrical engineering had a lasting influence on the modern world.

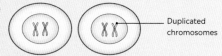

△ Hiram Maxim

1883
THE MAXIM GUN
American-born inventor Hiram Maxim developed the first effective single-barrel machine gun. It used the force of recoil of each bullet to eject the used cartridge and automatically load the next one. Capable of firing over 600 rounds per minute, the Maxim gun was soon licensed to armies across Europe. Its extensive use in World War I claimed millions of lives.

The Pasteur-Chamberland filter paved the way for large-scale filtration of drinking water in cities

1884
DISEASE FILTERS

French microbiologists Charles Chamberland and Louis Pasteur invented a device that improved water quality by removing microorganisms using a filter made from unglazed porcelain. This allowed water to pass through but excluded many pathogens such as bacteria. It was later discovered that some toxins and viruses were small enough to pass through the filter, leading to advances in the science of virology.

△ **Pasteur-Chamberland** water filter

1884
BACTERIAL STAINS

Danish bacteriologist Hans Christian Gram developed a staining technique to render bacteria more visible. Gram staining is used to identify two large groups of bacteria: gram-positive bacteria have thick cell walls and stain purple or blue; gram-negative bacteria, with thinner cell walls, stain pink.

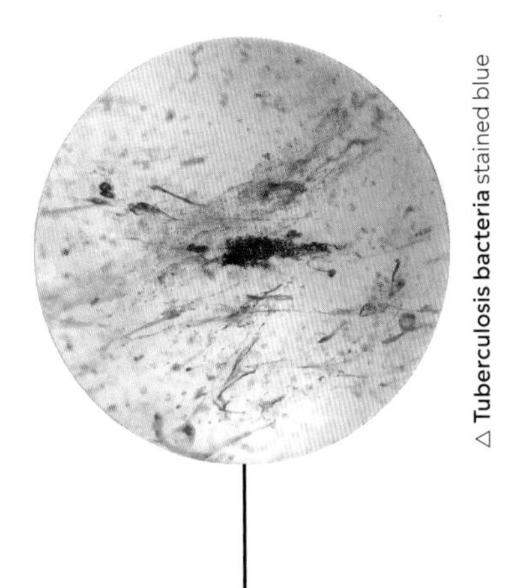

△ **Tuberculosis bacteria** stained blue

1884

1884 The prime meridian passing through Greenwich, England is named as Longitude 0 degrees, becoming the centre of world time

1884 Anglo-Irish engineer Charles Parsons patents a steam turbine for use in power stations and steam ships

VIRUSES

A virus is a tiny package of genetic material, either DNA or RNA, coated in a package of proteins. It acts like a parasite, using the machinery of an organism's host cells to replicate itself. Some viruses are harmless, or even beneficial, but others cause fatal diseases in their hosts. A virus is not generally considered to be alive, partly because it requires a living host in order to replicate.

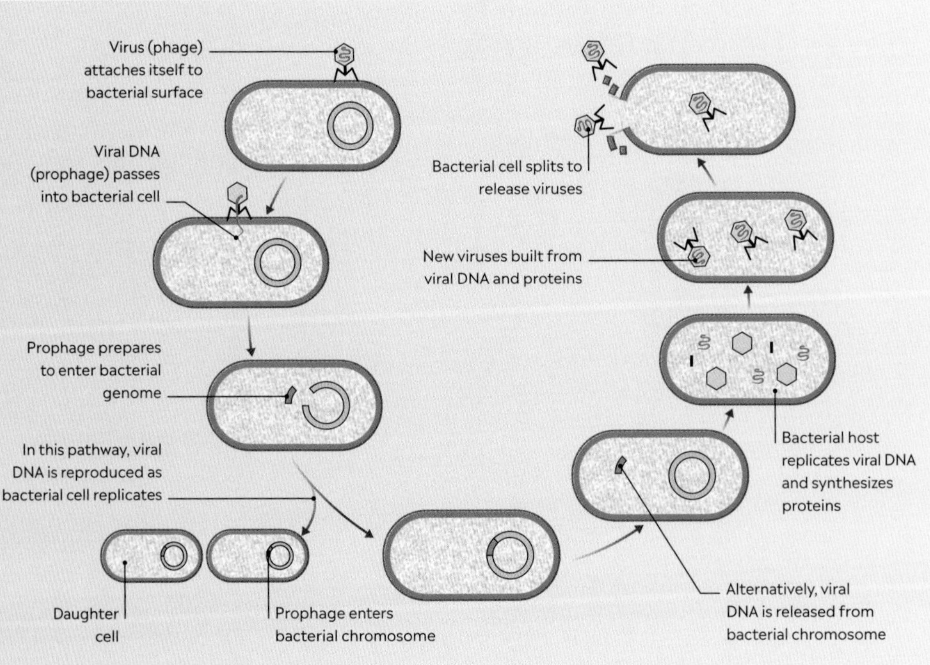

Viral replication
Viruses enter host cells to reproduce. They use the machinery and raw components of the host to replicate their own DNA and proteins thousands of times.

△ Svante August Arrhenius

1884

IONIC DISSOCIATION

Swedish physicist and chemist Svante August Arrhenius formulated his theory of electrolytic dissociation after concluding that the conductivity of a solution is a result of the presence of ions in the solution. He demonstrated that, in solution, the molecules of an electrolyte dissociate into two types of charged particle: positively charged ions (cations) and negatively charged ions (anions).

1885

EARLY BIOMETRICS

Francis Galton was the first to establish the scientific use of a biometric identifier – fingerprints – to identify individuals. He collected over 8,000 sets of such prints, which he studied in precise detail and from which he was able to provide an effective print classification system. He proposed that an individual's fingerprints are formed before birth and remain largely unchanged throughout life.

◁ Fingerprint

1885

1885 A bright new star appears in the Andromeda Nebula; it is later recognized to be a distant supernova

1885 English physicist Lord Rayleigh predicts surface earthquake waves, later named Rayleigh waves

1885 English engineer Horace Darwin invents the rocking microtome, an instrument that cuts thin slices of biological material

1885 German chemist Clemens Winkler discovers germanium, another of the missing elements from Mendeleev's periodic table

1885

BENZ AUTOMOBILE

The first practical motor car was the Benz Patent Motorwagen, created by German engineer Carl Benz in 1885 and patented the following year. Benz's car was a tricycle powered by a single cylinder four-stroke internal combustion engine. A chain drove the rear wheels, and a simple toothed steering column turned the single front wheel. Altogether, more than 20 of the cars were built. Benz's wife, Bertha, famously demonstrated the car's practicality by completing a 106-km (66-mile) drive across Germany from Mannheim to Pforzheim.

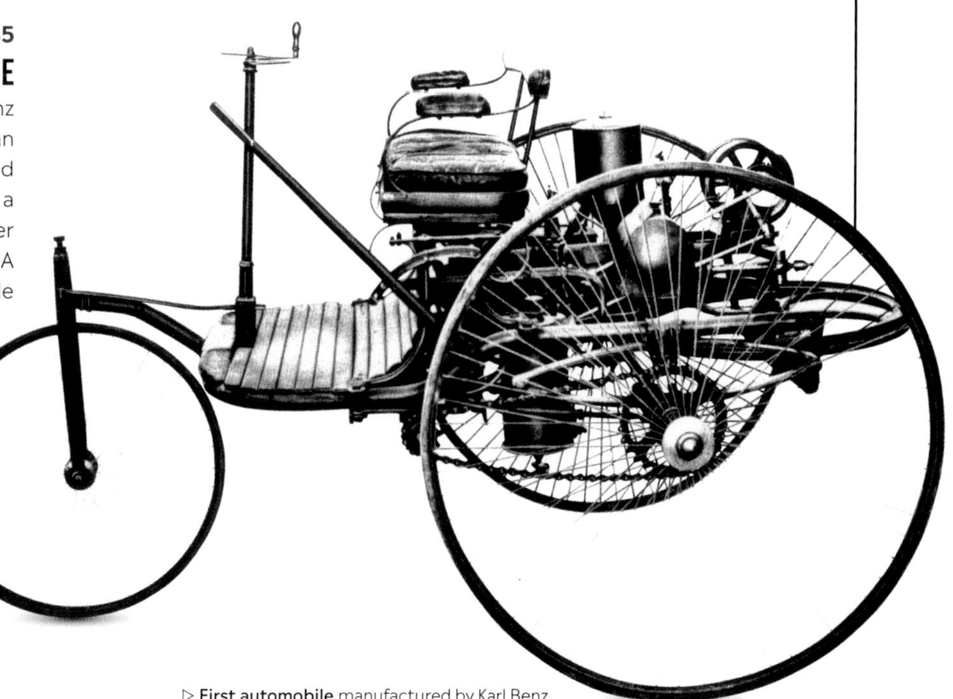

▷ **First automobile** manufactured by Karl Benz

1886

NITROGEN FIXING IN LEGUMES

German chemists Hermann Hellriegel and Hermann Wilfarth studied the ability of leguminous plants (those in the pea family) to assimilate nitrogen from the air. They discovered that the root nodules of legumes contained bacteria that could "fix" atmospheric nitrogen, converting it into ammonia or related compounds that could be used by the plants.

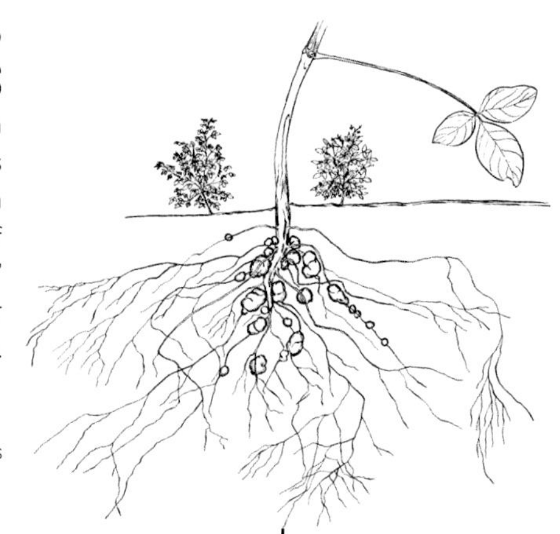

▷ **Nitrogen fixing** nodules on legume roots

By the end of the 19th century, 83 elements had been discovered; today 118 elements are known

1886

1886 The periodic table fills out as predicted elements such as dysprosium, gadolinium, fluorine, and germanium are discovered

1886 Eugen Goldstein discovers canal rays, analogous to cathode rays but moving in the opposite direction

1887 Austrian physicist Ernst Mach establishes the ratio of an object's speed to the speed of sound – the Mach number

1886

ALUMINIUM PRODUCTION

Scientists had produced small samples of aluminium since the 1820s, but the metal remained as expensive as silver until US chemist Charles Martin Hall and French scientist Paul Héroult independently discovered a way of producing aluminium cheaply using electric current.

▷ **Washington monument** being capped with aluminium

1887

LARGEST REFRACTING TELESCOPE

Construction finished on the world's largest refracting (lens-based) telescope at the Lick Observatory in California, US. With a 90-cm (36-in) objective lens and 17.4-m (57-ft) main tube, it remained one of the world's most powerful telescopes for several decades and was used to discover Jupiter's fifth moon and photograph the sky in unprecedented detail.

△ **The Lick** refracting telescope

1857–94
HEINRICH HERTZ
Born in Germany, Heinrich Hertz studied science and engineering before becoming a physics professor. He carried out research in several areas, devoting much of his career to the study of radio waves. The unit of frequency, the hertz, is named after him.

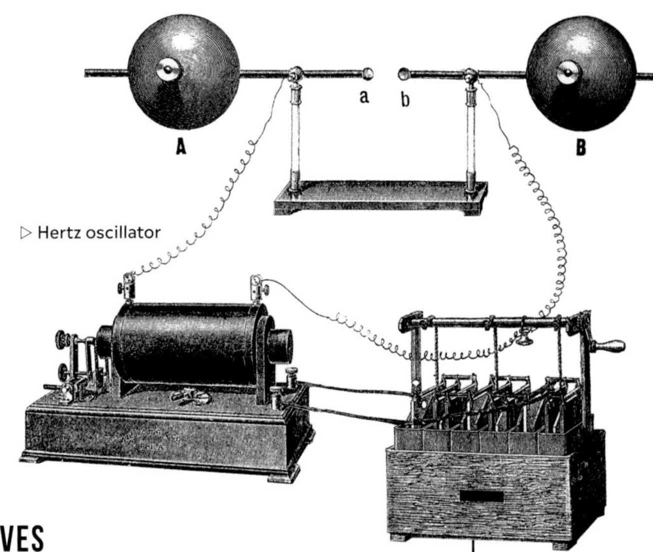

▷ Hertz oscillator

1888
RADIO WAVES
Throughout the 1880s, German physicist Heinrich Hertz tried to produce and detect radio waves, the existence of which had been predicted two decades earlier. In 1888, he succeeding in generating radio waves by creating regular sparks between two metal rods. The waves could be detected by a receiver with a smaller spark gap between its two rods.

1888

1887 Albert Michelson and fellow American Edward Morley conduct a more accurate version of their 1881 experiment, with the same result

1887 The photoelectric effect, by which a polished metal emits electrons when hit by visible or ultra-violet light, is discovered by Heinrich Hertz

1888 French chemist Henry-Louis Le Chatelier finds how the temperature, pressure, and concentration in chemical reactions can be used for efficient syntheses

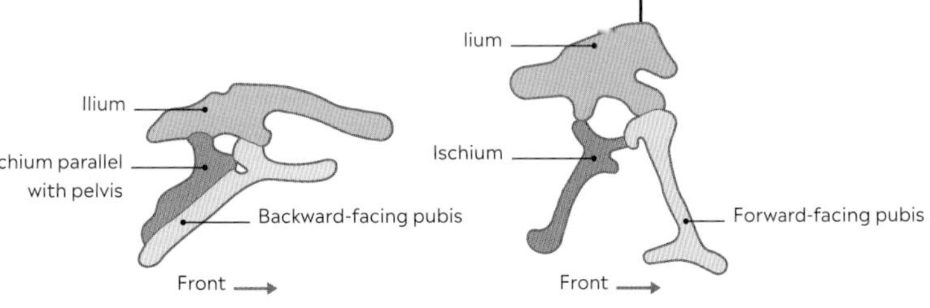

1887
DINOSAUR CLASSIFICATION
The study of dinosaurs was advanced when Harry Govier Seeley, an English palaeontologist, proposed that these extinct reptiles could be divided into two groups, based on differences in the structure of their hips: the "bird-hipped" ornithischia and the "lizard-hipped" saurischia. This proved a useful division, although it has since been challenged.

Pelvis shapes
In the ornithischian pelvis (hip), the pubis points backwards, parallel with the ischium; in the saurischian pelvis, it points forwards and downwards.

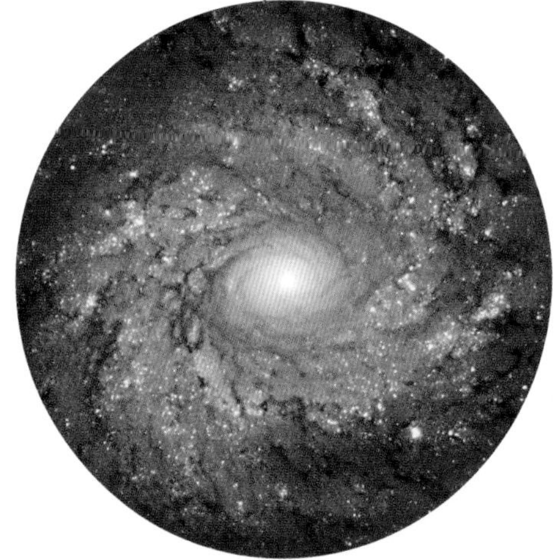
△ **Spiral galaxy** NGC 3982

1888
THE NGC CATALOGUE
Danish astronomer John Louis Emil Dreyer published his *New General Catalogue* of galaxies, star clusters, and cosmic gas clouds. Building on the work of William, Caroline, and John Herschel, it became the standard catalogue of so-called deep-sky objects.

1889
ISOSTASY

US geologist Clarence Dutton realized that Earth's varied topography results from a process of gravitational equilibrium that he called isostasy. Earth's surface crustal rocks "float" upon the denser, viscous mantle rock below. Crustal rocks adjust their buoyancy and height according to their density. The light, less dense continental rocks rise higher than the heavier and denser crustal rocks of the ocean floor.

Maintaining equilibrium
Just as blocks of the same density floating on water rise according to their size (like an iceberg) to maintain their equilibrium, Earth's crustal rocks, which "float" upon the denser mobile mantle, adjust their height above sea level according to their size and density.

Floating blocks all of the same density
Larger blocks rise higher than small ones
Line of equilibrium
Water denser than the blocks
BLOCKS IN WATER

Continental crust
Ocean crust
L ine of equilibrium
Denser viscous mantle rock
ROCKS OF EARTH'S CRUST FLOAT ON MANTLE

1889

1889 German physician Oskar Minkowski discovers the role of the pancreas in regulating blood sugar levels and links it to diabetes

1889 George Eastman, US founder of Eastman Kodak, revolutionizes photography by inventing flexible photographic film with a clear base

1889 The Eiffel Tower in Paris is completed and becomes the world's tallest structure

1889
SYNAPSES DISCOVERED

While studying the nervous system's structure, Spanish neuroscientist Santiago Ramón y Cajal discovered gaps (synapses) between individual nerve cells and showed that neurons function as independent cells, rather than making up a single network. He and Camillo Golgi recieved a Nobel Prize for their work on neurology in 1906.

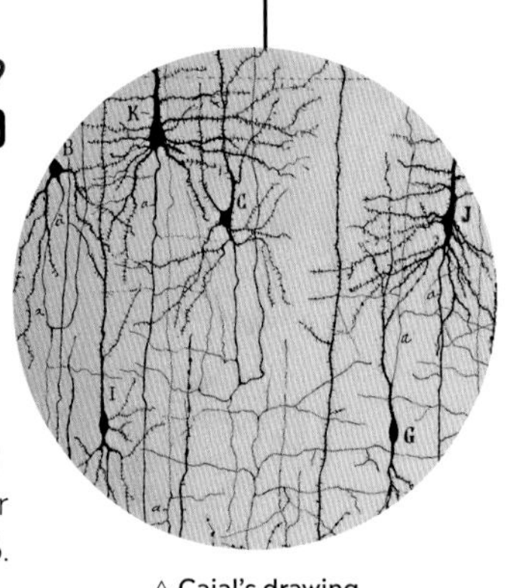

△ **Cajal's drawing** of brain structures

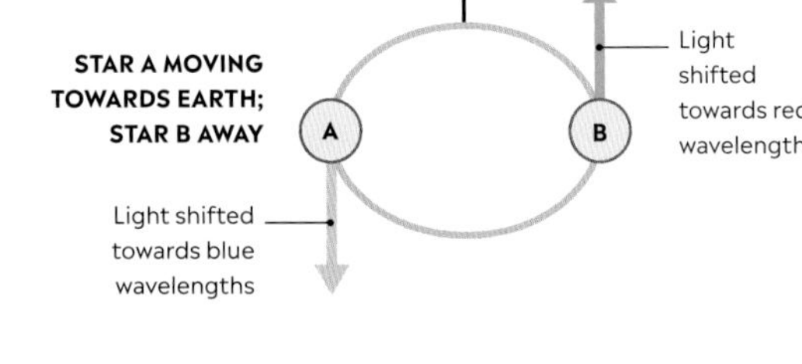

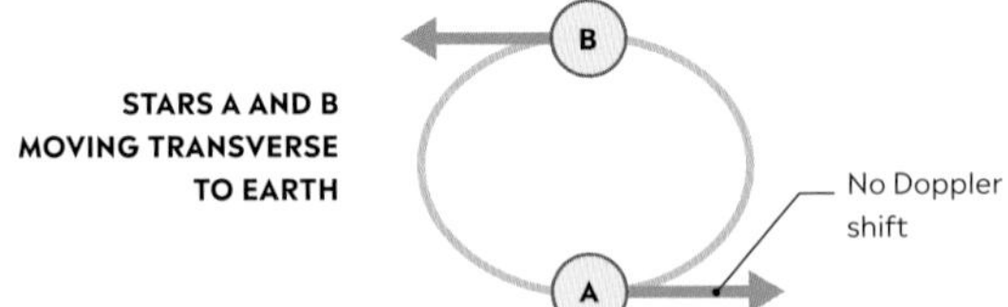

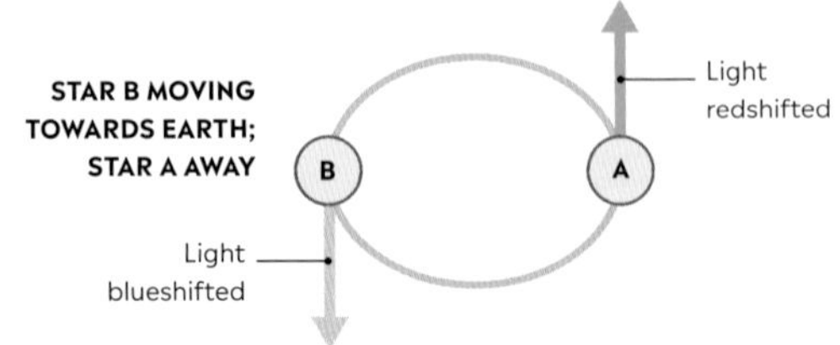

1889
SPECTROSCOPIC BINARY STARS

Germany's Carl Vogel and US astronomer Edward Charles Pickering studied the quality of emitted light to show that many stars are in fact binary pairs. The orbits of binary stars are such that as one move towards Earth, the other moves away. The apparent wavelength of light from each star is shifted according to the Doppler effect (*see p.130*). These shifts allow astronomers to calculate the speeds and relative masses of the stars.

1889

ACTIVATION ENERGY

Swedish scientist Svante Arrhenius investigated how increasing a reaction's temperature increased the rate at which it progressed. He proposed that an "activation energy" must be reached before chemicals will react. It helped to explain why some reactions, such as in fireworks, quickly generate enough energy to explode.

△ Fireworks

1890

ANIMAL TROPISMS

Inspired by earlier work on plant tropisms (growth in response to certain stimuli), German-US biologist Jacques Loeb proposed that animals exhibit tropisms comparable to those of plants. He suggested, for example, that caterpillars responded to light. Loeb also believed that some of his ideas would one day provide the basis for a mathematical theory of human conduct.

△ **A caterpillar** climbs towards light

1890

1890 German physiologist Emil von Behring creates antitoxins for the life-threatening toxin-producing bacterial infections diptheria and tetanus

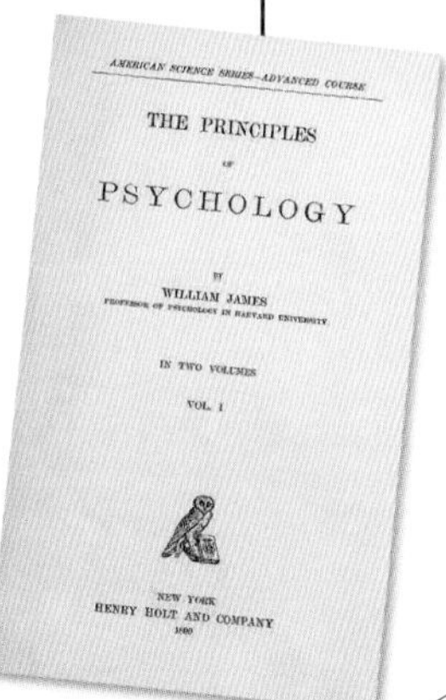

AMERICAN SCIENCE SERIES—ADVANCED COURSE

THE PRINCIPLES OF PSYCHOLOGY

BY WILLIAM JAMES

PROFESSOR OF PSYCHOLOGY IN HARVARD UNIVERSITY

IN TWO VOLUMES

VOL. I

NEW YORK

HENRY HOLT AND COMPANY

1890

◁ *The Principles of Psychology*

1890

THE PRINCIPLES OF PSYCHOLOGY

US psychologist and philosopher William James outlined his ideas underlying the study of psychology in his book *The Principles of Psychology*. In it, he identified four major aspects of psychology: stream of consciousness, emotion, habit, and will. His work was highly influential and helped to establish the study of psychology as a scientific discipline.

1843–1910 WILLIAM JAMES

A leading 19th-century US thinker, William James abandoned his training as a physician to focus on psychology and philosophy. He laid down the principles of psychology and established pragmatism and radical empiricism as philosophies.

> "The great thing, then, in all education, is to make our nervous system our ally instead of our enemy."
>
> *WILLIAM JAMES,* THE PRINCIPLES OF PSYCHOLOGY, *1890*

1891

RISING SAP IN PLANTS

It had long been known that plants take in water and transport it up into their branches, twigs, and leaves, but the route and mechanism behind this was a mystery. Polish-German botanist Eduard Strasburger demonstrated that the sap within cell lines known as the xylem forms continuous columns from the roots to the leaves, and that the rise of the sap is not a result of air or root pressure but of some other underlying mechanism. This mechanism was later found to be capillary action.

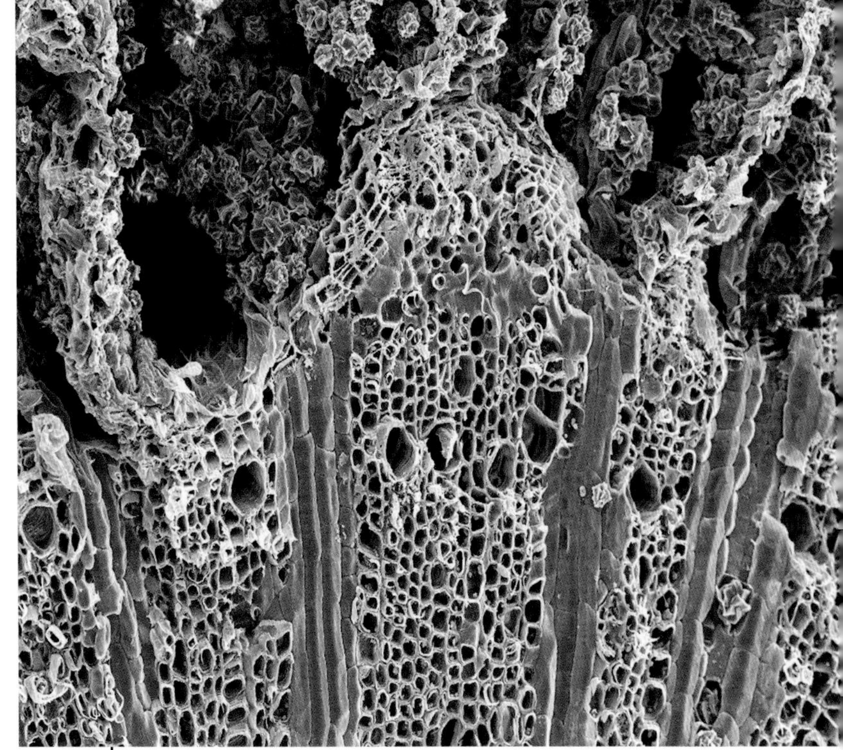

▷ **Xylem vessels** carry water

1891

1891 German chemist Agnes Pockels publishes her first work on the properties of liquid and solid surfaces, helping to establish surface science

1891 George Stoney proposes the name "electron" for the basic unit of electrical charge

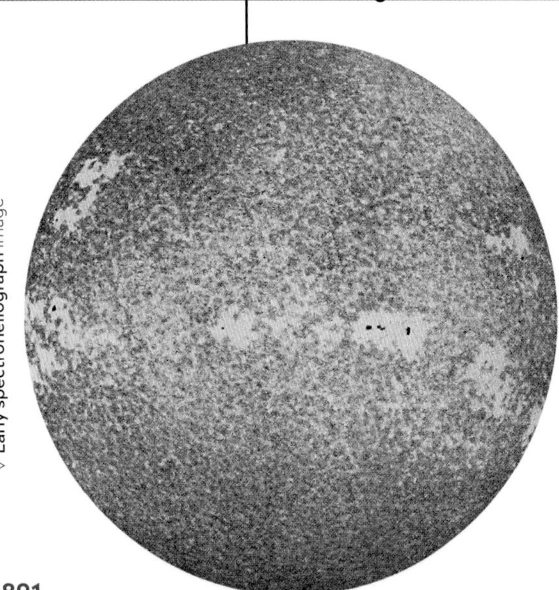

▷ **Early spectroheliograph** image

1891

IMAGING THE SUN

George Ellery Hale in the US and Henri-Alexandre Deslandres in France independently invented the spectroheliograph, an instrument that projects sunlight of a single narrow wavelength on to a photographic plate. By discarding other wavelengths, spectroheliograph images allow the Sun's surface detail to be seen, revealing granulated patterns in its visible photosphere.

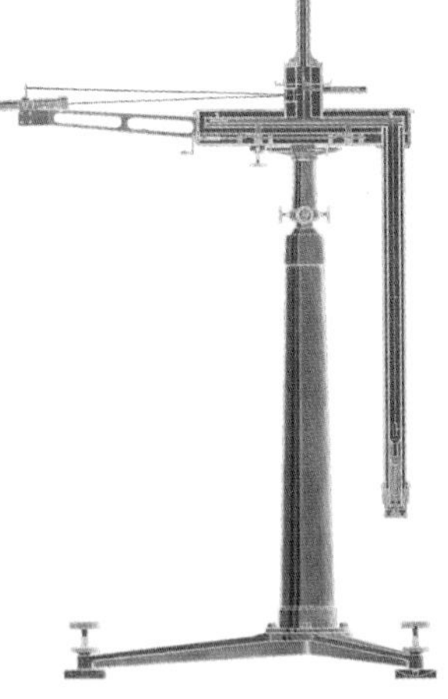

△ **Eötvös'** torsion balance

1891

MASS AND GRAVITY

Scientists had long used the term "mass" in relation to both the weight of objects due to gravity and the inertia of objects (resistance to changes in motion). However, there was no reason why these two forms of mass should be considered the same. In 1891, Hungarian physicist Loránd Eötvös designed an experiment, using an early version of his torsion balance, that proved inertial and gravitational mass to be identical to at least 1 part in 20 million. His work helped to pave the way for Albert Einstein's general theory of relativity.

▽ **Film frames** from *Fred Ott's Sneeze*

1891

MOTION PICTURES

Following his success with the phonograph, Thomas Edison commissioned his employee, British inventor William Dickson, to make a motion picture camera. The kinetograph captured moving images at 40 frames per second. Dickson also invented a device on which to display the moving images, the kinetoscope, and left behind one of the oldest surviving motion pictures, *Fred Ott's Sneeze*.

"I am experimenting upon an instrument which does for the eye what the phonograph does for the ear."

THOMAS EDISON, IN A NOTE TO HIS LAWYER, 1888

1892 German biologist August Weismann distinguishes "germ plasm", the cells of ovaries and testes that transmit heritable information, from somatic cells

1892 Irish physicist George Fitzgerald suggests that the length of a moving object is shorter than that of a stationary one

1893 British astronomer E.W. Maunder discovers the Maunder Minimum, a period of very low sunspot activity around the late 1600s

1893 Austrians Josef Breuer and Sigmund Freud publish *On the Psychical Mechanism of Hysterical Phenomena*, helping launch the study of psychoanalysis

1893

▽ **Early vacuum** flasks

1892

THE DEWAR FLASK

While experimenting with very cold temperatures, Scottish chemist and physicist James Dewar invented a double-walled flask with a vacuum between the walls, designed to prevent heat flow in or out. His design became popular for keeping drinks hot or cold.

△ **A red-orange** tungsten filament

1893

LINKING TEMPERATURE AND WAVELENGTH

German physicist Wilhelm Wien deduced a relationship between the temperature of an object and the peak wavelength of the spectrum of the radiation it emits. Wien's Law could be used to work out the temperature of objects emitting radiation.

1895

THE FIRST X-RAY

While experimenting with cathode rays, German physicist Wilhelm Röntgen detected invisible radiation that penetrated a thick piece of black cardboard and caused a fluorescent screen to illuminate. He named his discovery X-rays and within a few weeks, he used it to take a photograph of the bones in his wife's hand.

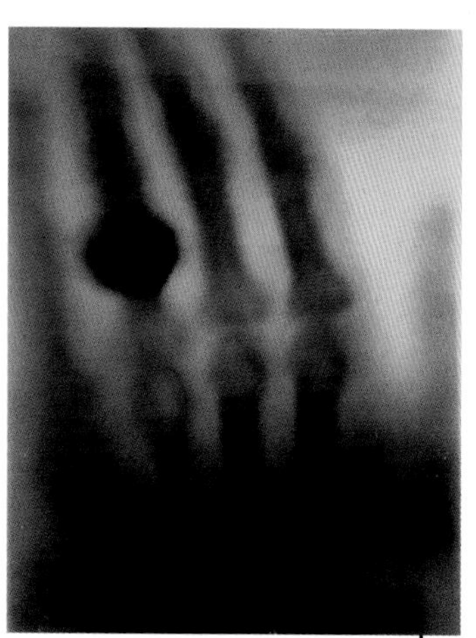

△ **First X-ray** taken by Röntgen

1894

MALARIA AND MOSQUITOS

After discovering that mosquitos were involved in spreading filariasis, a disease caused by a parasitic roundworm, Scottish parasitologist Patrick Manson turned his attention to malaria, finding that this disease is also spread by mosquitos. He published his work in the article "On the nature and significance of the crescentic and flagellated bodies in malarial blood".

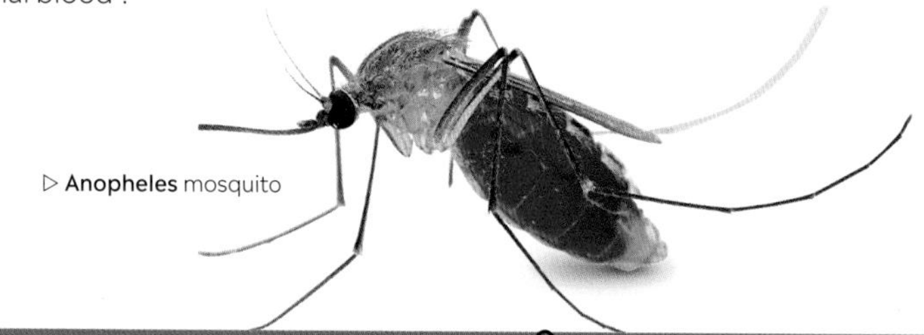

▷ **Anopheles** mosquito

1894

1894 German inventor Rudolf Diesel develops the efficient and powerful engine named after him

1894 British chemists John Strutt and William Ramsay discover the element argon; Ramsay goes on to isolate helium

1894 Dutch palaeoanthropologist Eugène Dubois publishes his discovery of *Pithecanthropus erectus*, an early hominid fossil

1894 British physiologists Albert Sharpey-Schafer and George Oliver describe the effects of an adrenal substance later identified as epinephrine (adrenaline)

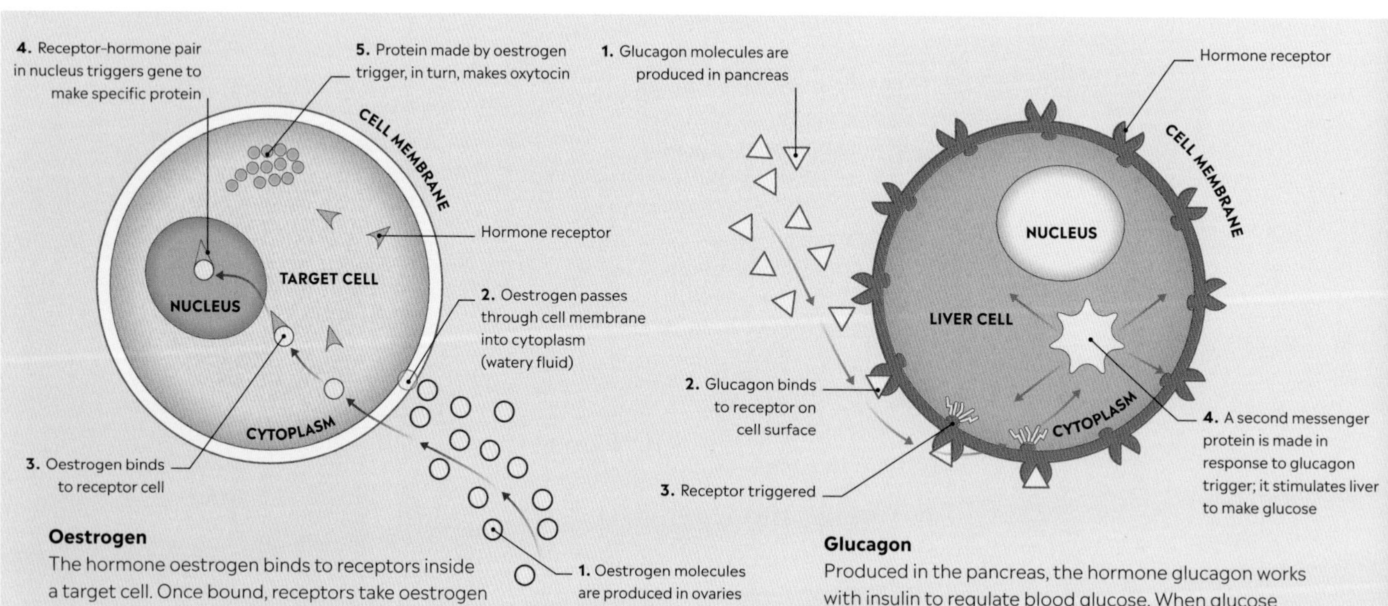

4. Receptor-hormone pair in nucleus triggers gene to make specific protein
5. Protein made by oestrogen trigger, in turn, makes oxytocin
CELL MEMBRANE
Hormone receptor
TARGET CELL
NUCLEUS
2. Oestrogen passes through cell membrane into cytoplasm (watery fluid)
CYTOPLASM
3. Oestrogen binds to receptor cell
1. Oestrogen molecules are produced in ovaries
1. Glucagon molecules are produced in pancreas
Hormone receptor
CELL MEMBRANE
NUCLEUS
LIVER CELL
2. Glucagon binds to receptor on cell surface
CYTOPLASM
3. Receptor triggered
4. A second messenger protein is made in response to glucagon trigger; it stimulates liver to make glucose

Oestrogen
The hormone oestrogen binds to receptors inside a target cell. Once bound, receptors take oestrogen into the nucleus, triggering protein production. This protein in turn triggers production of other proteins.

Glucagon
Produced in the pancreas, the hormone glucagon works with insulin to regulate blood glucose. When glucose levels are low, glucagon binds to receptors on the surface of liver cells, causing the release of stored glucose.

△ **Factories** pumping out carbon dioxide and other pollutants, c. 1900

1896
CARBON DIOXIDE AND CLIMATE
Swedish scientist and father of climate change science Svante Arrhenius used the basic principles of physical chemistry to estimate how changes in the amount of carbon dioxide in the atmosphere affect Earth's surface temperature. He calculated that if carbon dioxide levels halved, Earth's surface temperature would fall by 4–5 °C (7.2–9 °F). Conversely, if carbon dioxide levels doubled, they would trigger a rise of about 5–6 °C (9–10.8 °F). This figure is now estimated to be 2–3 °C (3.6–5.4 °F).

1896 German biochemist Eduard Buchner discovers that an enzyme, zymase, drives fermentation, producing carbon dioxide and alcohol

1896

1895 Dutch physicist Hendrik Lorentz correctly proposes that mass increases with speed, becoming infinite at the speed of light

1896 Dutch physician Christiaan Eijkman's studies of chickens suffering from a beriberi-like disease uncovers dietary deficiency; this leads to the identification of thiamine (vitamin B)

HORMONES
The chemical messengers found in animals, plants, and fungi, hormones regulate growth and development. Some animal hormones affect behaviour – adrenaline, for example, increases the heartbeat and causes restlessness. The hormone insulin controls blood-sugar levels by instructing cells to take up glucose from the bloodstream. Some hormones regulate other hormones – for example, adrenaline inhibits insulin secretion. In plants, hormones called auxins encourage shoots and stems to grow towards sunlight. Auxins are also involved in mutually beneficial interactions between plants and fungi.

▽ **Henri** Becquerel

1896
RADIOACTIVITY
French physicist Henri Becquerel discovered that phosphorescent (glow-in-the-dark) uranium compounds emit penetrating radiation similar to X-rays. Finding that non-phosphorescent uranium compounds did the same, he realized that the radiation was coming from the uranium itself. He had discovered radioactivity.

A material's radioactivity may be measured in becquerels or curies

RADIOACTIVITY

Uranium
Radioactive isotopes of uranium can be extracted from natural ores, such as torberite (above).

The nucleus of an atom consists of protons, which carry a positive charge, and neutrons, which have a similar mass to protons but lack electrical charge. An atom's nucleus is considered stable when the number of protons is in balance with the number of neutrons. This balance is not a simple equivalence in number because, for most atoms, more neutrons than protons are present – for example, an atom of copper has 29 protons and 34 neutrons in its nucleus.

Over time, atomic nuclei that are unstable will disintegrate, giving up energy or particles (*see below, left*) to achieve a more stable configuration. This emission of energy or particles is called radioactivity. It is a physical rather than a chemical process that depends on the configuration of the nucleus. Some elements exist as both stable and unstable versions – for example, carbon-12 has 6 protons and 8 neutrons and is stable, while carbon-14 has 6 protons and 8 neutrons and is radioactive. These different versions are called isotopes.

Radioactivity occurs naturally, but it can also be induced. Some radioactive elements present at the formation of our planet persist because they have long half-lives (*see below, right*), and other radioactive nuclei are formed when nonradioactive atoms are hit by cosmic rays. Radioactive substances can also be made by bombarding nuclei with energy or particles in either a nuclear reactor (*see pp.224–25*) or a particle accelerator.

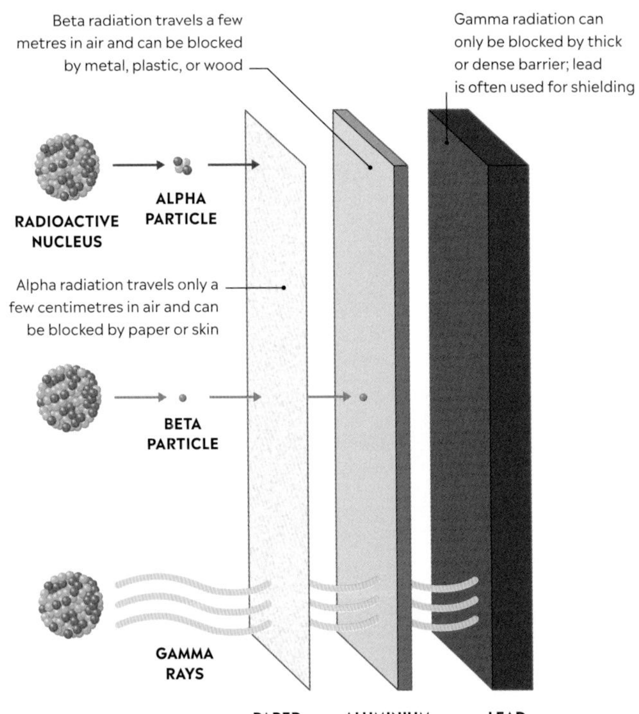

Beta radiation travels a few metres in air and can be blocked by metal, plastic, or wood
Gamma radiation can only be blocked by thick or dense barrier; lead is often used for shielding
RADIOACTIVE NUCLEUS
ALPHA PARTICLE
Alpha radiation travels only a few centimetres in air and can be blocked by paper or skin
BETA PARTICLE
GAMMA RAYS
PAPER
ALUMINIUM
LEAD

TYPES OF RADIATION

Unstable nuclei emit excess energy or mass as particles or radiation. A nucleus with too much mass emits an alpha particle (two protons and two neutrons); if it has too many neutrons it emits a beta particle, changing one neutron into a proton. A nucleus with too much energy emits gamma rays, which carry great energy without changing any of the particles.

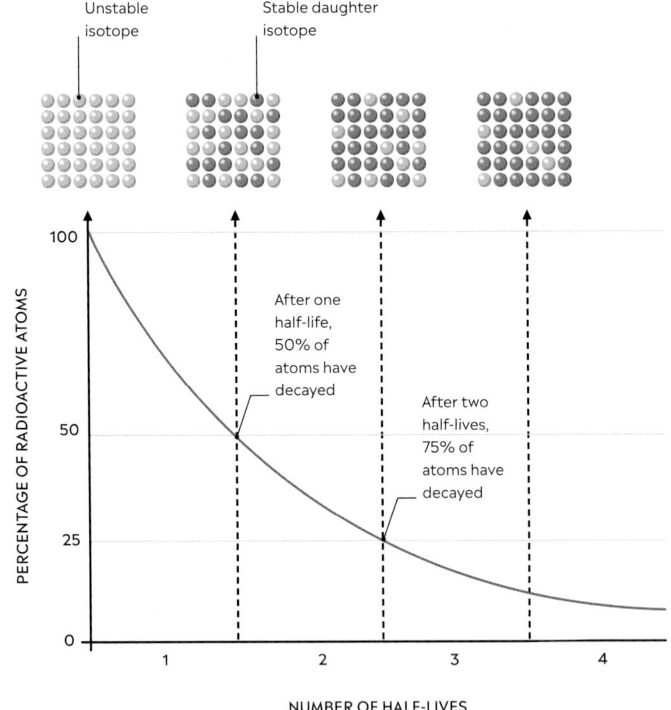

Unstable isotope
Stable daughter isotope
100
50
25
0
PERCENTAGE OF RADIOACTIVE ATOMS
After one half-life, 50% of atoms have decayed
After two half-lives, 75% of atoms have decayed
1
2
3
4
NUMBER OF HALF-LIVES

MEASURING RADIOACTIVITY AND HALF-LIFE

Each radioactive element has a characteristic half-life – the time taken for half of the atoms to disintegrate (or decay). Half-lives range from millionths of a second to billions of years. The radioactivity of a sample of an element is measured by counting how many atoms decay every second. The SI unit is the becquerel (Bq): one Bq is simply one disintegration per second.

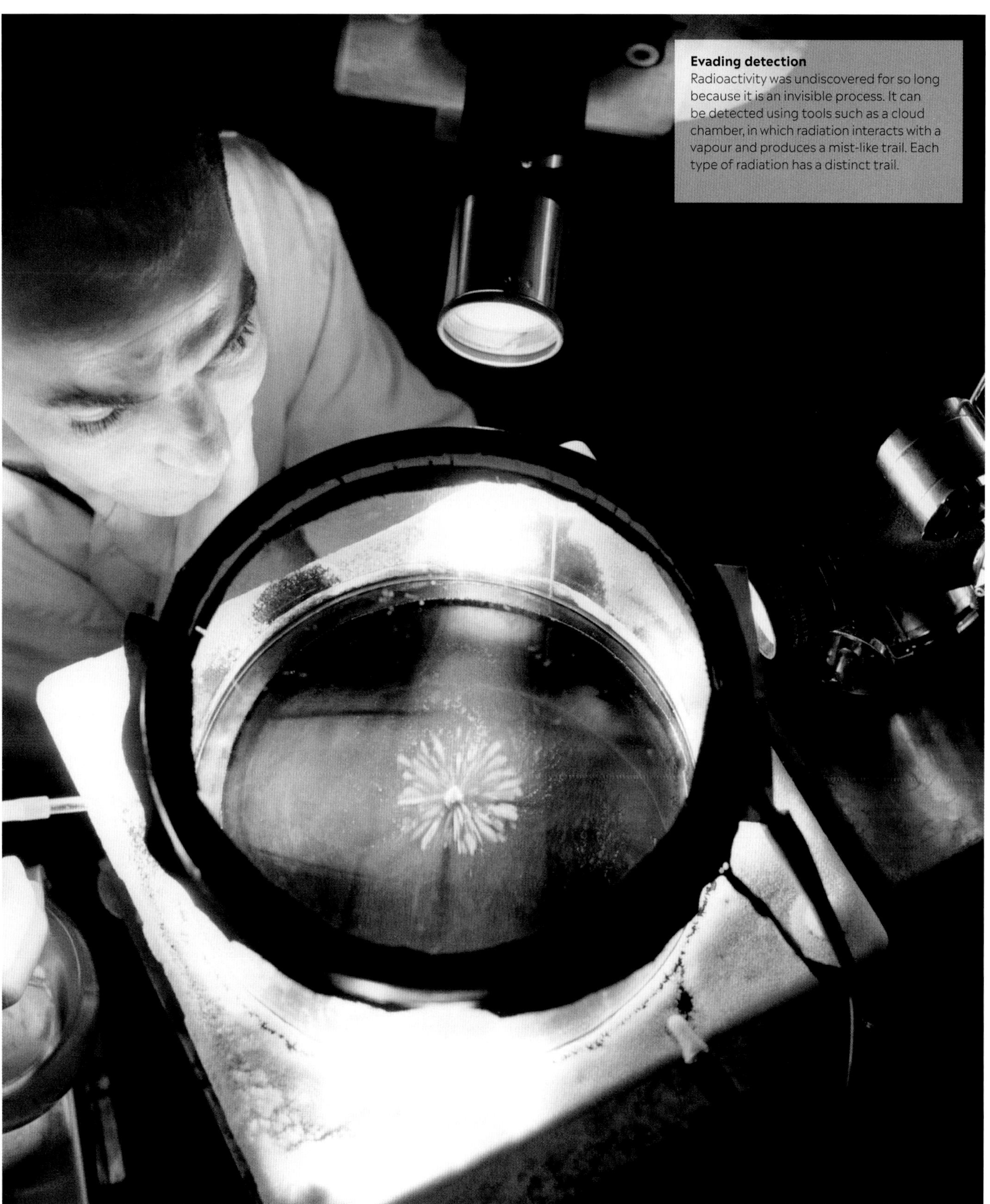

Evading detection
Radioactivity was undiscovered for so long because it is an invisible process. It can be detected using tools such as a cloud chamber, in which radiation interacts with a vapour and produces a mist-like trail. Each type of radiation has a distinct trail.

ELEMENTARY PARTICLES

The ultimate building blocks of the Universe, elementary particles combine to form atoms. There are two broad families of elementary particles: those that make up matter (fermions) and those that carry forces (bosons). Some subatomic particles, such as electrons and photons, can exist on their own, while some can only exist in combination with others – for example, quarks combine to form composite particles like protons and neutrons. The Standard Model, the most successful theory of particle physics, classifies all known elementary particles according to their properties.

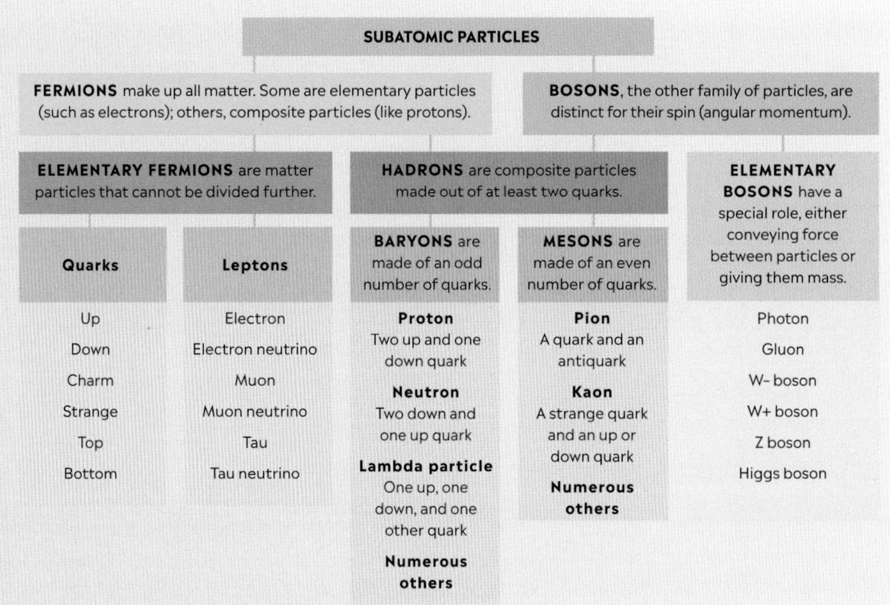

SUBATOMIC PARTICLES
FERMIONS make up all matter. Some are elementary particles (such as electrons); others, composite particles (like protons).
BOSONS, the other family of particles, are distinct for their spin (angular momentum).
ELEMENTARY FERMIONS are matter particles that cannot be divided further.
HADRONS are composite particles made out of at least two quarks.
ELEMENTARY BOSONS have a special role, either conveying force between particles or giving them mass.
Quarks
Leptons
BARYONS are made of an odd number of quarks.
MESONS are made of an even number of quarks.
Up
Down
Charm
Strange
Top
Bottom
Electron
Electron neutrino
Muon
Muon neutrino
Tau
Tau neutrino
Proton
Two up and one down quark
Neutron
Two down and one up quark
Lambda particle
One up, one down, and one other quark
Numerous others
Pion
A quark and an antiquark
Kaon
A strange quark and an up or down quark
Numerous others
Photon
Gluon
W- boson
W+ boson
Z boson
Higgs boson

1897

1897 French chemist Paul Sabatier discovers that nickel can be used as a catalyst to add hydrogen atoms to carbon compounds

1897 Karl Braun, a German physicist and engineer, invents the oscilloscope, a device for displaying electrical signals

1897 Ernest Rutherford, the New Zealand physicist, distinguishes between alpha and beta radiation

1897 A refracting telescope with a 1-m (39-in) lens is completed at Yerkes Observatory, USA

1897

THE ELECTRON

Throughout the 1890s, physicists debated the nature of cathode rays. In 1897, British physicist J.J. Thompson used an electric field inside a Crookes tube to deflect the rays, proving them to be streams of tiny particles, or"electrons". The electron was the first known subatomic particle.

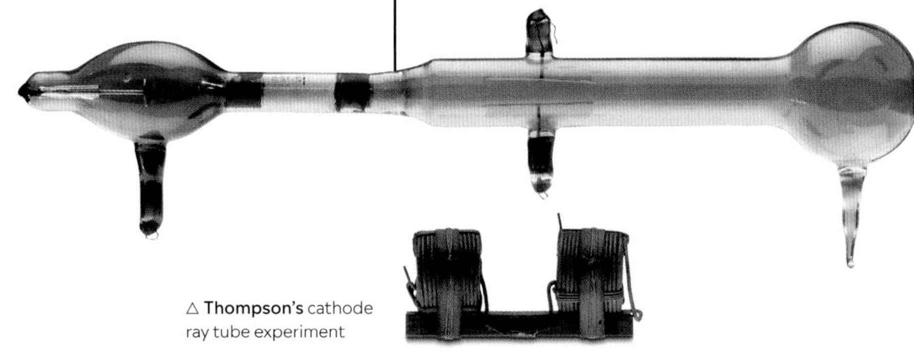

△ **Thompson's** cathode ray tube experiment

> "I can see no escape from the conclusion that they [cathode rays] are charges of negative electricity carried by particles of matter."
>
> *J.J. THOMPSON,* PHILOSOPHICAL MAGAZINE, *1897*

▽ **Mosaic virus** in tobacco plant

1898
VIRAL AGENTS

Dutch microbiologist Martinus Beijerinck published the results of his experiments using filtration, demonstrating that mosaic disease of tobacco plants is caused by an infectious agent even smaller than a bacterium. Although he was unable to isolate this agent, he regarded it as a "contagious living fluid" and named it a virus.

1867–1934
MARIE CURIE

Born Maria Skłodowska in Warsaw, Poland, Marie Curie moved to Paris, France and began studying physics and chemistry. A pioneer in radioactivity, her research, mostly carried out with her husband Pierre, earned her two Nobel prizes (in 1903 and 1911).

1898 Polish-French chemist Marie Curie and her husband, Pierre, discover the radioactive element polonium

1899

△ **Jagadish Chandra Bose**

1898 The inert gases krypton, neon, and xenon are found by Scottish chemist William Ramsay and British chemist Morris Travers

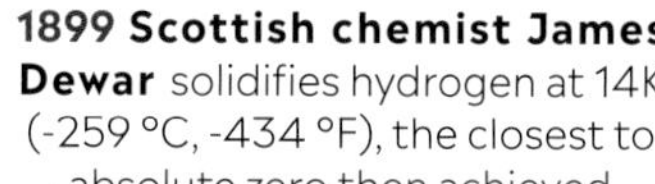

1899 Scottish chemist James Dewar solidifies hydrogen at 14K (-259 °C, -434 °F), the closest to absolute zero then achieved

1897
MICROWAVES

After two years of experimenting with short wavelength radio waves, now known as microwave radiation, Indian physicist and biologist Jagadish Chandra Bose demonstrated his discoveries at the Royal Institution, London. His apparatus used microwave radiation to ring a bell and explode gunpowder across the room. Microwave radiation is now used in telecommunications, including Wi-Fi.

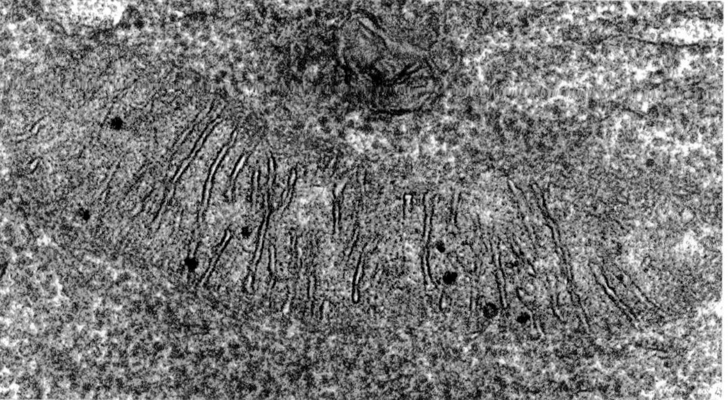
△ **Mitochondria** under an electron microscope

1898
MITOCHONDRIA OBSERVED

German microbiologist Carl Benda was one of the first to make microscopic studies of the internal structure of cells. Using a crystal violet stain to render cell components visible, Benda noticed that the cells contained certain small structures in the cytoplasm. He named these mitochondria, postulating that they may be involved in cell metabolism, which later proved to be correct.

1900

THE UNCONSCIOUS MIND

Austrian neurologist Sigmund Freud founded psychoanalysis, seeking to understand human personality. He studied medicine in Vienna and Paris and established a private practice in Vienna treating patients suffering from nervous disorders. In 1900, Freud published his major work, *The Interpretation of Dreams*. This book analyses dreams and explains their meaning in relation to unconscious experiences and desires.

△ Sigmund Freud

"The interpretation of dreams is the royal road to a knowledge of the unconscious activities of the mind."

SIGMUND FREUD, THE INTERPRETATION OF DREAMS, *1900*

1900

1900 British scientist Owen Richardson realizes that electric charge released from heated metals involves the release of electrons

1900 French physicist Paul Villard discovers a third kind of radiation emitted by radioactive substances – gamma rays

1900 French physicist Henri Becquerel finds that beta rays are made of particles with the same mass and charge as electrons

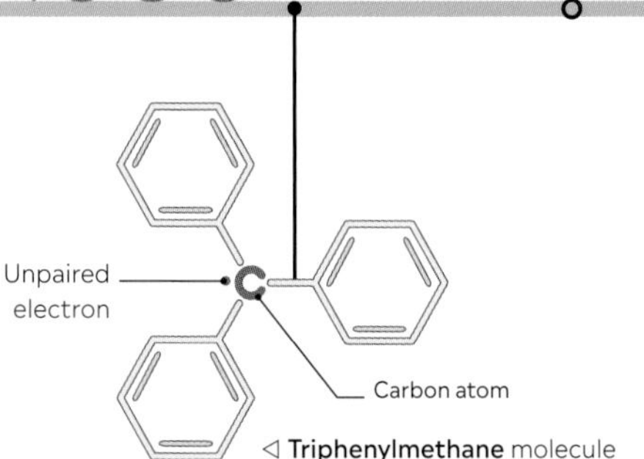

◁ **Triphenylmethane** molecule

1900

STABLE FREE RADICAL DISCOVERED

The Russian-born American chemist Moses Gomberg isolated the first stable organic free radical in the compound triphenylmethane. The molecule's central carbon atom makes just three bonds rather than the usual four, and has an unpaired electron. The discovery led to an understanding of how the free electron on a radical makes it highly reactive. Radicals are an extremely important part of many chemical reactions.

△ **The colour of light** emitted by hot coals depends on their temperature.

1900

QUANTA OF ENERGY

Max Planck formulated an equation to explain the spread of wavelengths emitted by objects called black bodies, which are ideal at absorbing and emitting radiation. Planck's equation relies on the existence of "packets" of energy called quanta, essential in quantum theory.

1858–1947

MAX PLANCK

Born in the German city of Kiel, Planck studied physics and maths in Munich, where his family had moved when he was nine years old. He spent most of his career as a professor of theoretical physics at the Friedrich-Wilhelms-Universität, Berlin.

1901
BLOOD GROUPS

Early blood transfusions, though sometimes successful, often led to complications and fatalities. Austrian physiologist Karl Landsteiner discovered the reason. He showed that there are different types of human red blood cells and that safe transfusions require the use of blood groups that are compatible.

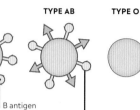

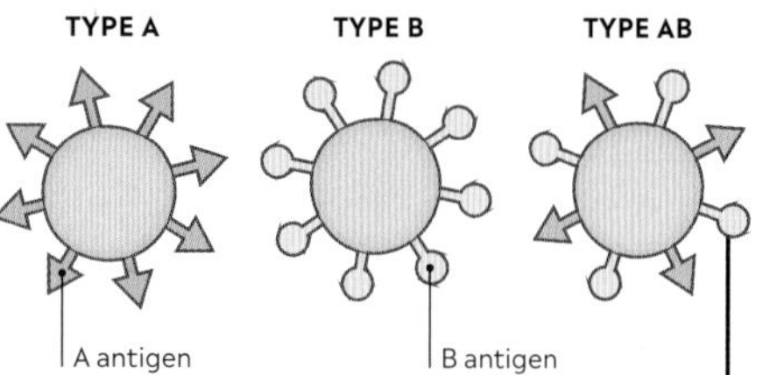

Blood groups
There are four main blood groups – A, B, AB, and O, defined by whether they carry A antigens, B antigens, both antigens, or no antigens. Each group may also be rhesus (Rh) positive or negative.

1900
MUTATION THEORY

The process of "mutationism" was proposed by Dutch geneticist Hugo de Vries in 1901 as a way of explaining evolution. It conflicted with Charles Darwin's theory of gradual evolution by natural selection because it invoked sudden jumps – changes in organisms that created new species. De Vries's studies of the evening primrose showed that it seemed to suddenly produce new forms that he named mutations. This led him to suggest that new species also came about suddenly. Later studies discredited mutation theory and gave further support to Darwin's ideas.

◁ **Evening primrose** leaves studied by de Vries

1901 Magnesium-based compounds that promote useful organic reactions are discovered by French chemist Victor Grignard

1901

1900 The noble gas radon is discovered by German chemist Friedrich Dorn; it is highly radioactive

1900 English biologist William Bateson is the first to use the term "genetics" and establish it as a science

1901 Italian inventor Gugliemo Marconi claims to have sent radio signals across the Atlantic

1901
THE OKAPI

The okapi (*Okapia johnstoni*), a relative of the giraffe, is found in forests at an altitude of 500–1500 m (1,640–4,920 ft) in the Democratic Republic of the Congo and formerly also in Uganda. It resembles a giraffe, but with a much shorter neck, a mainly dark brown body, and zebra-like stripes on its legs. Previously unknown except in tales and legends, it was first officially described and named in 1901 by English zoologist Philip Lutley Sclater.

▷ **Okapi**

Quantum computing
Certain quantum phenomena, such as the ability of a particle to exist in two states simultaneously, can be harnessed for use in computing. Quantum computers can be used to solve certain problems that are too complex for standard computers.

QUANTUM PHYSICS

At the scale of atoms and subatomic particles, the behaviour with which we are familiar in everyday life is overtaken by behaviour that seems to defy common sense. This is the world of quantum physics. At this scale, matter can behave like waves, and light can behave like particles, due to a phenomenon known as wave-particle duality *(see p.84)*.

Central to quantum physics is the concept of quantization. If something is quantized, it has a minimum possible quantity (this is known as a quantum), and all quantities of it can only be whole-number multiples of that quantum, meaning it is discrete. Light, for example, comes in quanta known as photons.

Another important feature of quantum physics is uncertainty. Classical physics describes a clockwork-like Universe in which anything can, in theory, be predicted with perfect accuracy. This is impossible in the quantum world, because it is not deterministic but probabilistic. In other words, it is impossible to calculate the precise location of an electron; it is possible only to calculate the probability that it is within a certain area. Until it is measured, it is simultaneously in every possible place. Even when it is measured, the more precisely its position is known, the less precisely its momentum is known. This concept is known as the uncertainty principle.

The quantum world is filled with curious phenomena like these. However, they are not always discernible at the larger scale for which classical physics is particularly effective at describing behaviours.

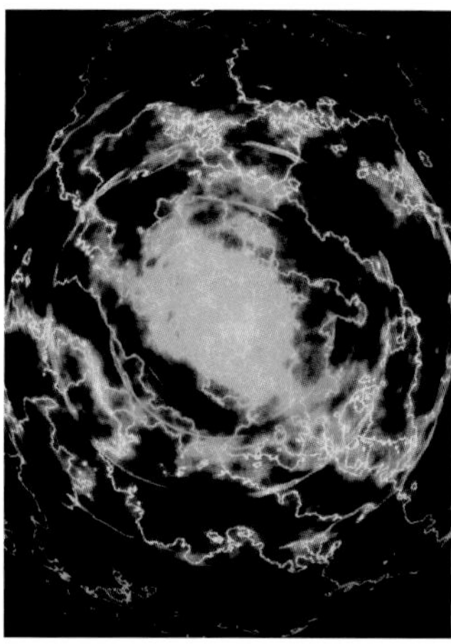

Electron cloud
The probability of finding the electrons of a helium atom in various locations can be depicted through the use of a quantum cloud.

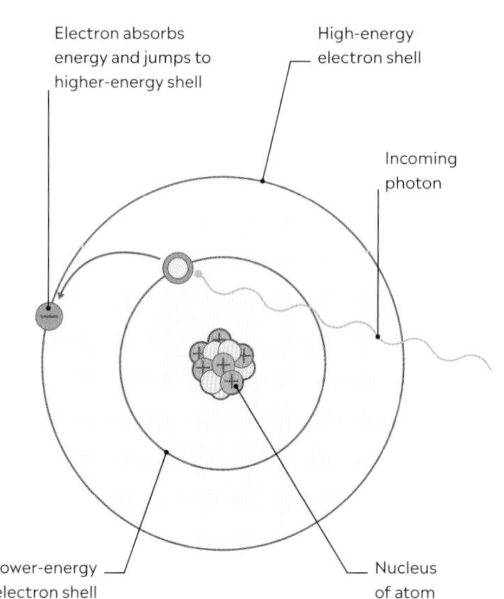

PACKETS OF ENERGY

Quantization means that there is a minimum possible amount of any physical quantity, which is known as a quantum. For example, light comes in quanta known as photons, which cannot be divided any further. When an electron in orbit around a nucleus absorbs a photon, it may gain enough energy to jump to a higher energy level.

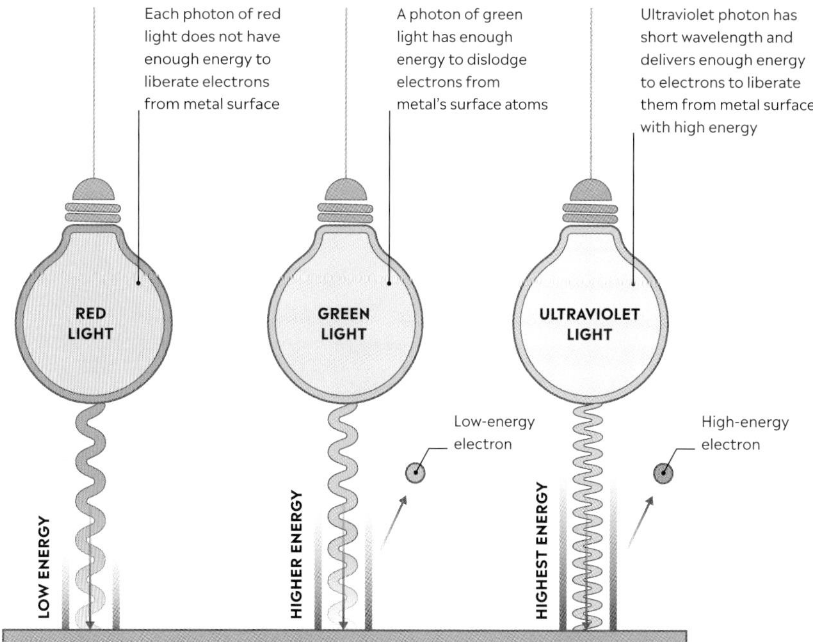

THE PHOTOELECTRIC EFFECT

Electrons can be emitted from a material exposed to light, producing an electric current. This is known as the photoelectric effect. When photons are absorbed by electrons trapped in orbit around an atom in the material, the electron may gain sufficient energy to escape. However, this requires the light to have a minimum frequency.

▽ Philippe Lenard

1902
PHOTOELECTRIC EFFECT
Ultraviolet light causes metals to eject negatively charged electrons. This is the so-called photoelectric effect. In 1902, German physicist Philippe Lenard found that the energy of the ejected electrons increases with the frequency of the light - a fact that is at odds with the wave theory of light. Einstein would explain this phenomenon using quantum theory.

1902
PROBING THE ATMOSPHERE
French meteorologist Teisserenc de Bort pioneered the use of instrument-carrying balloons to investigate the upper atmosphere. He discovered that air temperature decreased to -51 °C (-60 °F) at 17km (11 miles) high, but remained the same above this. He named the lower layer the troposphere and the upper one the stratosphere.

△ **Weather balloon** at launch

1902

1902 British physiologists Ernest Starling and William Bayliss discover secretin, the first known hormone

1902 Ernest Rutherford and British chemist Frederick Soddy propose that when radioactive elements disintegrate, they transform into other elements

STRUCTURE OF THE ATMOSPHERE
Data gathered from balloons, aeroplanes, and satellites show that Earth's atmosphere has a layered structure. The lowermost layer, the troposphere, supports life with its abundant oxygen and changing weather. Above this are four more layers - the stratosphere, mesosphere, thermosphere, and exosphere - of varying temperatures and composition. The stratosphere's ozone layer prevents solar radiation from harming life in the troposphere below. While solar radiation warms the upper stratosphere, temperatures in the mesosphere fall to around -90°C (-130°F), before rising to 1,500°C (2,700°F) in the thermosphere above. The outermost exosphere contains very little gas and merges into space.

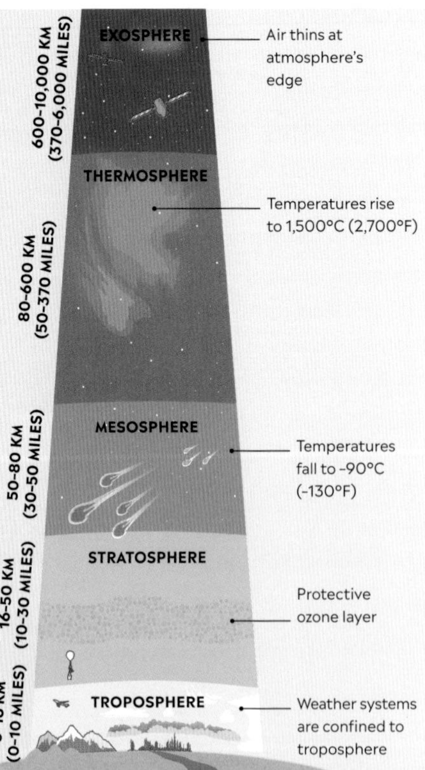

1871-1937
ERNEST RUTHERFORD
Born in Nelson, New Zealand, Rutherford carried out pioneering research in the UK and Canada, contributing to our understanding of the structure and behaviour of atoms. He was awarded the 1908 Nobel Prize in Chemistry.

△ **The Wright Flyer** at Kitty Hawk, 17 December 1903

1903

POWERED FLIGHT

On 17 December 1903, brothers Wilbur and Orville Wright made the first successful heavier-than-air flight at Kitty Hawk, North Carolina. The 12-second flight was the culmination of years of research in which they developed innovative aerodynamic techniques such as wing warping and a moveable rudder.

1903 Dutch physician Willem Einthoven develops an early electrocardiograph that could record the heart's electrical activity

1903 Austrian physicist Ernst Mach questions the notion of absolute space, helping to provide inspiration for Einstein's theories of relativity

1903

△ **Ivan Pavlov** experimenting with conditioning

1903

CONDITIONED REFLEX

Russian physiologist Ivan Petrovich Pavlov noticed that his dogs would start to produce saliva as soon as they saw a person bringing them food. By combining feeding with a particular sound, such as a metronome, the dogs would salivate on hearing the sound, even in the absence of food, in what he called a conditioned or conditional reflex.

1903

CHROMOSOME THEORY OF INHERITANCE

Walter Sutton, an American geneticist, established that the laws of inheritance were reflected in the behaviour of chromosomes, the carriers of genetic information. A similar conclusion was reached by German zoologist Theodor Boveri.

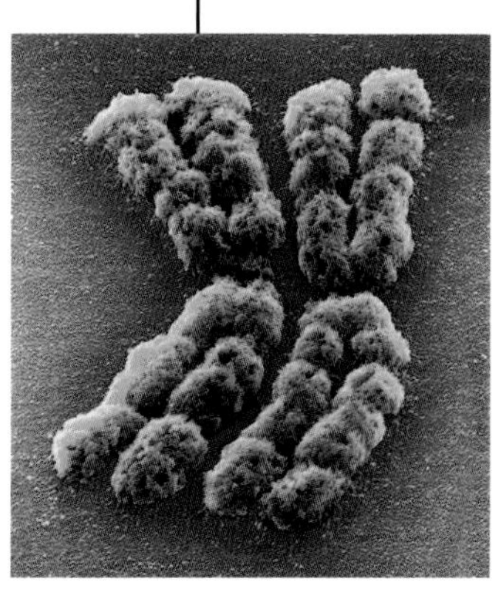

▷ **Chromosomes** carry genetic information

"Learn, compare, collect the facts!"

IVAN PAVLOV, WRITING ON "GRADUALISM"

1904
FLEMING'S VALVE

British physicist John Fleming developed an evacuated glass tube containing a heated wire and a metal plate. Making use of thermionic emission (the release of electrons from metal) it is called a valve because it allows current to flow in one direction only. It makes possible many basic electronic devices that will prove crucial to the development of radio.

▽ **Synaptic signals** in the cerebral cortex

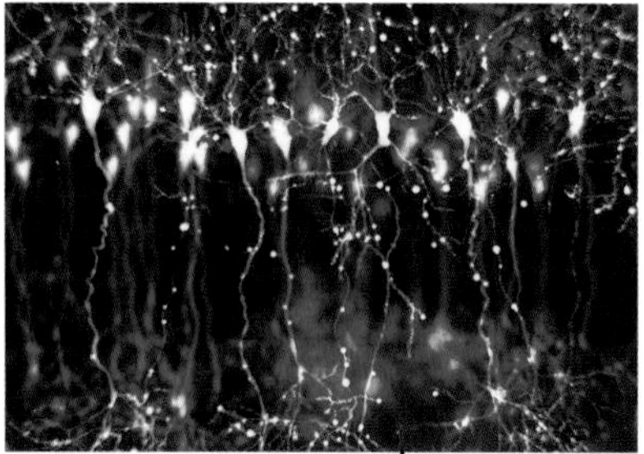

1904
NEURON THEORY

Spanish scientist Santiago Ramón y Cajal is widely considered the father of neuroscience. Through microscope studies of brain and spinal cord tissues he demonstrated how nerve cells (neurons) connect with each other, rather than being continuous. In 1906, he won a Nobel Prize with Italian Camillo Golgi.

△ **Fleming tube** or thermionic valve

1904

1904 Dutch astronomer Jacobus Kapteyn identifies two streams of stars, moving in opposite directions; this is later seen as evidence for the rotation of the Milky Way

1904 British physicist J.J. Thomson publishes his "plum pudding" model of the atom, later shown to be inaccurate.

1904 British physicist Charles Barkla proves that X-rays are a form of electromagnetic radiation, like light, infrared, and radio waves

1904
INTERSTELLAR MATTER

While studying the spectrum of the binary star system Mintaka in Orion, German astronomer Johannes Franz Hartmann noted that most of its spectral absorption lines shifted back and forth in a regular pattern caused by the stars' orbit; however, lines associated with calcium did not. He suggested that an unseen cloud containing calcium lay between the star and Earth. This was the first evidence for the sparse "interstellar medium" that is now known to fill much of the Milky Way.

◁ **Mintaka** (top left) lies in Orion's belt

1905

EINSTEIN'S FOUR LANDMARK PAPERS

Albert Einstein published four transformative papers in what was called his "annus mirabilis". The first concerned the photoelectric effect, proving that light energy is delivered in quanta. The second, on Brownian motion (*see below*), proved the existence of atoms, then still doubted by some. The third introduced the Special Theory of Relativity, and the fourth featured the equation $E = mc^2$, a consequence of special relativity.

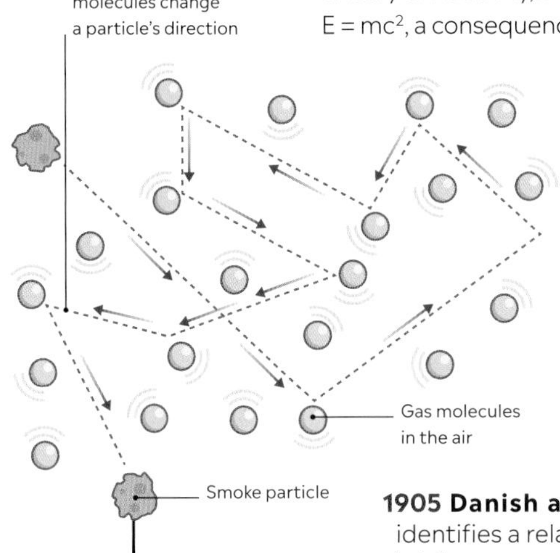

Brownian motion
Brownian motion can be observed under the microscope as the random, jerky movement of smoke particles. Einstein proved that it is the result of invisible atoms and molecules bumping into the larger particles.

1879–1955
ALBERT EINSTEIN

Albert Einstein was born in Ulm, Germany. His Special and General Theories of Relativity, along with his contributions in quantum theory, laid the foundations of modern physics and made him one of the most celebrated scientists of all time.

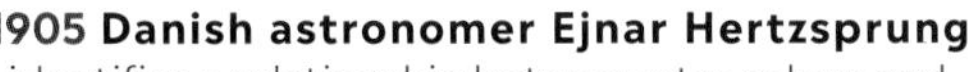

1905 Danish astronomer Ejnar Hertzsprung identifies a relationship between star colour and brightness, and distinguishes between classes of stars with different luminosities

1906 German chemist Walther Nernst states the third law of thermodynamics

1906

1905 American geneticists Nettie Stevens and Edmund Wilson discover chromosomes that determine sex in mammals

1906 J.J. Thomson proposes that an atom has a number of electrons equal to that element's atomic number

▷ **Intelligence test** for children

1905

IQ TESTING

Intelligence quotient (IQ) testing is a method of seeking to assess and compare "intelligence". French psychologist Alfred Binet invented the first such test, with colleague Théodore Simon, to establish which school children might require extra assistance.

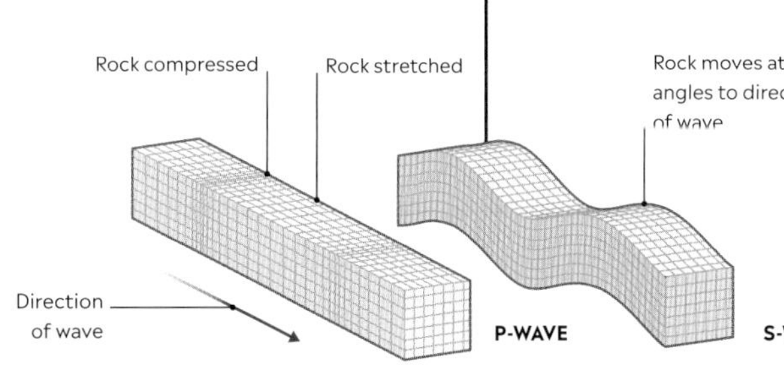

1906

EARTH'S CORE

By studying earthquake shock waves, the British physicist R.D. Oldham found differences in the properties of primary or compressional (P) waves and secondary or shear (S) waves. P waves travel through Earth, arriving at the opposite side later than they should if they passed straight through. He concluded that Earth's composition must change at depth with a core of denser material that slows the waves down.

P and S waves
Shock waves travel through Earth as either Primary, compressional, waves or Secondary, shear, waves. While P and S waves travel through solid rock, S waves do not travel through liquid.

(38)

Ersetzt man in der Bewegungsgleichung die gestrichenen Grössen durch die ungestrichenen, so erhält man zunächst

$$\frac{m\frac{dq}{dt}}{\left(1-\frac{q^2}{c^2}\right)^{\frac{3}{2}}} = \varepsilon \mathfrak{e}_x \quad \cdots \quad (26).$$

Berücksichtigt man, dass

$$\frac{\frac{dq}{dt}}{\left(1-\frac{q^2}{c^2}\right)^{\frac{3}{2}}} = \frac{d}{dt}\left\{\frac{q}{\sqrt{1-\frac{q^2}{c^2}}}\right\}$$

ist, und dass die rechte Seite von (26) nach einer Anmerkung des § 2 als die auf den materiellen Punkt wirkende Kraft $\mathfrak{K}$ aufzufassen ist, so nimmt (26) die Form an

$$\frac{d}{dt}\left\{\frac{mq}{\sqrt{1-\frac{q^2}{c^2}}}\right\} = \mathfrak{K}_x$$

Soll also in der Relativitätstheorie der Impulssatz aufrecht erhalten werden, so müssen wir den in der geschweiften Klammer stehenden Ausdruck als den Impuls des materiellen Punktes auffassen. Hieraus schliessen wir verallgemeinernd, dass $\frac{mq}{\sqrt{1-\frac{q^2}{c^2}}}$ dem Impulsvektor eines beliebig bewegten materiellen Punktes gleich ist. Soll also der Impulssatz in der Relativitätstheorie aufrecht erhalten und die Grundlage der Lorentz'schen Elektrodynamik beibehalten werden, so muss die Vektorgleichung der Bewegung des materiellen Punktes unter der Einwirkung der beliebigen Kraft $\mathfrak{K}$ lauten

$$\frac{d}{dt}\left\{\frac{mq}{\sqrt{1-\frac{q^2}{c^2}}}\right\} = \mathfrak{K} \quad \cdots \quad (27)$$

Ist die einzige auf den materiellen Punkt wirkende Kraft elektrodynamischer Natur, so ist hiebei $\mathfrak{K} = \varepsilon\left\{\mathfrak{e} + \left[\frac{q}{c}, \mathfrak{h}\right]\right\}$ zu setzen.

Es ist leicht zu zeigen, dass (27) auch dem Energiesatz gerecht wird, wenn $\mathfrak{K}q$ als Ausdruck für die pro Zeiteinheit an den materiellen Punkt geleistete Arbeit beibehalten wird. Man erhält nämlich

$$\mathfrak{K}q = q\frac{d}{dt}\left\{\frac{mq}{\sqrt{1-\frac{q^2}{c^2}}}\right\} = \frac{d}{dt}\left\{\frac{mq^2}{\sqrt{1-\frac{q^2}{c^2}}}\right\} - \frac{mq\dot{q}}{\sqrt{}} = \frac{d}{dt}\left\{\frac{mq^2}{\sqrt{1-\frac{q^2}{c^2}}} + mc^2\sqrt{1-\frac{q^2}{c^2}}\right\}$$

oder

$$\mathfrak{K}q = \frac{d}{dt}\left\{\frac{mc^2}{\sqrt{1-\frac{q^2}{c^2}}}\right\}. \quad \cdots \quad (27a)$$

Der Ausdruck unter der Klammer rechts spielt die Rolle der Energie des bewegten Massenpunktes. Dieser Ausdruck

$$E = \frac{mc^2}{\sqrt{1-\frac{q^2}{c^2}}} \quad (28)$$

Einstein's notes
Albert Einstein proposed his special theory of relativity in 1905. It was intended as an update to mechanics when classical mechanics was found to be incompatible with James Clerk Maxwell's equations of electromagnetism.

THE SPECIAL THEORY OF RELATIVITY

Mass and energy
Nuclear weapons work on the principle that mass is concentrated energy; a colossal amount of energy can be contained within a tiny mass.

Proposed in 1905, Albert Einstein's special theory of relativity describes how an object's speed affects mass, time, and space. Light always travels through a vacuum at the same speed, no matter how quickly its source is moving. It transpires that light is special: it has a universal speed limit. As an object approaches the speed of light, its mass increases towards infinity, as does the energy that is required to move it, making it impossible to reach the speed of light. Combined with the principle that the laws of physics are the same in all inertial frames of reference - that is, physics is the same whether someone is standing still or on a train moving at a constant speed - this has many interesting consequences.

The theory states that time and space are not absolute but relative, so they can warp when the speed of light is neared. (Something moving fast enough to experience this is travelling at a "relativistic speed".) This means that two people experience different lengths of time passing if one remains on Earth and the other travels at relativistic speed in space. These phenomena are known as time dilation and length contraction.

Furthermore, mass and energy are equivalent, the two connected by the speed of light. The effects of Einstein's theory seem counterintuitive but have been confirmed with experiments examining objects at relativistic speed. These effects even need to be accounted for in everyday technologies such as satellite navigation.

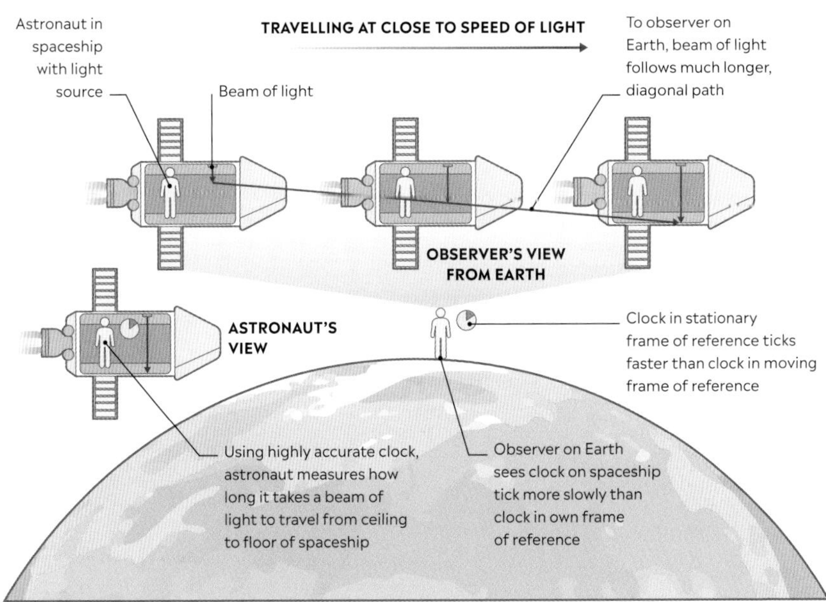

TIME DILATION

The special theory of relativity says that time is relative, meaning that it passes at different rates for observers who are moving at different speeds. If a clock were to move through space at near the speed of light, it would appear to an observer standing on Earth to tick noticeably slower. The faster the clock moves, the greater the time dilation.

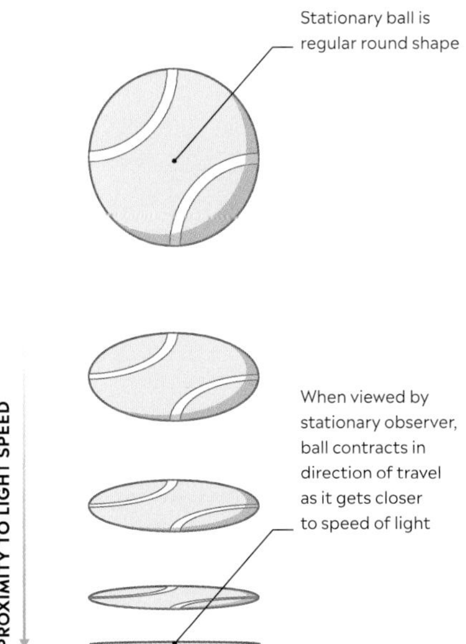

LENGTH CONTRACTION

Einstein's theory predicts an effect known as length contraction. This means the space around a moving object (including the object and any measuring devices) contracts from the point of view of an observer. This becomes more extreme as the object nears light speed.

1907

DATING ROCKS

American chemist Bertram Boltwood discovers that, over millions of years, radioactive uranium decays to lead. He measured the proportion of the two elements in minerals such as zircon to calculate the age of the mineral.

△ **Zircon** crystal (dark crystal in centre)

1908

COLOUR PHOTOGRAPHY

Gabriel Lippmann, a French physicist, won a Nobel Prize for producing the first colour photographic plate. Rather than relying on dyes to reproduce colour, Lippmann treated his sensitized plates with mercury to form a mirror. When viewed, the mirrored plate reflected light that interfered with incident light to reproduce the original colours. The images were technically difficult to achieve and the process did not come into widespread use.

1907

1908 Ejnar Hertzsprung suggests ranking the brightness of stars by their calculated appearance at a standard distance

1907 The first synthetic peptide is created from 18 amino acids by German chemist Emil Fischer, unlocking research into protein structures

1908 German physicist Hans Geiger develops a technique for detecting alpha particles, and so creates the Geiger counter for measuring radiation

1908 The International Conference on Electric Units and Standards formalizes the ampere and the ohm as international units

RADIOMETRIC DATING

Over the past century, radiometric dating has revolutionized Earth science, making available a reliable timeline of Earth's formation and history. Some rocks contain minerals that include unstable radioactive elements - such as uranium and potassium - that decay at known rates to more stable elements. Measuring the ratio of the radioactive elements to their decay products in a sample of rock reveals how much time has passed since the rock formed.

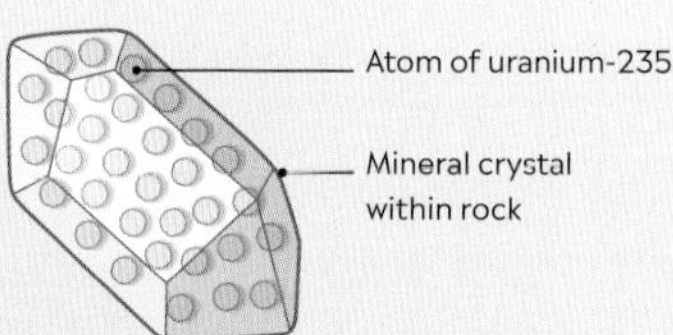

1. Newly formed rock
As a rock forms, minerals grow within cooling magma, and some contain radioactive elements such as uranium-235.

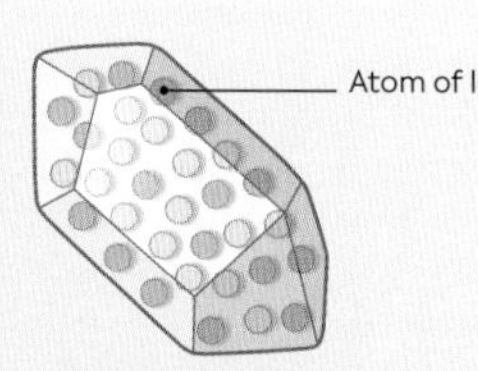

2. After 704 million years
The concentration of the uranium-235 atoms has been halved by decay to a different element, lead-207.

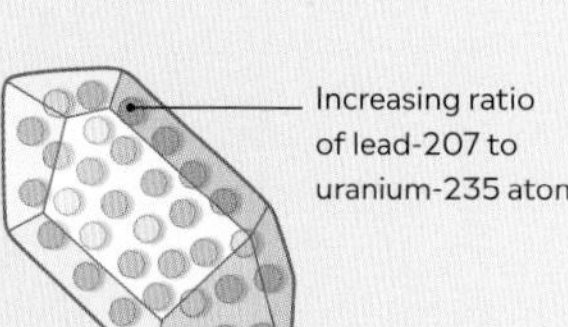

3. After 1.406 billion years
The concentration of uranium-235 atoms has halved again, with the stable element lead-207 now being dominant.

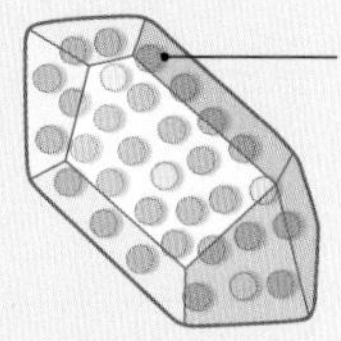

4. Present day
If a sample of rock has a lead-207: uranium-235 ratio of 7:1, it indicates that the rock is 2.812 billion years old.

How uranium-lead dating works
When a mineral containing uranium-235 crystallizes, the atoms of uranium-235 trapped within the crystal begin to decay into atoms of lead-207 at a specific fixed rate.

◁ **Early colour photograph** by Gabriel Lippmann

Bakelite is an excellent insulator, so was ideal for making electrical devices

1909

SYNTHETIC PLASTIC

Belgian-born American chemist Leo Hendrik Baekeland invented the first synthetic resin, which he named Bakelite. The substance was made commercially using the chemicals phenol and formaldehyde. It was a tough, water- and chemical-resistant plastic that could be set into any shape and used to make objects such as telephones and radios.

◁ **Bakelite telephone** from the early 1920s

1909 Croation geophysicist Andrija Mohorovičić finds that earthquake wave speeds change with rock composition

1910

1910 J.J. Thomson and British chemist Francis Aston measure the masses of different forms, or isotopes, of elements

△ **Fritz Haber** in his laboratory

1908

INDUSTRIAL PRODUCTION OF AMMONIA

German chemist Fritz Haber discovered an effective way to convert the gases nitrogen and hydrogen into ammonia (NH_3) using high pressure and catalysts. The process enabled nitrogen-based fertilizers to be produced industrially, which increased crop yields to feed growing populations. Ammonia also creates nitrate compounds that were used as powerful explosives during World War I.

▷ ***Drosophila*** eye colour variants

1910

SEX-LINKED TRAITS

American embryologist Thomas Hunt Morgan studied inheritance of eye colour in fruit flies (*Drosophila melanogaster*). Most had red eyes, but a few had white. All the white-eyed flies were male, so demonstrating that a feature of an animal could be linked to its sex.

1911
COSMIC RAYS

An electroscope is an instrument with two thin metal leaves. Ionizing radiation charges the leaves and causes them to repel one another. Austrian-American physicist Victor Hess investigated the phenomenon, carrying an electroscope to high altitudes in a balloon. He found that the radiation originates in space and so discovered cosmic rays.

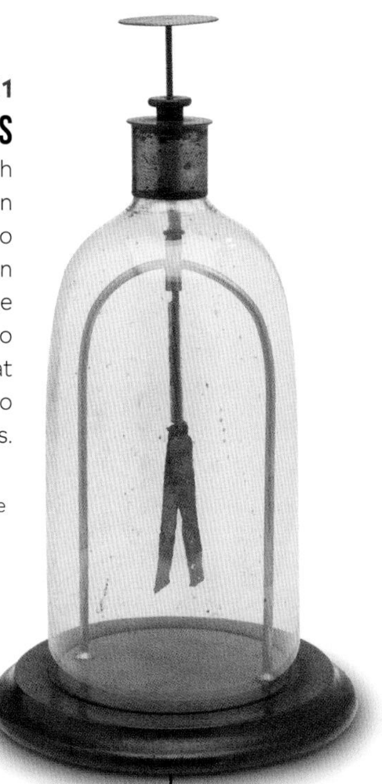

▷ **Gold leaf** electroscope

1868–1921
HENRIETTA SWAN LEAVITT

Born in Lancaster, Massachusetts in the US, Leavett studied at Oberlin College and then at the "Harvard Annex", where she volunteered in the observatory. Her studies in astronomy provided the tools used to map stars and the distances between them.

1911

1911 German physician Alois Alzheimer finds abnormalities in the brains of patients suffering degenerative symptoms

1911 Dutch physicist Heike Onnes discovers superconductivity in mercury wire at very low temperature

1912 German physicist Max Von Laue experiments with X-ray diffraction, providing information on the atomic structure of crystals

△ **Tree rings** in felled tree

1911
CLIMATE RECORD IN TREES

The number and thickness of a tree's annual growth rings record its age and changes in climate as the tree grows. By matching the pattern of annual growth rings in long-lived trees, American astronomer Andrew Douglass was able to build the first historic record of climate-linked tree growth, called dendrochronology.

Alpha scattering experiment
Alpha particles from a radioactive source were directed towards gold foil. Many particles passed through, but some were deflected.

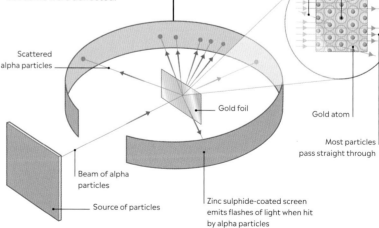

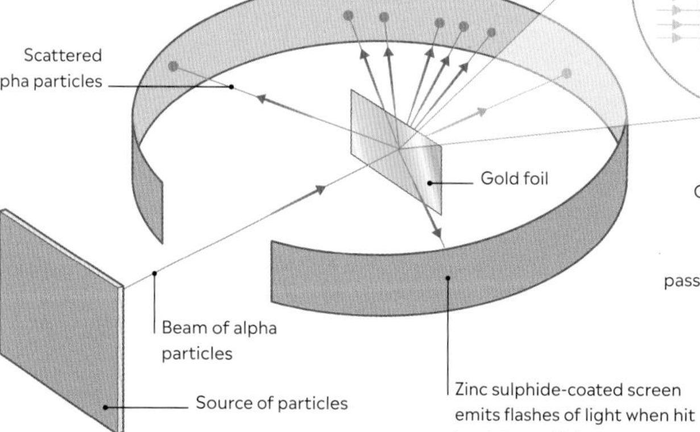
A small proportion of particles are deflected when they hit the dense nucleus
Nucleus of gold atoms
Scattered alpha particles
Gold foil
Gold atom
Most particles pass straight through
Beam of alpha particles
Source of particles
Zinc sulphide-coated screen emits flashes of light when hit by alpha particles

1911
CONSTITUTION OF MATTER

Starting in 1908, Ernest Rutherford and two colleagues began a series of experiments to investigate how matter deflects alpha particles (particles consisting of two protons and two neutrons bound together). In one, they sent alpha particles through thin gold foil. By analysing the deflections of the particles, they concluded that the positive charge in atoms is concentrated at the atomic centre, discovering the atomic nucleus.

PLATE TECTONICS

Earth's outermost layers are broken into several plates that are constantly moving. Over the course of Earth's history, these plates have pushed against one another to form mountains and moved apart to create oceans in a process known as plate tectonics.

Plate motion

The movement of plates is driven by Earth's internal temperature. Rising heat generates volcanoes and breaks plates apart. Where two plates collide, the heavier plate sinks below the other into the mantle.

Plate carried along by movement of upper mantle
Plates forced apart by upward mantle plume
CRUST
Plate sinks, pulled downwards partly by its own weight
Hot material from lower mantle rises towards surface
UPPER MANTLE
LOWER MANTLE
Plate consists of brittle crust fused to top layer of mantle
CORE

1912

CONTINENTAL DRIFT

German scientist Alfred Wegener pieced together the shapes of the world's landmasses and speculated that they once formed a single supercontinent, which was consistent with geological structures and fossils that had since "drifted" apart. His work was largely ignored at first.

△ Alfred Wegener

1912 Henrietta Swan Leavitt discovers the relationship between the luminosity and the period of variability of fluctuating stars called Cepheids, providing a way to measure distances of stars in other galaxies

1912 German chemist Friedrich Bergius develops a process to convert coal dust and hydrogen into petrol

1912

1912 Polish-born biochemist Casimir Funk isolates a substance from rice that cures a disease in pigeons, naming it a "vitamine"

1912

PILTDOWN MAN

Charles Dawson, a British amateur archaeologist, found part of a human-like skull in gravel beds near the village of Piltdown in Sussex, England. The find was initially taken to be 500,000 years old – a missing link between apes and humans. However later technology showed the remains came from two species: the braincase from a modern human and the jawbones from an ape. Piltdown Man was a scientific fraud.

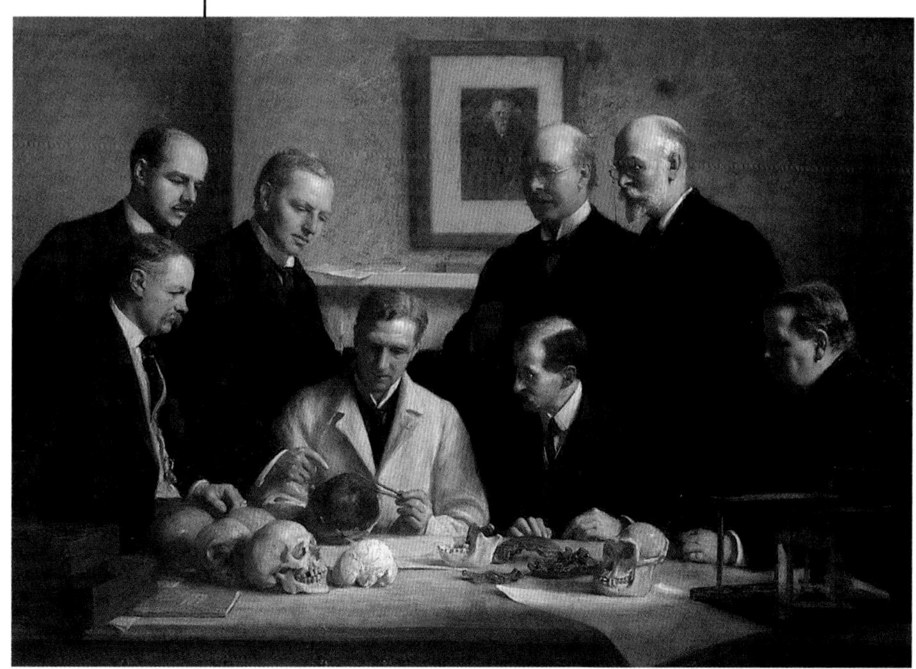

△ **Examination of the** Piltdown skull

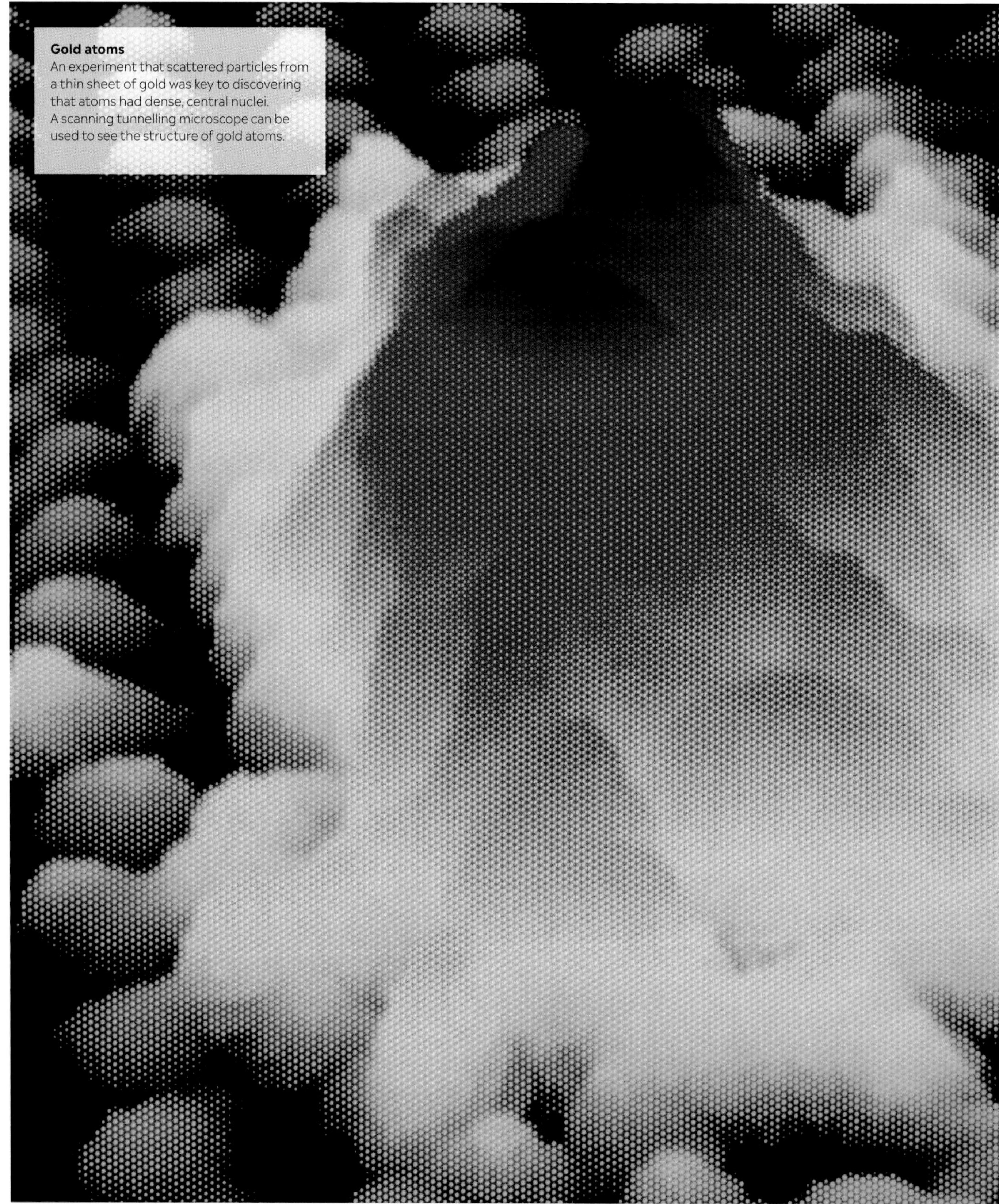

Gold atoms
An experiment that scattered particles from a thin sheet of gold was key to discovering that atoms had dense, central nuclei. A scanning tunnelling microscope can be used to see the structure of gold atoms.

ATOMIC STRUCTURE

Atoms were long thought to be indivisible, but they are in fact made up of smaller subatomic particles: protons, neutrons, and electrons. Different atoms have different numbers of constituent particles, but all have the same basic structure of a central nucleus surrounded by orbiting electrons.

An atom's nucleus contains positively charged protons and electrically neutral neutrons, with the exception of a hydrogen atom; in its most common form, hydrogen contains just a single proton.

Almost all of an atom's mass - more than 99.9 per cent - is found in the nucleus. However, the nucleus is tiny in comparison with the atom as a whole. To give a sense of perspective, if an atom was the size of a sports stadium, its nucleus would be the size of a pea.

Protons and neutrons have approximately the same mass. The number of protons in an atom (known as the atomic number) determines what chemical element the atom is - for example, any atom with 26 protons is iron, no matter how many neutrons and electrons it has. The number of neutrons determines what isotope (particular type of the atom) it is.

Around the nucleus are tiny electrons arranged in shells. They are held in place by their electrical attraction to protons in the nucleus, having equal and opposite (negative) charge. The closer an electron is to the nucleus, the more strongly it is attracted and the harder it is to strip away from the atom. If the number of protons and electrons is the same, the atom is electrically neutral; but if they are different, the atom is known as an ion.

The Cavendish Laboratory
Sir Ernest Rutherford was a leading figure in understanding atomic structure. At Cambridge University's Cavendish Laboratory, he proposed a Solar System-like model of the atom.

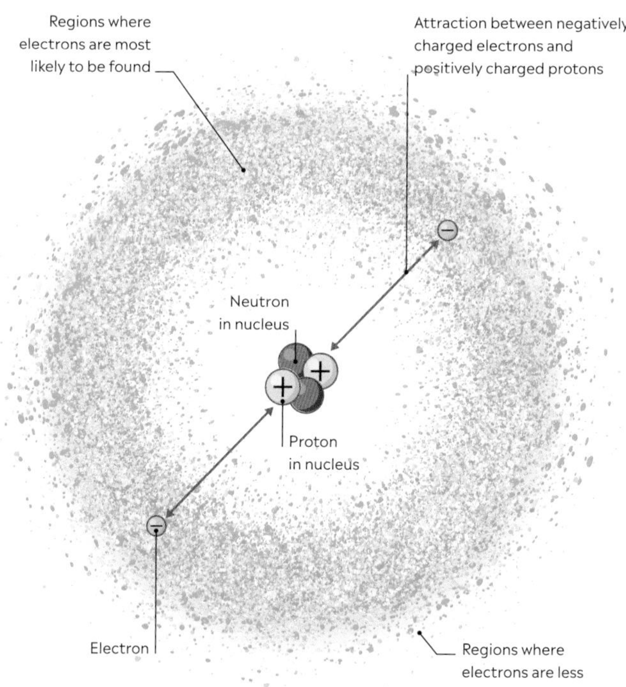

STRUCTURE OF A HELIUM ATOM

Helium is the second-lightest element after hydrogen. Electrically neutral, a helium atom contains just two protons and (usually) two neutrons in its nucleus, and two electrons surrounding it. The locations of the electrons can be represented by a cloud that shows the probability of finding an electron in a given region, with higher density meaning a higher probability.

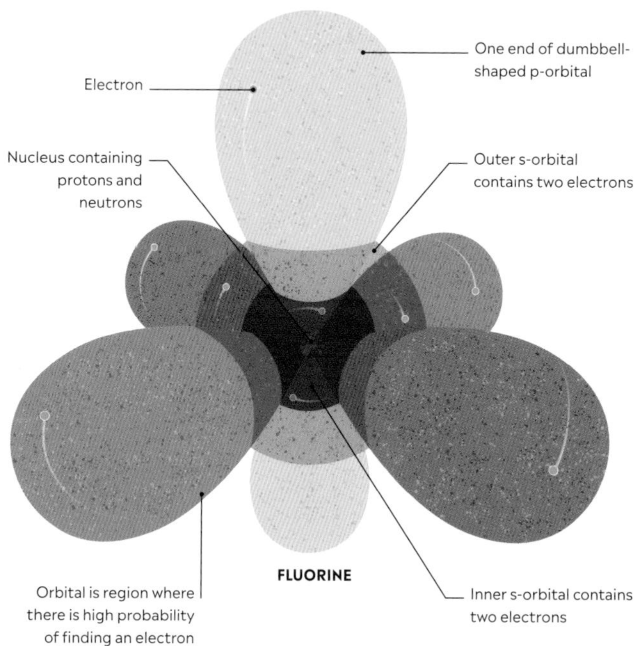

ELECTRON ORBITALS

Electrons do not orbit nuclei in the way planets orbit the Sun. Due to their quantum nature (*see pp.180–81*), it is impossible to pinpoint their locations. Instead, electrons exist in orbitals - that is, regions surrounding the nucleus. Orbitals can hold up to two electrons each, and they fill up starting with the one closest to the nucleus.

1913

HERZSPRUNG-RUSSELL DIAGRAM

Danish astronomer and chemist Ejnar Hertzsprung and American astronomer Henry Norris Russell independently investigated patterns in the colour and brightness of stars. In 1911, Hertzsprung published a diagram showing that more luminous stars in the Pleiades cluster were hotter and bluer than fainter stars. Two years later, Russell produced an expanded diagram covering a much wider range of stars and showing the same pattern.

◁ Henry Norris Russell

STELLAR EVOLUTION

Over billions of years, stars are born, age, and eventually die. They spend most of their lives shining by fusing hydrogen nuclei to make helium in their cores. The rate of this process depends on both the mass of the star and its core temperature. Once the hydrogen in a star's core is exhausted, fusion expands into the rest of the star, but the exhausted core may later reignite to fuse heavier elements. These changes cause the star to brighten and swell enormously. Once the core is totally exhausted, the star's life may end in gradual cooling, the gentle expulsion of its outer layers, or a violent supernova explosion.

1913

1913 Danish physicist Niels Bohr uses quantum theory to propose that electrons orbit the nucleus in "allowed" orbits

1913 French physicist Charles Fabry finds that an atmospheric ozone layer filters out solar ultraviolet radiation

1913 German biochemist Leonor Michaelis and Canadian physician Maud Menten develop an equation describing rates of enzyme reactions

Isotopes of carbon

Two varieties or isotopes of carbon show that the number of protons (atomic number) is constant, while the number of neutrons varies, resulting in a difference in overall atomic mass.

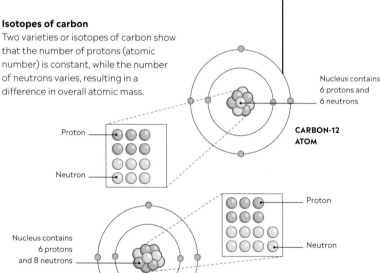

Nucleus contains 6 protons and 6 neutrons

Proton

Neutron

CARBON-12 ATOM

Proton

Neutron

Nucleus contains 6 protons and 8 neutrons

CARBON-14 ATOM

1878–1958

J.B. WATSON

John Broadus Watson grew up in South Carolina, US. He studied psychology at the University of Chicago before teaching at Johns Hopkins University in Baltimore, Maryland. He is famous for promoting the approach known as behaviourism.

1913

THEORY OF ISOTOPES

While studying radioactivity, Frederick Soddy noted that some elements, which were the products of decay, had more than one atomic mass but the same atomic number. He coined the term isotope - meaning "at the same place" (in the periodic table) - after a suggestion from his friend Margaret Todd, a Scottish physician.

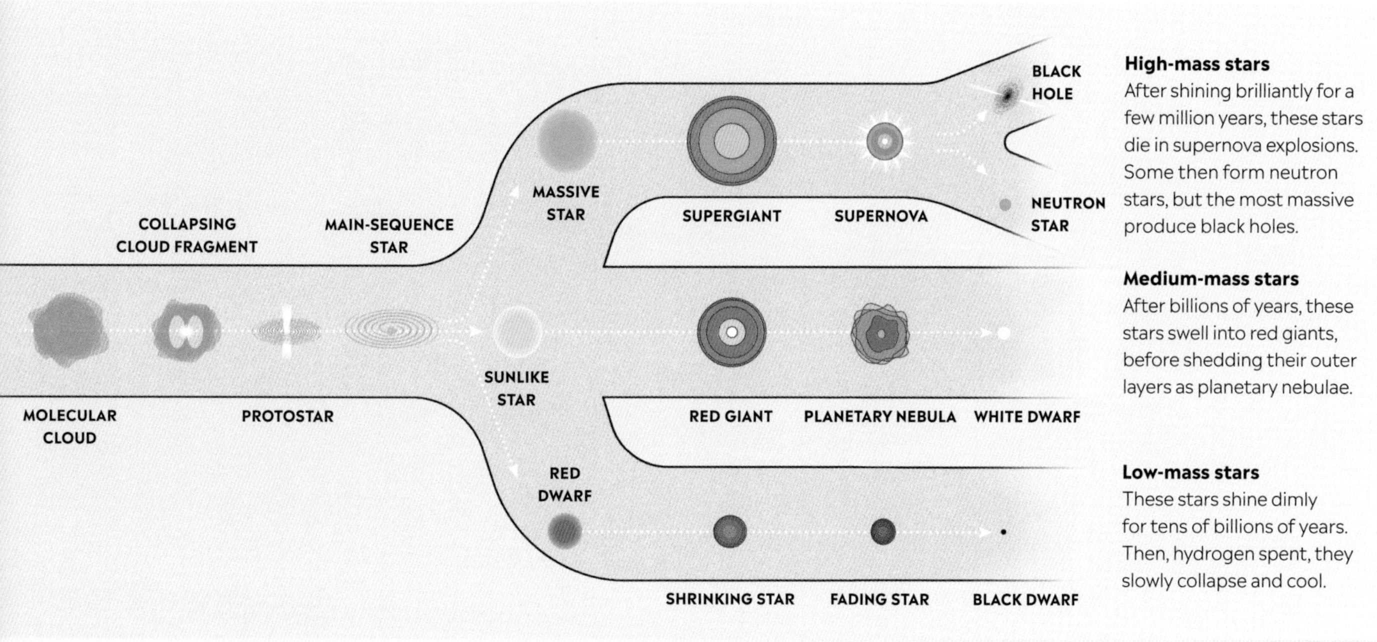

High-mass stars
After shining brilliantly for a few million years, these stars die in supernova explosions. Some then form neutron stars, but the most massive produce black holes.

Medium-mass stars
After billions of years, these stars swell into red giants, before shedding their outer layers as planetary nebulae.

Low-mass stars
These stars shine dimly for tens of billions of years. Then, hydrogen spent, they slowly collapse and cool.

1913 US physicist Robert Millikan publishes the results of his "oil drop" experiment, with which he calculated the electric charge of the electron

1914 Thyroxine, a hormone produced by the thyroid gland, is isolated by American chemist Edward Calvin Kendall

1914

1914 British physicist Henry Moseley determines that atomic number (the number of protons in an atom's nucleus) is a real quantity, not just an ordering of the periodic table

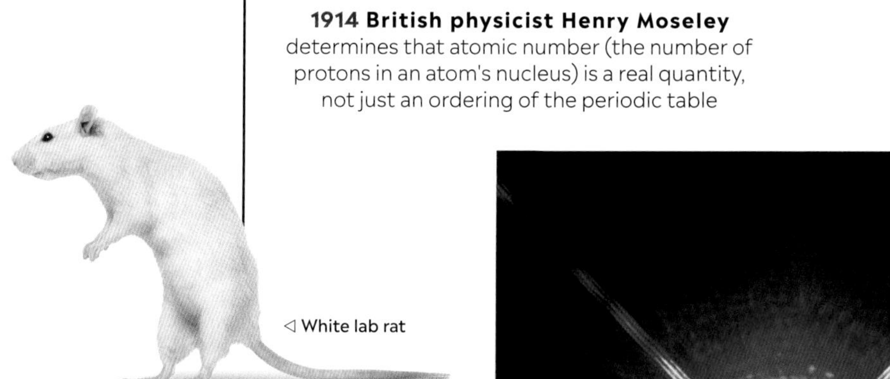
◁ White lab rat

1913
BEHAVIOURISM

US psychologist J.B. Watson and others sought to bring objectivity to the study of behaviour. He outlined his theories at Columbia University in a lecture titled *Psychology as the Behaviourist Views it*. He sought to probe behaviour experimentally and explain it based on reflexes, associations, and the effects of reinforcers. Watson favoured the use of lab rats as test animals.

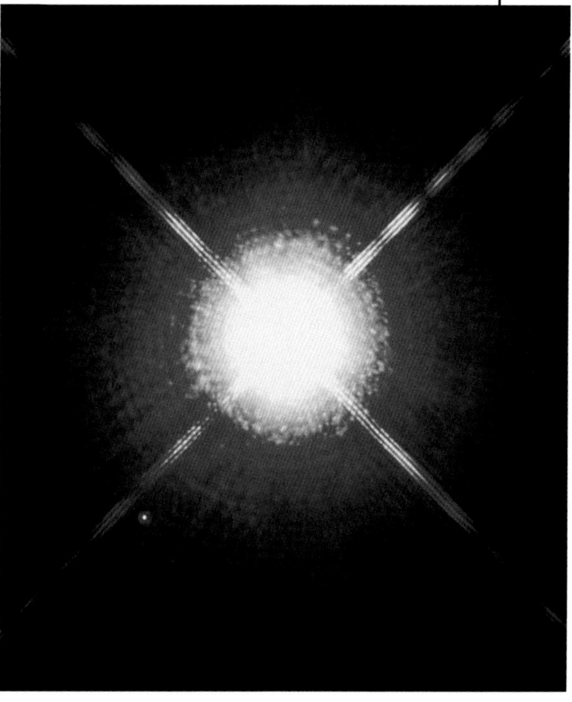

1914
WHITE DWARFS

US astronomer Walter Sydney Adams confirmed that the faint but nearby star 40 Eridani B had a hot white surface normally associated with much more luminous stars, and could be a new class of stellar object. In 1915, he showed that Sirius B, the faint companion of the brightest star in the sky, had similar properties. Stars of this type, now known as white dwarfs, were later shown to be the collapsed cores of Sun-like stars.

◁ **Sirius B** is the tiny dot to the lower left of the much brighter star Sirius

1916
SCHWARZCHILD RADIUS

Einstein's General Theory of Relativity described how the presence of mass or energy in space distorts, or curves, spacetime, the four-dimensional continuum of the Universe. The curvature can be calculated using a set of formulae called the field equations. In 1916, German physicist Karl Schwarzschild used the equations to work out the gravitational field around a spherical object. He discovered that, at a certain radius from a very dense object's centre, the field is strong enough that not even light can escape; he thus found a mathematical expression for a black hole. The "Schwarzchild radius" determines the position of the black hole's event horizon.

Two black holes colliding

Gravitational waves radiate out at the speed of light

Black holes

When huge stars come to the end of their lives, gravity causes them to collapse down to below their Schwarzschild radius, and become black holes. So strong is the gravitational influence of black holes that when they collide, they produce ripples in spacetime, called gravitational waves.

1915

1915 The first transatlantic wireless call is made from Arlington, Virginia, USA, to the Eiffel Tower in Paris, France

1915 Proxima Centauri, the Sun's closest neighbouring star, is discovered by British astronomer Robert Innes

▽ **Rock salt** crystal

◁ **Sodium chloride** crystal structure

1915
X-RAY CRYSTALLOGRAPHY

English father and son physicists Henry and Lawrence Bragg published a book detailing their advances in the practical use of X-ray diffraction to determine the arrangement of atoms in crystals. Their first results emerged from their identification of the crystal structures of several compounds, including rock salt (sodium chloride) and diamond.

1915
EVOLUTION BY JUMPS

British geneticist Reginald Punnet studied the evolution of mimicry in butterflies. In the Mormon swallowtail butterfly, for example, he noted that some females resembled an unrelated species that was not palatable to predators. Punnet suggested that this change came about through evolutionary jumps, rather than gradually as Darwin's theory generally proposed.

△ **Mormon swallowtail** butterfly

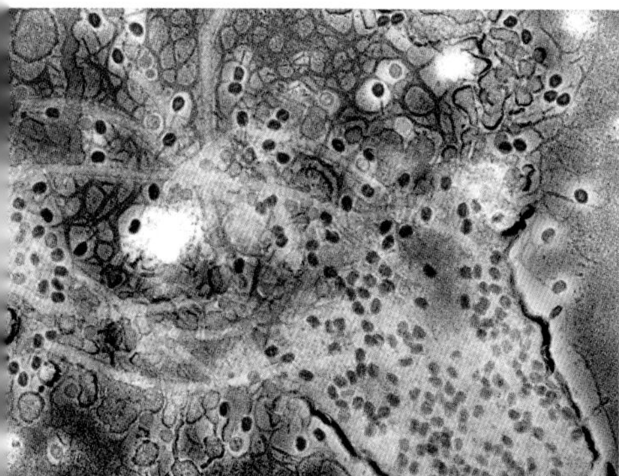

1916

BACTERIA-EATERS

French-Canadian microbiologist Félix d'Herelle discovered a type of virus, later called bacteriophage ("bacteria-eater"), which infects and destroys bacteria. In doing so, the virus takes over the bacterial cell, forcing it to create new viruses.

◁ **Bacteriophages** infecting *E. coli*

1918

MASS SPECTROMETRY

British physicist Francis Aston built the first practical mass spectrometer. The instrument can separate atoms and isotopes that have different masses. An electric current ionizes (imparts a charge to) atoms of a substance inside a glass globe. Electric and magnetic fields accelerate the resulting ions, which follow different trajectories according to their masses.

◁ **Mass spectrometer** from 1919

1917 The Hooker telescope enters service in California, US. It is the largest telescope in the world until 1949

1918

1916 Albert Einstein publishes the final version of his groundbreaking general theory of relativity (*see below*)

1918 US astronomer Harlow Shapley shows that the Milky Way is far larger than previously thought and that the Sun is far from its centre

GENERAL THEORY OF RELATIVITY

The general theory of relativity describes gravity not as an attractive force between massive objects – as Isaac Newton had proposed – but as a geometric property of space-time produced by these objects. Massive objects such as stars cause the "fabric" of space-time to warp. These distortions manifest as gravity. General relativity accounts for observations that could not be explained using Newton's theory, and it was used to predict phenomena that have been observed since – for example, gravitational waves in 2015, a century after the theory was proposed.

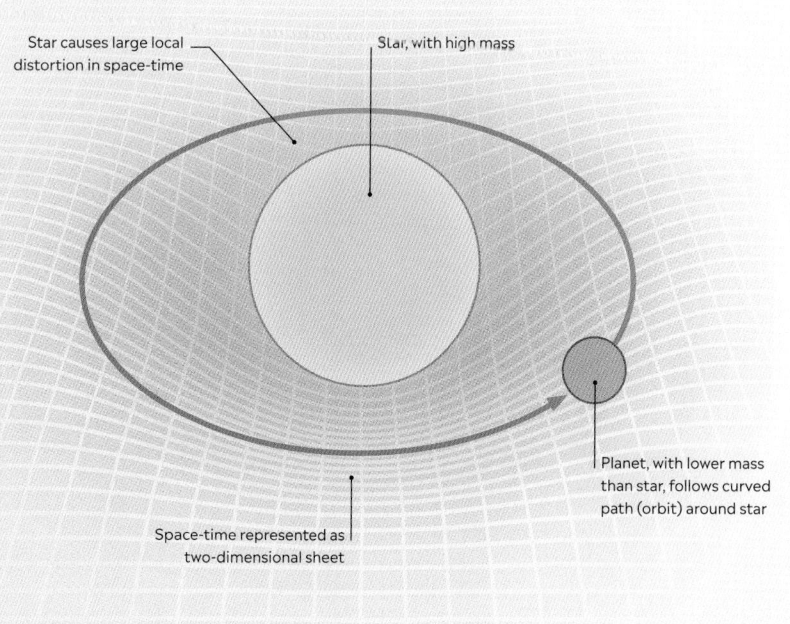
Star causes large local distortion in space-time
Star, with high mass
Planet, with lower mass than star, follows curved path (orbit) around star
Space-time represented as two-dimensional sheet

Gravity and space-time
Einstein's theory models the three dimensions of space and the one dimension of time into four-dimensional space-time. Space-time is shaped by the presence of mass.

CHEMICAL BONDS

The properties of metals
Metals owe many of their properties to their metallic bonds. They have lots of free electrons, making them good conductors of heat and electricity. They are also malleable when heated.

A chemical bond is an interaction between atoms that creates substances made of molecules, ions, metals, or crystals that we encounter in everyday life. The subatomic particle that makes chemical bonds possible is the negatively charged electron.

An atom has a central nucleus that contains positively charged protons, surrounded by an equal number of electrons arranged in shells around the nucleus. The number of electrons in the outer shell (valence shell) determines what type of bonds - and how many - are formed.

Molecules are formed by nonmetal atoms (including carbon, oxygen, and nitrogen), which bond by sharing pairs of electrons to complete the outer shells of electrons. Known as covalent bonds (*see below*), these can occur between atoms of either the same or different elements. Some atoms can share more than one pair of electrons to form multiple bonds, as well as double or triple bonds.

Metal atoms can donate the electrons in their outer shells to nonmetal atoms, creating positive metal ions (anions) and negative nonmetal ions (cations). These form ionic bonds (*see below*) using electrostatic forces - crystals are formed via these bonds, for example. Metal atoms can also give up their outer-shell electrons to form metallic bonding, a lattice of metal ions surrounded by shared free electrons.

Weaker electrostatic chemical bonds also occur between molecules. One example of this is the hydrogen bonds in water that make it a liquid at room temperature.

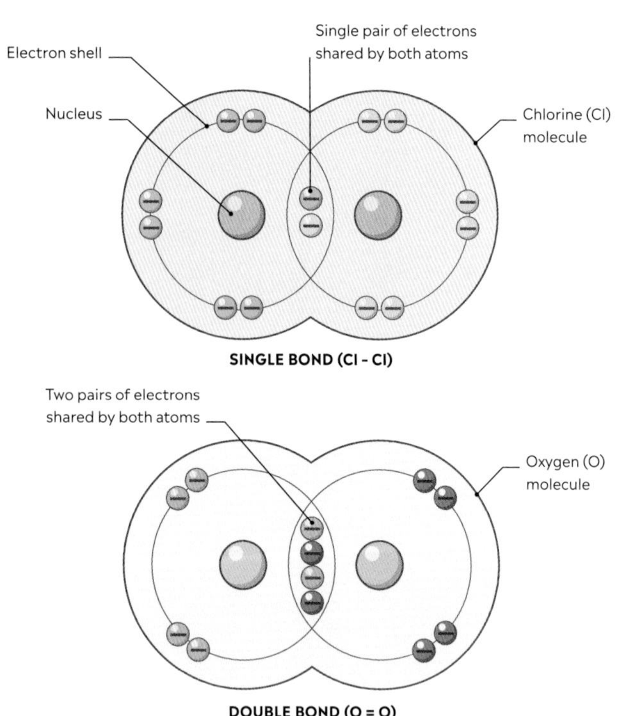

COVALENT BONDS

A covalent bond forms when two nonmetal atoms share a pair of electrons. Sharing electrons allows the atoms in the molecule to fill their outer electron shells, making them more stable. The number of electrons available to pair is called the valency, and this can be used to work out how different elements combine.

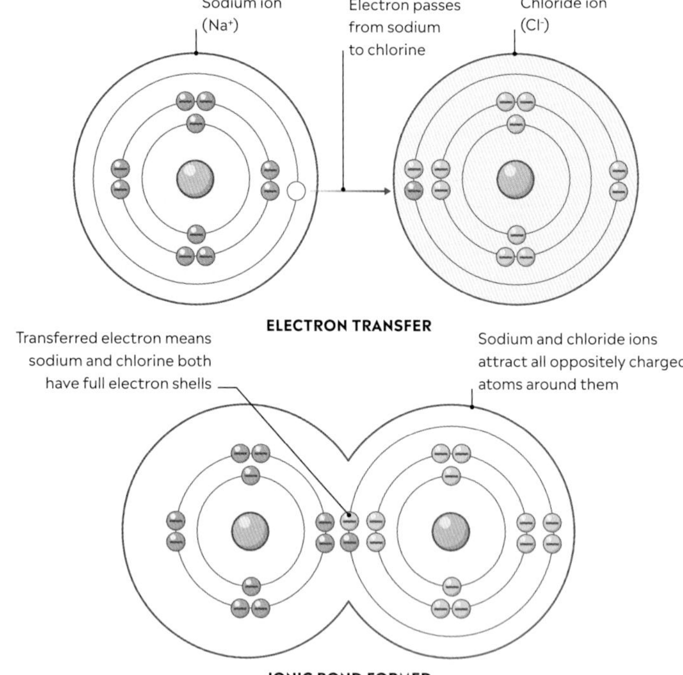

IONIC BONDS

An ionic bond forms when electrons transfer from the atom of one element (usually a metal) to the atom of another element (usually a nonmetal). The outer electron shell of each atom is filled, and the positive and negative ions that are created bond as a result of the electrostatic attraction they have to one another.

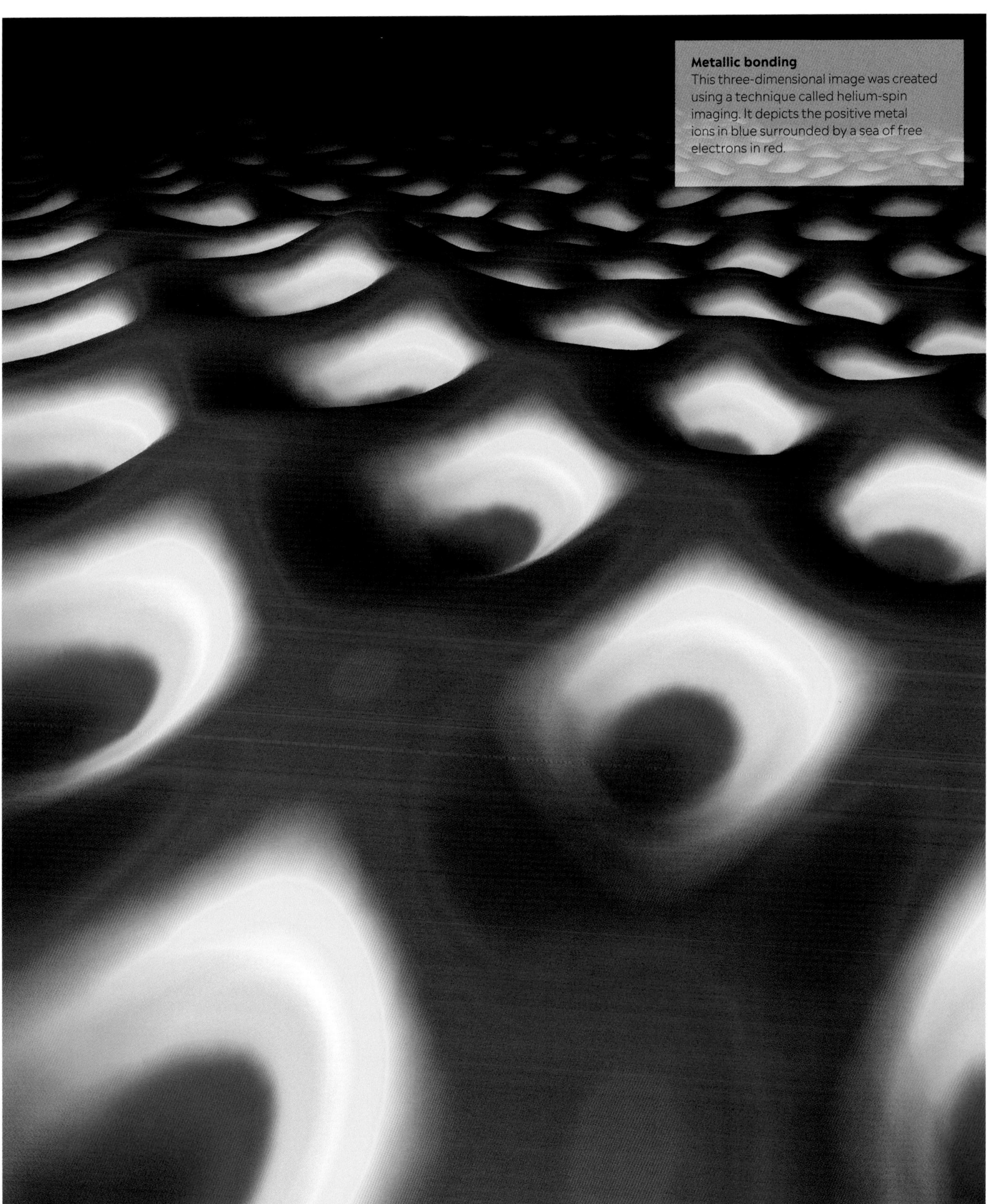

Metallic bonding
This three-dimensional image was created using a technique called helium-spin imaging. It depicts the positive metal ions in blue surrounded by a sea of free electrons in red.

▽ **Total eclipse** of the Sun

1919
CONFIRMATION OF EINSTEIN'S THEORY

Einstein's General Theory of Relativity suggested that the presence of mass would distort spacetime and so alter the course of light that would otherwise have travelled in a straight line. His equations enabled scientists to calculate the amount by which space would be distorted. A total solar eclipse on 29 May offered an opportunity to test the theory. Scientists in two separate expeditions took photographs of the eclipsed Sun. Analysis showed that stars adjacent to the Sun in the sky had moved their positions by exactly the amount predicted by Einstein's theory.

1879–1958
MILUTIN MILANKOVIC

Born in Serbia, then part of Austro-Hungary, Milanković studied and practised engineering in Vienna. He later taught mathematics and became interested in the astronomical causes of Ice Ages, studying differences in climate over time.

1919

1919 American chemist Irving Langmuir explains how atoms can form covalent bonds; he is awarded a Nobel prize in 1932

1919 German physicist Albert Betz publishes his law relating to the maximum energy a wind turbine can extract from moving air

1919 Ernest Rutherford discovers that elements, such as nitrogen, can be "disintegrated" by bombarding them with alpha particles, which causes them to emit protons

1919
BEE DANCE

Austrian biologist Karl von Frisch studied the lives of honey bees (*Apis mellifera*), explaining how they transmit information using dance-like movements. A bee dancing in a tight circle informs the other bees that food is nearby, while a more complex waggle-dance transfers information about more distant sources - both direction and distance. Direction is indicated by the angle of the dance to the vertical, and distance by the time the bee takes during its waggle at the centre of the dance. Partly to honour this work, von Frisch received the Nobel Prize in 1973.

◁ **Bees** dance in a hive

1920

MILANKOVIC CYCLES

Serbian astrophysicist Milutin Milanković developed a theory concerning the predictable changes in Earth's motion around the Sun. He suggested that they control long-term climate change, such as the advance and retreat of ice sheets, and demonstrated that three aspects of orbital motion are at work: variations in the shape of Earth's orbit around the Sun, called orbital eccentricity; changes in the angle of Earth's axis of rotation, called obliquity or tilt; and changes in the direction of the axis of rotation, called precession. As Earth travels around the Sun, regular variations in these three aspects of the Earth–Sun relationship combine to vary the amount of solar (heat) energy that reaches Earth. Together they are known as the Milanković cycles.

Milankovitch cycles
Over 100,000 years Earth's orbit varies from circular to elliptical (eccentricity). Over 42,000 years the tilt of Earth's axis changes, and over 28,200 years the axis wobbles (precession). Together and over time, these changes affect Earth's surface temperature and correspond to changes in climate, such as the onset and ending of ice ages.

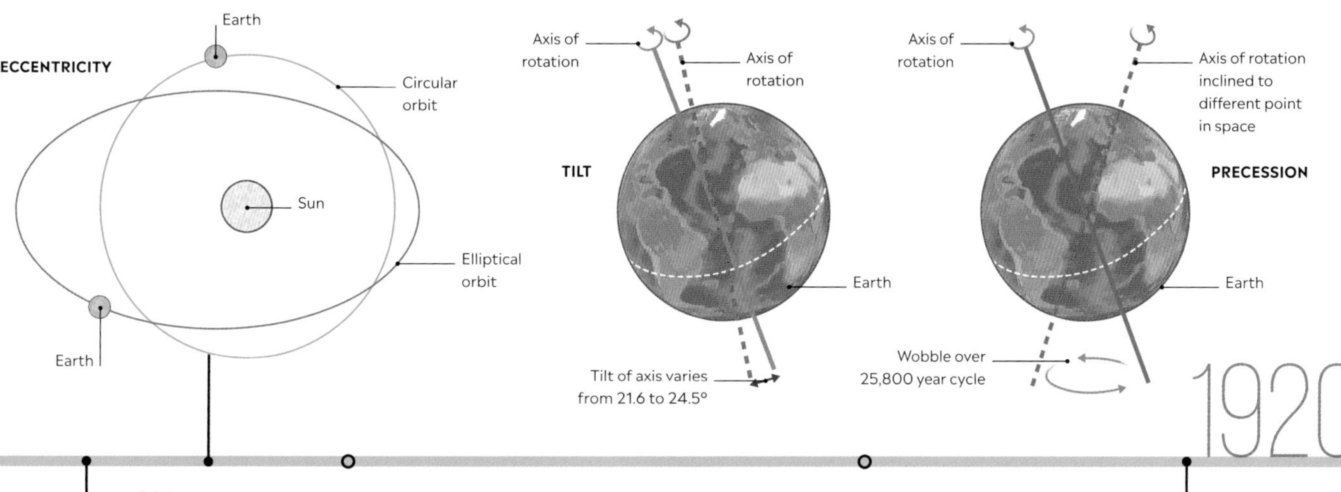

1920

1920 US astronomers Albert Michelson and Francis Pease measure the diameter of the star Betelgeuse. They find it is about 300 times bigger than the Sun

1920 Norwegian meteorologist Jacob Bjerknes discovers that weather fronts forming along an undulating line in the Atlantic grow into atmospheric waves known as cyclones

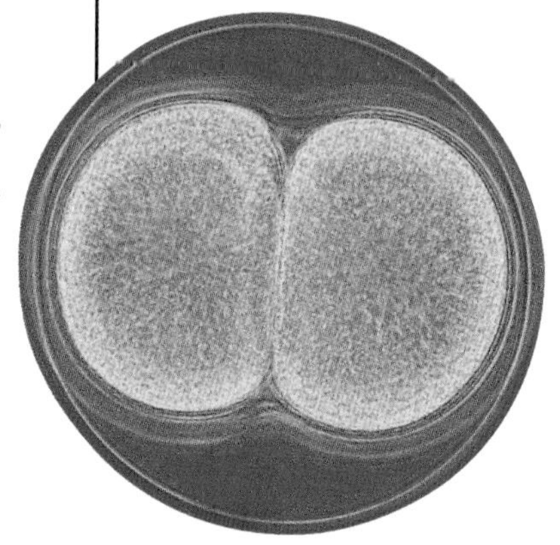
▷ Two-cell embryo stage

1920s

ANAEMIA

Anaemia is a condition in which the number of red blood cells circulating through the body is too low. American physiologist George Hoyt Whipple suggested methods of stimulating the formation of new red blood cells through diet. He established that foods such as meat, especially liver, help to alleviate the problem.

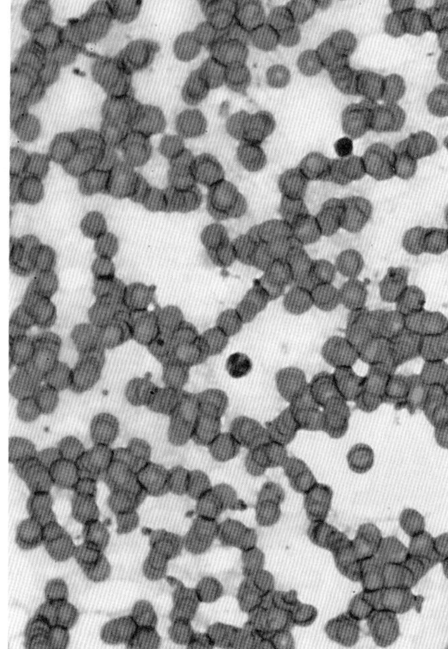
△ **Red blood cells,** micrograph

1920

EMBRYO ORGANIZERS

In their research into amphibian embryos, German scientists Hans Spemann and Hilde Mangold made a discovery that changed the focus and direction of investigations in developmental biology. They established the existence of a cluster of cells that induces the development of the central nervous system. Induction is the key process by which certain cells influence the development of surrounding cells.

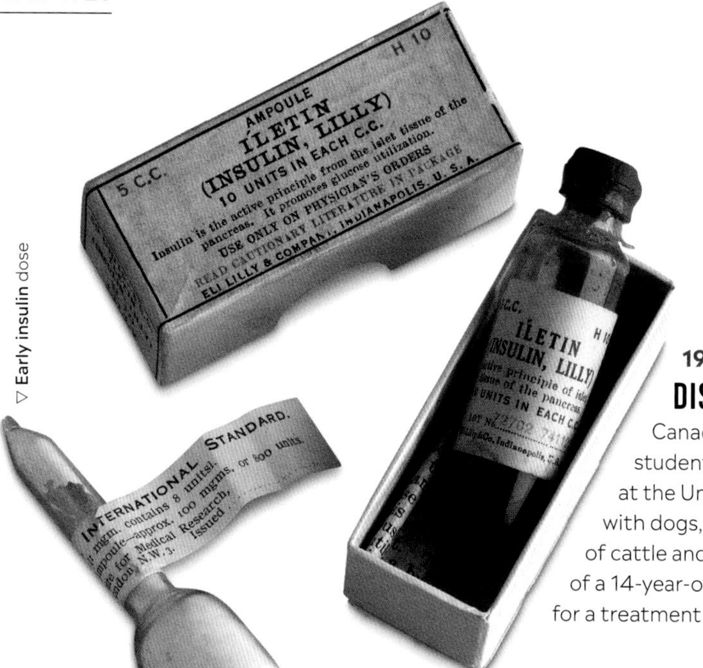

▷ Early insulin dose

The hormone insulin helps to regulate the amount of glucose in the bloodstream

1921

DISCOVERY OF INSULIN

Canadian surgeon Frederick Banting and medical student Charles Best isolated the hormone insulin at the University of Toronto, Canada. After initial trials with dogs, they extracted insulin from the pancreases of cattle and successfully improved the condition of a 14-year-old diabetic boy, thus paving the way for a treatment for this previously fatal disease.

1921

1921 German biologist Otto Loewi proves that nerves act via chemical transmitters

1921 Scottish biologist Alexander Fleming discovers a substance that kills bacteria; he names it lysozyme

1922 Norwegian mineralogist Victor Goldschmidt classifies chemical elements according to their associations with various earth materials

1921

VITAMINS AND RICKETS

American biochemist Elmer Verner McCollum helped to discover a number of key dietary substances through his feeding experiments with rats. He found that a substance from eggs and butter was essential for healthy growth, and named it factor A (later vitamin A) and also helped to identify vitamin B. Rickets is a condition that affects the development of bone. McCollum established that rats affected by the condition recovered when fed cod-liver oil, naming the key substance involved vitamin D. He and colleagues found that sunlight also protected the rats from rickets.

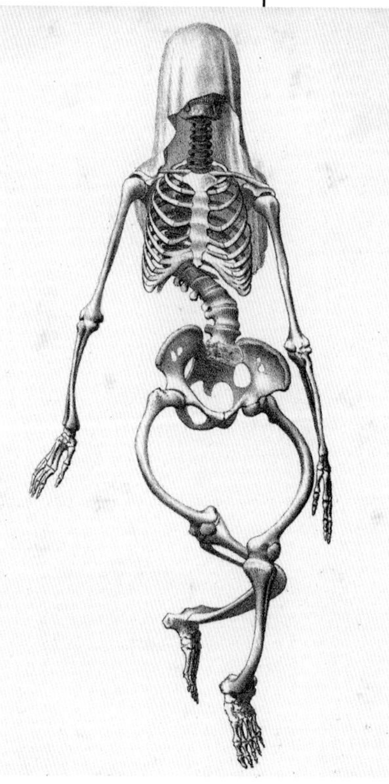

▷ **Skeleton** of human rickets victim

VITAMINS

Essential for our body's growth, vitality, and wellbeing, vitamins are nutrients that, although only needed in tiny amounts, are required for making use of other nutrients. Vitamins are made by plants and animals, whereas minerals come from soil and water. The body is able to store any vitamins that dissolve in fat, whereas water-soluble vitamins cannot be stored and must therefore be replenished more frequently.

Vitamin functions
There are many different types of vitamin, both fat-soluble and water-soluble. They can be found in various foods and have different functions in the body.

1922
PARTICLE SPIN

In their landmark experiment, German physicists Otto Stern and Walther Gerlach sent silver atoms through a magnetic field, establishing that particles' spin is quantized – it can only have certain values. This helps to explain the arrangement of electrons in atoms.

Fields and particle spin induction
The spin of a particle is analogous to physical rotation, and it gives subatomic particles an associated magnetic field.

△ Louis de Broglie

1923
PARTICLES AND WAVES

US physicist Arthur Compton established the particle nature of X-rays, previously thought of as waves. In the same year, French scientist Louis de Broglie showed that electrons, previously thought of as particles, also behaved as waves. These two advances helped establish wave-particle duality.

1923

1922 Applying the equations of general relativity, Russian physicist Alexander Friedmann suggests that space might be expanding

1923 The ultracentrifuge is invented by Swedish chemist Theodore Svedberg; it can separate small particles, including proteins, of differing masses

1923 Acids and bases are defined as chemical species tending to lose (acid) or gain (base) a proton

FAT-SOLUBLE

VITAMIN A Needed for vision, growth, and development; deficiency can affect sight.

VITAMIN D Improves uptake of some minerals, including calcium; low levels can cause rickets.

VITAMIN E Maintains healthy skin, eyes; deficiency can impair the immune system.

VITAMIN K Needed to help blood clot; deficiency can lead to bleeding and bruising.

WATER-SOLUBLE

VITAMIN B1 Releases energy from food and improves muscle and nerve function.

VITAMIN B2 Promotes healthy skin, eyes, and nervous system; deficiency leads to anaemia.

VITAMIN B3 (niacin) Maintains the nervous system, cardiovascular system, blood, and skin.

VITAMIN B5 Releases energy from food and helps break down fats; deficiency is rare.

VITAMIN B6 Improves nerve function and more; deficiency can affect mental health.

VITAMIN B7 (biotin) Needed for healthy bones and hair; deficiency causes dermatitis and muscle pain.

VITAMIN B9 (folic acid) Deficiency in pregnant women increases risk of spina bifida in their babies.

VITAMIN B12 Involved in red-blood-cell production; deficiency can lead to blood disorders.

VITAMIN C Promotes healthy skin, blood vessels, bones, and cartilage; deficiency can cause scurvy.

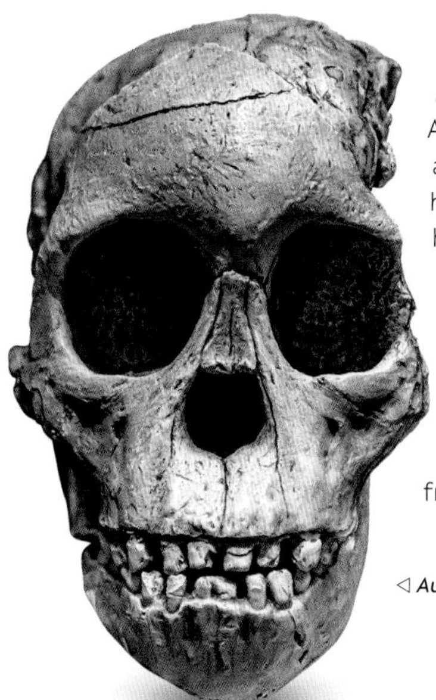

◁ *Australopithecus africanus* replica skull

1924
AUSTRALOPITHECUS AFRICANUS

Australian anthropologist Raymond Dart made a discovery that changed how we think about human evolution. While working in South Africa, he received fossils unearthed near the town of Taung. One was the skull of a young ape-like creature with a human-like jaw and teeth and a modest-sized brain. Dart named his find *Australopithecus africanus* ("southern ape of Africa"), recognizing it as important evidence of the evolution of early humans from their ape-like ancestors.

1926
WAVE MECHANICS

Erwin Schrödinger developed a new approach to quantum physics. In "wave mechanics", the behaviour of a quantum system (such as an atom) is described by a "wave function", which represents the probability of the system being in any of a number of allowed states. The Schrödinger equation can be used to work out the wave function.

▷ Erwin Shrödinger

1924

1924 Drawing on the work of Henrietta Swan Leavitt, US astronomer Edwin Hubble uses variable stars to show that so-called "spiral nebulae" are in fact distant galaxies

1924 Indian physicist Satyendra Nath Bose establishes a method to describe the behaviour of subatomic particles known as bosons

1924 German embryologists Spemann and Mangold identify organizer cells that direct embryonic development

1925 German chemist Carl Bosch develops a process for hydrogen manufacture on an industrial scale

1924
ELECTRON QUANTUM NUMBERS

Until now, the state of an electron in an atom had been described by three quantum numbers, representing its energy, angular momentum, and the orientation of its orbit. In 1924, Wolfgang Pauli suggested a fourth (later called "spin"). His exclusion principle states that no two electrons can have the same set of quantum numbers, helping explain the arrangement of electrons in atoms. In 1940, he extended the principle to all "half-integer spin" particles (fermions).

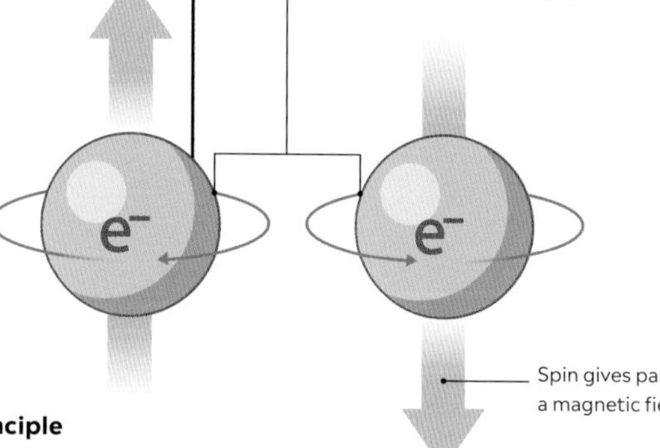
Two electrons at identical energy level, sharing three quantum numbers, but with opposing spin
e⁻
e⁻
Spin gives particles a magnetic field

Pauli exclusion principle
In any atom, paired electrons share the same three quantum numbers, but have opposing spin quantum numbers.

> "Physics... is much too hard for me anyway."
>
> *WOLFGANG PAULI, LETTER TO R. KRONIG, 1925*

1926
ADVANCES IN ROCKETRY

American engineer Robert Goddard launched the first liquid-fuelled rocket. The potential usefulness of liquid rocket propellants (in contrast to traditional powder fuels) for space flight had been identified by Russian teacher Konstantin Tsiolkovsky in 1903, but Goddard's brief flight, which lasted just 2.5 seconds, was the first practical demonstration of the concept. It kickstarted interest in rocketry and space exploration in several countries.

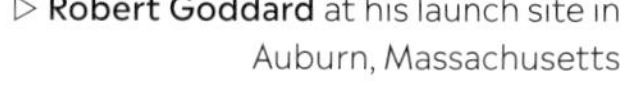

▷ **Robert Goddard** at his launch site in Auburn, Massachusetts

1901-54
ENRICO FERMI

Born in Rome, Fermi was a brilliant physicist. He developed quantum statistics to describe energy states, pioneered the production of heavy elements not found in nature, and, after moving to the USA, led a team that built the first nuclear reactor.

1926

1926 Enrico Fermi establishes a mathematical method to describe the behaviour of "fermions": subatomic particles that do obey the Pauli exclusion principle

1926
TELEVISION

Scottish inventor John Logie Baird gave the first public demonstration of his "televisor", a crude but impressive television system, in 1926. The device uses a Nipkow disk – one with holes arranged in a spiral that allows objects to be scanned several times per second. In 1927, he transmitted colour pictures, and in 1928, he sent images more than 700 km (435 miles) along a telephone line. Baird's system was superseded by all-electronic television.

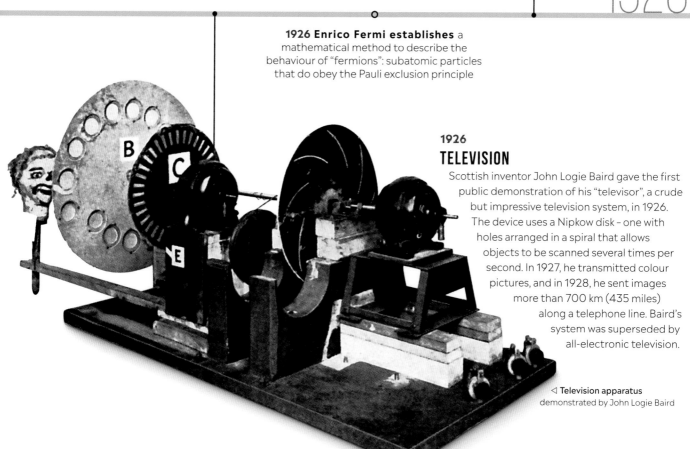

◁ **Television apparatus** demonstrated by John Logie Baird

HUMAN EVOLUTION

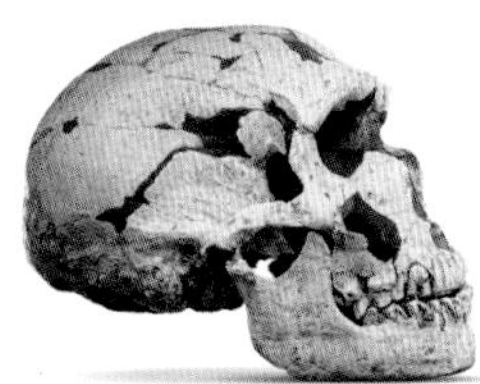

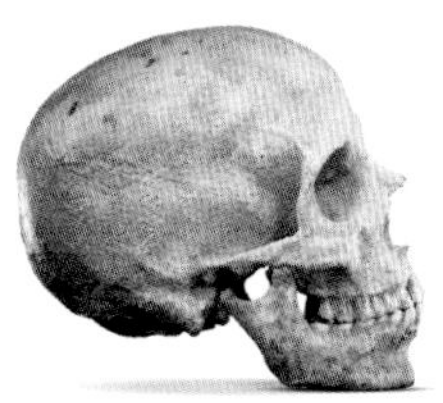

Coexisting species
Neanderthals lived between 430,000 and 40,000 years ago, coexisting with human ancestors. Genomics reveal that modern Europeans have about 1–2 per cent of Neanderthal DNA.

The study of fossils, DNA, and archaeological sites continues to help researchers understand how modern humans (*Homo sapiens*) evolved.

Studies of human and ape DNA suggest humans and chimpanzees shared a common ancestor about 6 million years ago. At that point, a branch in the evolutionary tree split off, leading to the so-called Hominini tribe. *Homo sapiens* is the sole survivor of this tribe, which included several species in the genera *Australopithecus* and *Homo*, among others. The earliest hominin species resembled chimpanzees but appears to have been able to use tools. (To add to the confusion, so can some chimps.)

For many years, it was thought that humans evolved in East Africa about 200,000 years ago and migrated across Eurasia 90,000–45,000 years ago. However, modern human remains dating back to at least 315,000 years ago have been discovered in northwest Africa, and some dating back to 200,000 years ago have appeared in the eastern Mediterranean. Far more recent hominins are still being discovered.

Twenty other members of the human family tree have been discovered since Neanderthals were recognized in 1864. For example, among recent discoveries, in 2003, a partial skeleton of a small hominin species, about 1 m (3 ft 3 in) tall, was excavated from a cave on the Indonesian island of Flores. In 2019, further small hominin remains were found in the Philippine island of Luzon. These were not the *Homo floresiensis* species discovered in 2003, but a new species, *Homo luzonensis*, dated to 67,000 years ago.

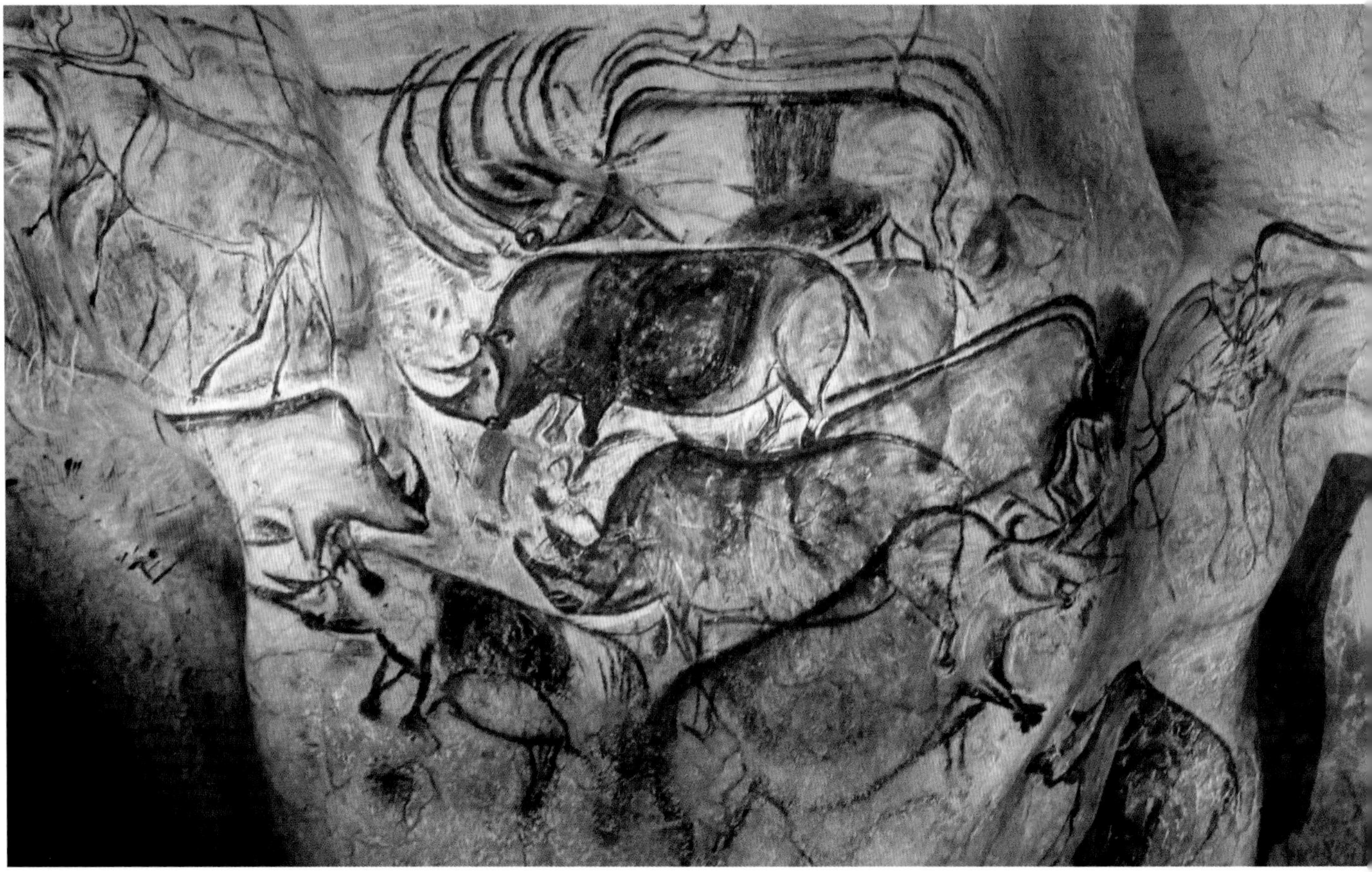

TIMELINE OF HUMAN EVOLUTION

The Hominini tribe includes numerous different species, genera, and families. The overall evolutionary lineage leading to *Homo sapiens* is clear, but the direct ancestral relationships less so. This timeline shows the periods at which different species existed but does not indicate how they relate to one another. Even so, the timeline is subject to change because new discoveries continue to be made.

KEY
Ardipithecus · *Australopithecus* · *Homo*

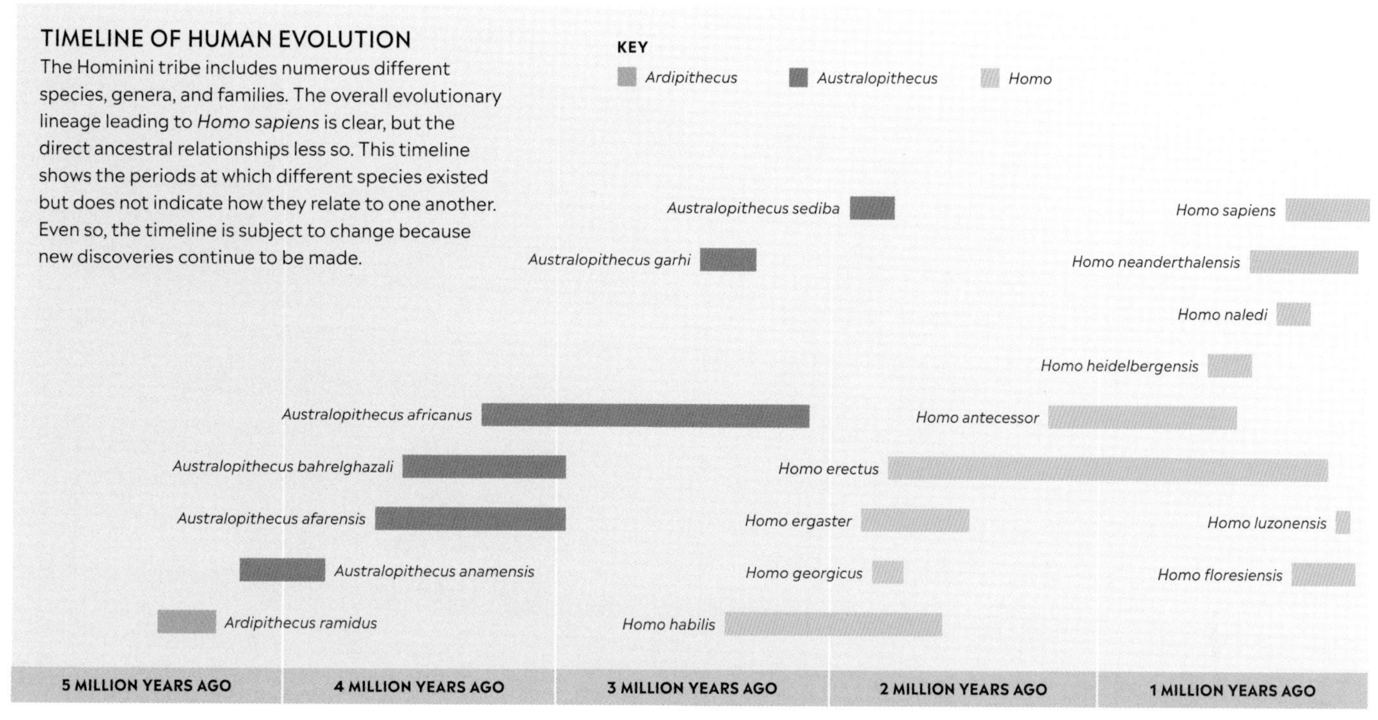

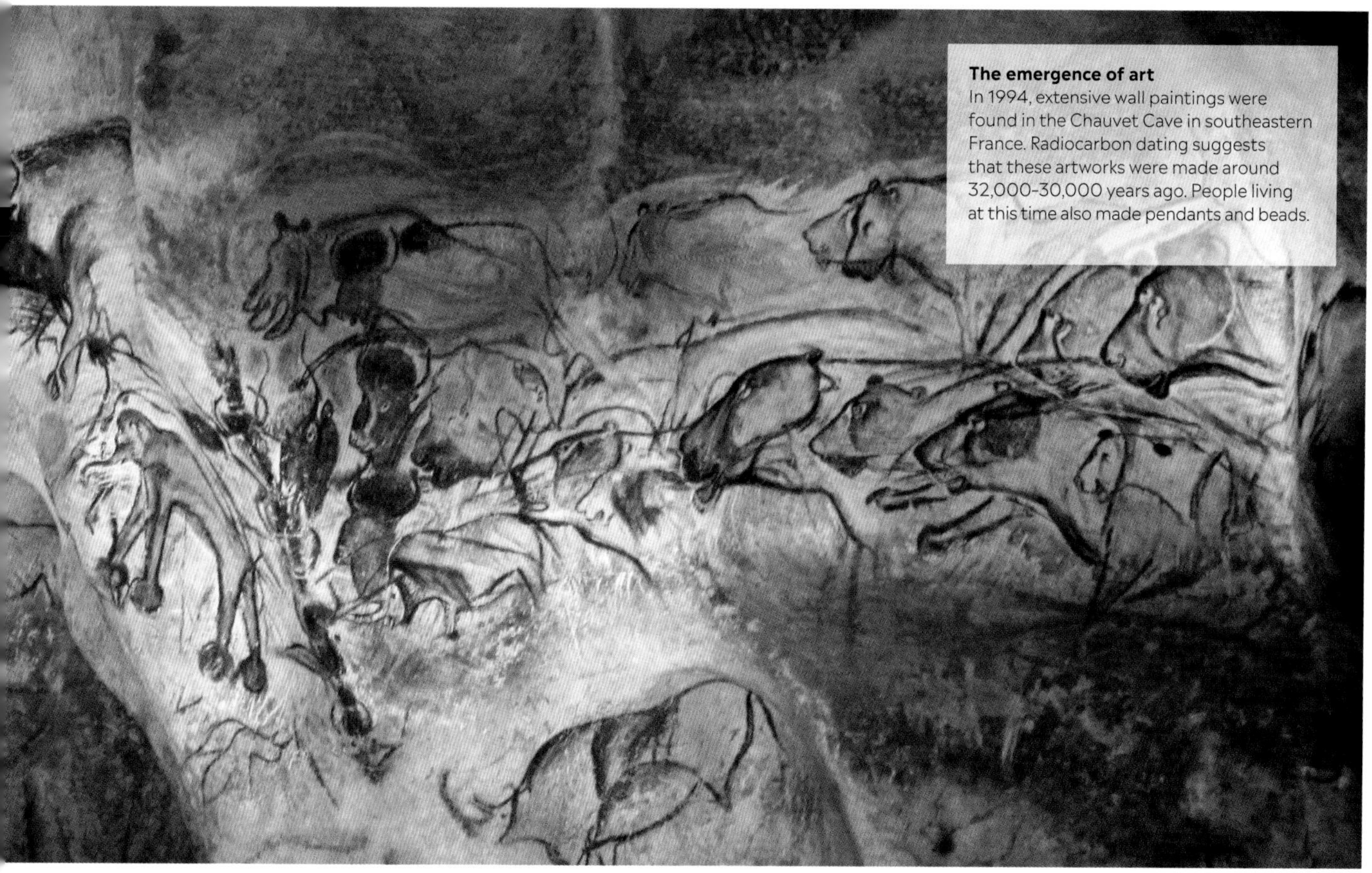

The emergence of art
In 1994, extensive wall paintings were found in the Chauvet Cave in southeastern France. Radiocarbon dating suggests that these artworks were made around 32,000–30,000 years ago. People living at this time also made pendants and beads.

1927

THE COSMIC EGG

Belgian physicist Georges Lemaître published a new analysis of Einstein's general relativity equations, arguing that they permitted a stable Universe that grows over time. In 1931, he traced the expansion backwards to a hot, dense origin he called the "primeval atom" – a forerunner of the Big Bang theory.

△ **Lemaître** was both a priest and physicist

△ **Werner Heisenberg**

1927

HEISENBERG UNCERTAINTY PRINCIPLE

In wave mechanics, the exact state of a particle is represented by a mathematical function. The wave nature of the function means that it is impossible to know precisely both the position and momentum of a particle. German physicist Werner Heisenberg showed that this limitation (which also applies to other pairs of variables) is a fundamental limitation in the Universe, not just a mathematical curiosity.

1927

1927 Electron diffraction is observed, confirming de Broglie's hypothesis that particles can behave as waves

1928 Hungarian biochemist Albert Szent-Györgi isolates hexuronic acid, later known as vitamin C, from the adrenal gland

1927 A German rocket society is founded, inspired by the writings of engineer Hermann Oberth. By 1930, it begins testing working rocket engines

> "I preferred to tell the truth that penicillin started as a chance observation."
>
> *ALEXANDER FLEMING, NOBEL LECTURE, 1945*

1928

PENICILLIN DISCOVERED

Scottish microbiologist Alexander Fleming discovered penicillin, the first true antibiotic. While checking Petri dishes that contained experimental colonies of *Staphylococcus* bacteria, he noticed something odd – a clear zone around a patch of mould. This suggested that the fungus (*Penicillium notatum*) had secreted something (later named penicillin) that stopped bacterial growth.

△ **Sir Alexander Fleming** in his laboratory

1929

THE ELECTROENCEPHALOGRAPH

German psychologist Hans Berger published his discovery of electroencephalography (EEG), a method of recording electrical activity in the brain, which he had first demonstrated in 1924. Berger was the first to describe different patterns of electric brain waves, such as alpha and the faster beta waves, as well as alterations in wave patterns in, for example, epilepsy, and how such patterns change with mental effort and attention.

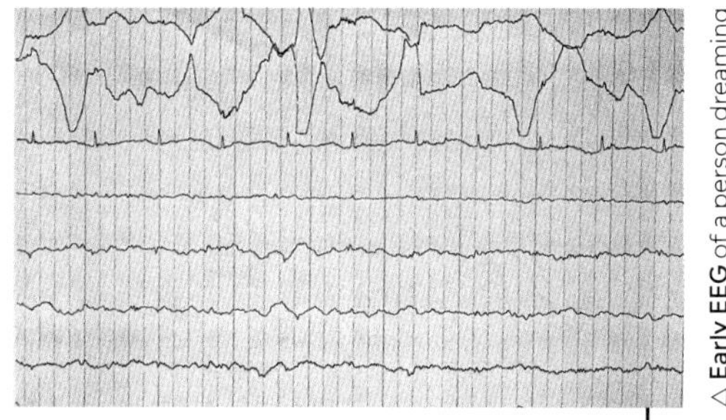

△ Early EEG of a person dreaming

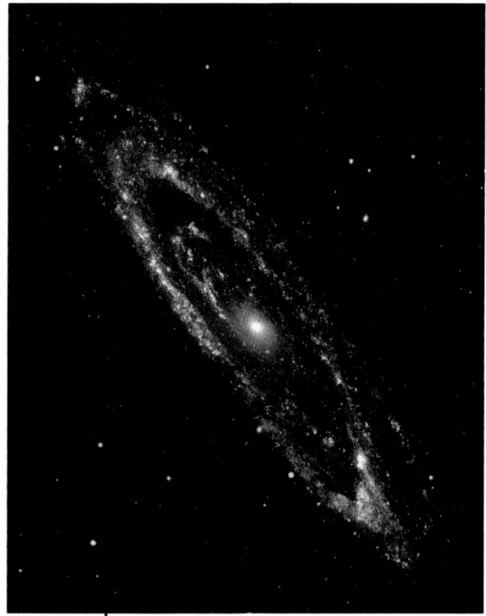

1929

EXPANDING UNIVERSE

After using the period-luminosity relationship in variable stars to estimate the distance of nearby galaxies such as Andromeda, American astronomer Edwin Hubble investigated other properties including Doppler shifts in their spectral lines. In 1929, he published evidence for a rule that, on average, more distant galaxies are receding faster from Earth, due to the overall expansion of the cosmos.

◁ Andromeda Galaxy

1928 British physicist Paul Dirac publishes an equation that combines relativity with quantum theory, and predicts the existence of antimatter

1929 Biochemists Adolf Butenandt from Germany and Edward Adelbert Doisy from the US independently isolate and purify oestrone, the first oestrogen to be identified

1929

THE BIG BANG

The modern Big Bang theory suggests that the Universe – all matter, energy, space, and time – originated in a vast explosion some 13.8 billion years ago. In the superhot and superdense conditions of the early Universe, matter and energy were interchangeable. However, as the Universe cooled, most energy became "locked up" in subatomic particles that came together to form the atoms, stars, and galaxies of the later Universe.

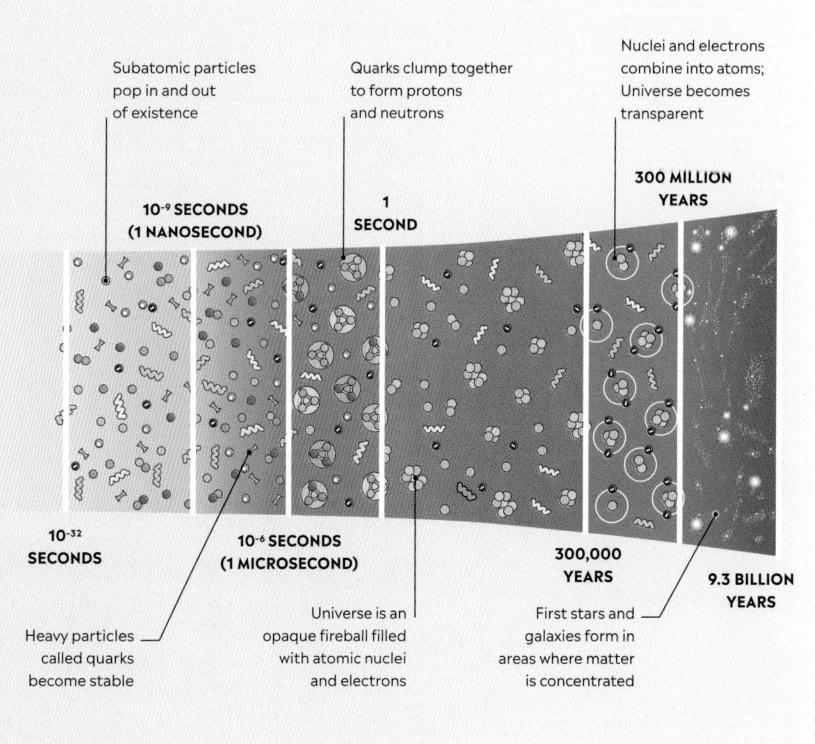
BIG BANG
Universe is infinitesimally small but already has all its mass and energy
0 SECONDS
10^{-35} SECONDS
An early burst of violent growth called inflation explains the patchy cosmic distribution of matter
10^{-32} SECONDS
Subatomic particles pop in and out of existence
10^{-9} SECONDS (1 NANOSECOND)
Heavy particles called quarks become stable
10^{-6} SECONDS (1 MICROSECOND)
Quarks clump together to form protons and neutrons
1 SECOND
Universe is an opaque fireball filled with atomic nuclei and electrons
300,000 YEARS
Nuclei and electrons combine into atoms; Universe becomes transparent
300 MILLION YEARS
First stars and galaxies form in areas where matter is concentrated
9.3 BILLION YEARS

Big Bang to first stars

As the Universe cooled rapidly while expanding from the original explosion, larger and more complex objects were able to form and remain stable.

1930

PLUTO DISCOVERED

While searching for a planet beyond Neptune, US astronomer Clyde Tombaugh discovered the dwarf planet Pluto. Although Pluto was initially classed as a planet in its own right, its small size led some astronomers to suspect it might be one of many icy bodies orbiting at the edge of the Solar System.

▷ **Pluto** photographed in 2015 by the US New Horizons space probe

1930

1930 US inventor Vannevar Bush invents the Intergraph, a mechanical computer

1930 British physicist Paul Dirac combines relativity and quantum physics to postulate the existence of antiparticles

1930 Swiss-born astronomer Robert Trumpler shows that dust grains in interstellar space dim the brightness of distant stars

1930 Austrian physicist Wolfgang Pauli hypothesizes the existence of the neutrino, a subatomic particle involved in radioactive decay

Pluto was named by an 11-year-old British schoolgirl, Venetia Burney

▷ **Beebe and Barton** with their bathysphere

1930

THE FIRST BATHYSPHERE

US marine biologist William Beebe and engineer Otis Barton built a pressure-resistant steel bathysphere connected by steel cable to a mother ship. Between 1930 and 1934 they made 35 dives off the coast of Bermuda and became the first people to observe life at depths down to 923 m (3,028 ft).

1930

TELESCOPE BREAKTHROUGHS

Estonian optician Bernhard Schmidt invented a revolutionary telescope that combined lenses and mirrors to produce sharp images of large areas of the sky. French astronomer Bernard Lyot perfected the coronograph, which blocks light from a bright source to reveal fainter objects.

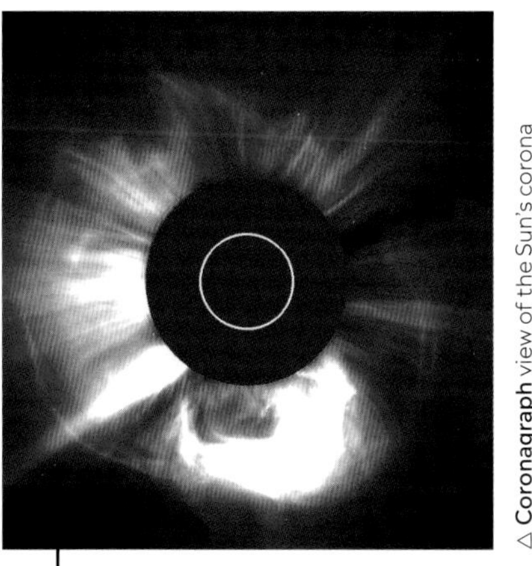

△ **Coronagraph** view of the Sun's corona

△ **Jansky** with his early radio telescope

1931

RADIO ASTRONOMY

While studying radio interference, US physicist Karl Jansky discovered radio signals coming from space. He determined that the strongest signals came from the direction of Sagittarius – the centre of our galaxy. His work sparked the development of radio astronomy.

1931 Deuterium is discovered by US scientist Harold Urey. It is a heavy isotope of hydrogen, with two neutrons rather than one

1931

1930 Frank Whittle, a British engineer, develops the first practical jet engine

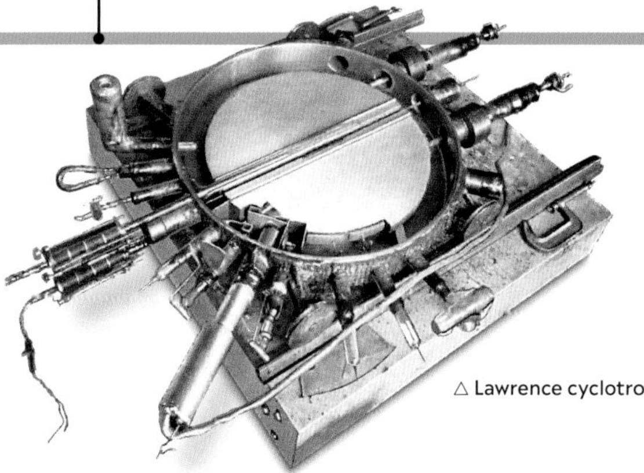

△ Lawrence cyclotron

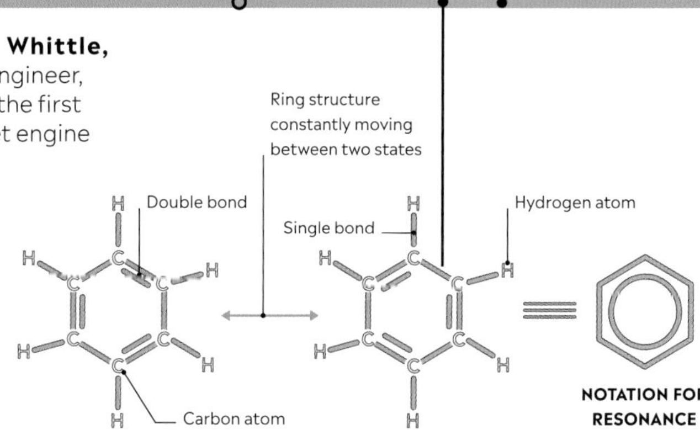

Benzene bonds

The bonds between carbon atoms in the benzene molecule resonate between single and double.

1930

THE CYCLOTRON

In the 1920s, physicists began developing devices that could accelerate ions (charged atoms) or other particles to high speeds to use as probes to investigate atomic structure. Early accelerators were linear, but in 1930, US physicist Ernest Lawrence built the first cyclotron – an accelerator that used a varying electric field to send particles around at ever-increasing speeds.

1931

BOND RESONANCE

To explain the properties of certain chemicals, and especially benzene, US chemist Linus Pauling developed the idea of electron resonance. As there were no isotopes of benzene, and the distance between carbon atoms is equal, he suggested that quantum mechanics could explain how bonding electrons moved between different states rather than maintaining three fixed double bonds.

1932

KLEIBER'S LAW

Swiss biologist Max Kleiber proposed that animals' body size and metabolic rate are related, but not linearly. Instead, metabolic rate scales to the ¾ power of body mass. Known as Kleiber's Law, it is now known to hold true from tiny bacteria right up to the largest animals.

△ **Rhinocerous** and bird

1932

DISCOVERY OF THE NEUTRON

In the 1920s, physicists had supposed that - in addition to positively charged protons - the atomic nucleus must contain another kind of particle, dubbed the neutron, which would have no electric charge. British physicist James Chadwick discovered the neutron in 1932.

▷ **Chadwick's neutron** detector (replica)

1932

1932 US astronomer Theodore Dunham identifies the gas carbon dioxide in the infrared spectrum of Venus

1932 In Cambridge, UK, physicists Ernest Walton and John Cockcroft split the atom, breaking lithium nuclei into helium nuclei

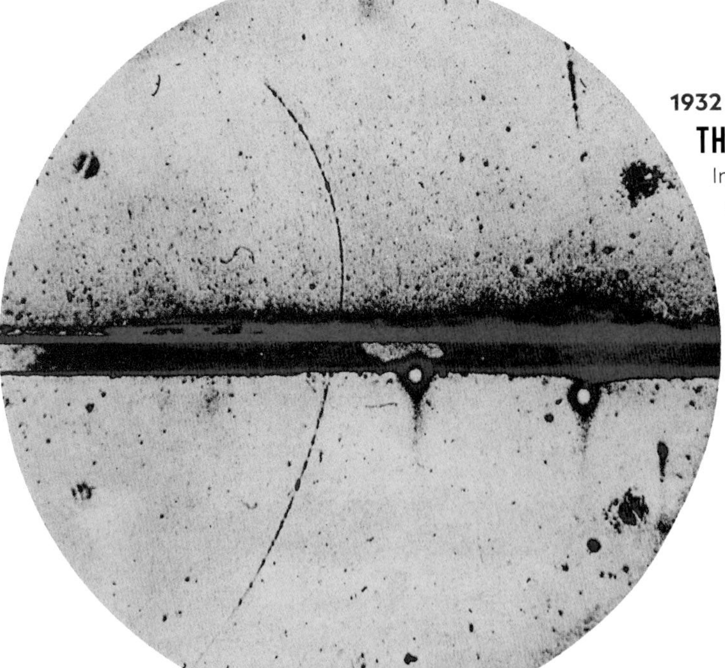

1932

THE POSITRON

In 1930, Paul Dirac had postulated the existence of an anti-electron - a particle with the same mass as the electron, but with an opposite (positive) electric charge. In 1932, US physicist Carl Anderson discovered the particle and named it the positron.

◁ **Anderson's cloud chamber** image of a positron showing its curving path

DARK MATTER

Most of the matter in the Universe is invisible to us. Observations now suggest there is about six times more dark matter than "normal" visible matter. However, dark matter's true nature remains puzzling. Studies have discounted the idea that it is simply normal matter in compact, hard-to-detect forms such as black holes. Instead, it seems more likely to consist of one or more undiscovered weakly interacting massive particles (WIMPs) that interact through gravity but not through electromagnetic radiation.

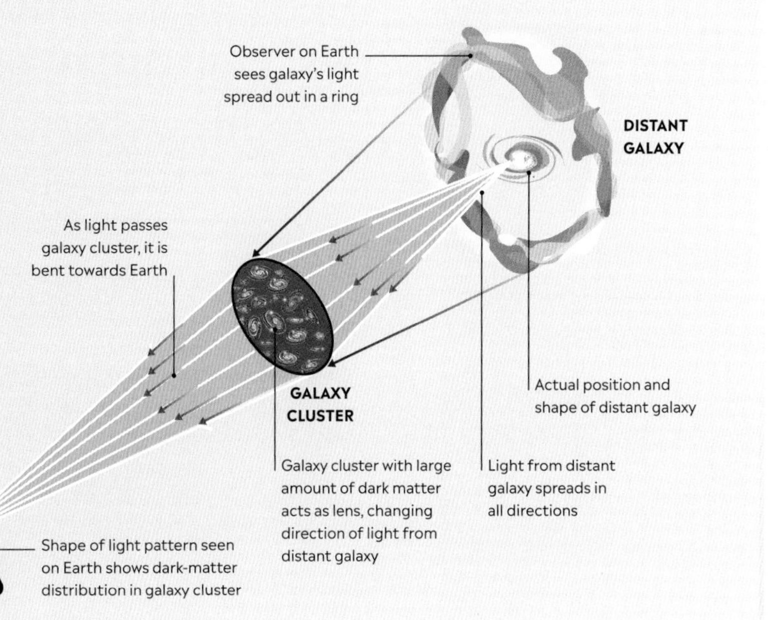
Observer on Earth sees galaxy's light spread out in a ring
DISTANT GALAXY
As light passes galaxy cluster, it is bent towards Earth
GALAXY CLUSTER
Actual position and shape of distant galaxy
EARTH
Galaxy cluster with large amount of dark matter acts as lens, changing direction of light from distant galaxy
Light from distant galaxy spreads in all directions
Shape of light pattern seen on Earth shows dark-matter distribution in galaxy cluster

Gravitational lensing
When light from a distant galaxy passes a massive galaxy cluster, it can be deflected to produce a distorted image visible from Earth. This "gravitational lensing" effect can provide valuable information about dark matter.

1933 Swiss astronomer Fritz Zwicky suggests the motions of galaxies are influenced by unseen dark matter

1933 Swiss chemist Tadeus Reichstein makes vitamin C in the laboratory, after the molecule is identified in 1928

1933

▽ **Ernst Ruska** at his electron microscope

1932
THE FIRST ELECTRON MICROSCOPE

German physicist Ernst Ruska had built the first prototype electron microscope in 1931, and the instrument produced its first images a year later. Capturing images using electrons rather than light, electron microscopes are capable of producing images with much higher magnification than a conventional microscope.

> The first electron microscope had a magnification of 400x; they can now reach 10,000,000x

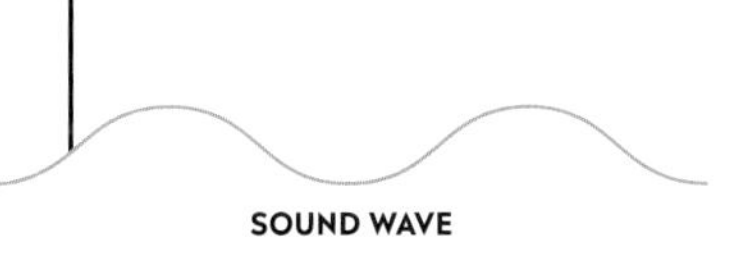
SOUND WAVE
AM
Wave amplitude is adjusted (modulated)
CARRIER WAVES
FM
Wave frequency is adjusted (modulated)

Modulation of a carrier wave
In amplitude modulation (AM), the sound wave modulates the carrier wave's amplitude (height); in frequency modulation (FM), it modulates the carrier wave's frequency.

1933
FM RADIO

Radio broadcasting is achieved by encoding the shapes of sound waves by modulating (changing) a "carrier" radio wave. In 1933, US engineer Edwin Armstrong invented frequency modulation (FM), in which the undulations of a sound wave are encoded as slight variations in the frequency of the carrier wave.

At its peak, a supernova will emit approximately as much light as an entire galaxy

1934

CHERENKOV RADIATION

Soviet physicist Pavel Cherenkov was the first to observe that water around a radioactive material glows blue. This Cherenkov radiation, visible in nuclear reactor coolant water, is a kind of shock wave, analogous to the sonic boom of supersonic aircraft, but with electromagnetic waves. It occurs when electrically charged particles travel faster than light can travel in the same medium.

◁ **The blue glow of** Cherenkov radiation in a nuclear reactor

1934

NOVAS AND SUPERNOVAS

Fritz Zwicky and German astronomer Walter Baade identified two classes of exploding star – novas and supernovas. They proposed that supernovas mark the transition of a normal star into a neutron star – Indian astronomer Subramanyan Chandrasekhar's theoretical superdense stellar remnant. Supernovas in fact mark the deaths of massive stars.

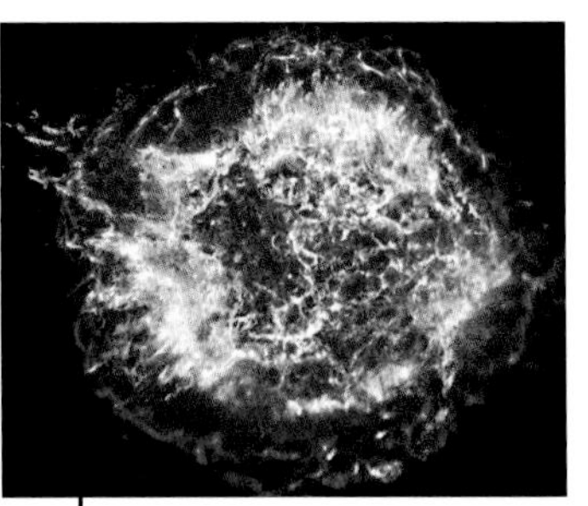

△ **Cassiopea A** supernova remnant

1934

1934 Irene and Frédéric Joliot-Curie produce the first artificial radioactive isotope by bombarding aluminium with alpha particles

1935 The synthetic polymer fibre nylon is developed by US chemist Wallace Carothers; it is the first to be sold commercially

1934

NEUTRON COLLISIONS

Soon after French chemists Irene and Frédéric Joliot-Curie reported having made elements radioactive by bombarding them with alpha particles, Italian physicist Enrico Fermi managed to do the same, but using neutrons instead. This proved easier because, having no electric charge, neutrons are not repelled from atomic nuclei and have more chance of hitting them. Fermi's research led the way for nuclear fission and the discovery of new, heavier elements.

◁ **Enrico Fermi** with a neutron counter

1935
THE EPR PARADOX

The EPR paradox was a thought experiment that questioned the reality of quantum physics. In quantum physics, the behaviour of all particles, or systems of particles, is entirely determined by a mathematical function called the wave function. The EPR paradox focussed on pairs of subatomic particles whose wave functions are co-dependent, or "entangled". According to the quantum physics of the time, entangled particles must be able to communicate faster than light, contradicting Einstein's proven special theory of relativity. The paradox has since been resolved.

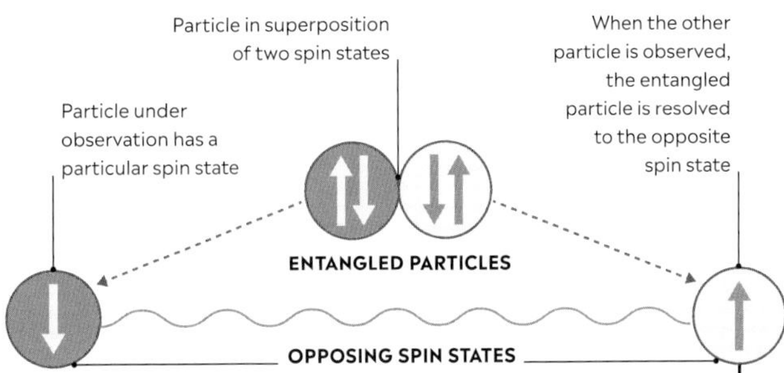
Particle in superposition of two spin states
When the other particle is observed, the entangled particle is resolved to the opposite spin state
Particle under observation has a particular spin state
ENTANGLED PARTICLES
OPPOSING SPIN STATES

A particle problem
The EPR paradox involves the quantum property of "spin", which can be "up" or "down". At first, two entangled particles are in a "superposition" of both states, but at a later time, observing one particle's state forces the other particle's wave function to "collapse" to the opposite state.

1935 US seismologist Charles Richter develops a numerical measure of earthquake activity

1935 Japanese physicist Hideki Yukawa proposes the force holding atomic nuclei together is carried by the exchange of particles called mesons

1935

1935 Swedish physiologist Ulf Svante von Euler isolates an active substance from seminal fluid, naming it prostaglandin

▷ **Konrad Lorenz** with his ducklings

1935
IMPRINTING

Austrian zoologist Konrad Lorenz concluded that instinct was important in animal behaviour. He established the concept of imprinting to describe, for example, the behaviour of ducklings and goslings, which instinctively form a bond with a parent (or with a human, as pictured above) soon after they hatch.

1935
RADAR

Scottish engineer Robert Watson-Watt demonstrated the concept behind radar technology, showing that aircraft could reflect radio waves. Using an existing radio broadcasting transmitter tower, he detected clear radio echoes from an aeroplane flying nearby. A system based on Watson-Watt's work provided vital early warnings during World War II.

▷ **Sir Robert Watson-Watt** statue

POLYMERS AND PLASTICS

The world's favourite toy
Millions of plastic construction kits are made by the Danish company Lego, who are developing bricks made only from recyclable plastic.

A polymer is a natural or synthetic substance made up of macromolecules - that is, large molecules that, in turn, are made from simpler chemical units called monomers. Essential for living organisms, polymers include proteins, cellulose, and nucleic acids. Synthetic polymers (including plastics) are abundant and important materials in the modern world. The first industrial plastic, Bakelite, was developed in 1907.

Thermoplastics are made from simple starting chemicals that react to create long-chain molecules that can be moulded or shaped using heat and pressure. By varying the initial chemicals, the properties of the plastic produced - including density, electrical conductivity, transparency, and strength - can be altered. This flexibility allows them to be made into a wide variety of products. Some are cheap to make and disposable, such as drinks bottles (from polyethylene terephthalate, or PET), flexible pipes (made of polyvinyl chloride, or PVC), lightweight packaging (made of foamed polystyrene), shatterproof windows (polymethyl methacrylate), or fabrics (such as nylon and spandex). Other plastics, such as polyurethane foam, are thermoset and, once made, cannot be remoulded because a chemical reaction has locked the polymer chains in place.

The problem of plastic waste has led to an increase in recycling. Biodegradable plastics that break down in nature have also been developed. Although disposable plastics cause concern, plastics do provide a lightweight alternative to metals - so, for example, cars containing plastic are much lighter and require less fuel.

MONOMERS

A monomer is a molecule that reacts, under the right conditions, to form long-chain polymers. Monomers can introduce a wide variety of different atoms into the long-chain, as well as the atoms or groups attached to it. These additions are used to alter the characteristics and properties of the polymer or plastic that is created.

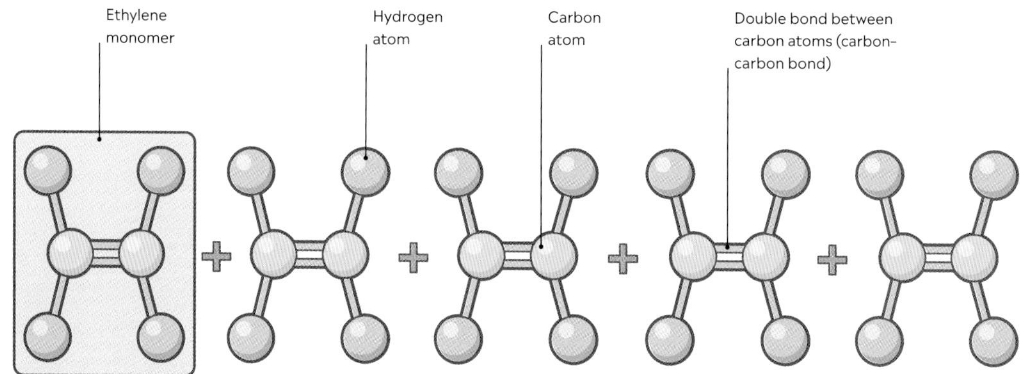

POLYMERS

A polymer is the product of controlled reactions of monomers. The long chains formed contain anything from 10,000 to 100,000 monomers. In this example, ethylene reacts to create the plastic polythene. By breaking the double bond, the reaction of ethylene adds monomers onto a chain that continues to add more monomers in a process called polymerization.

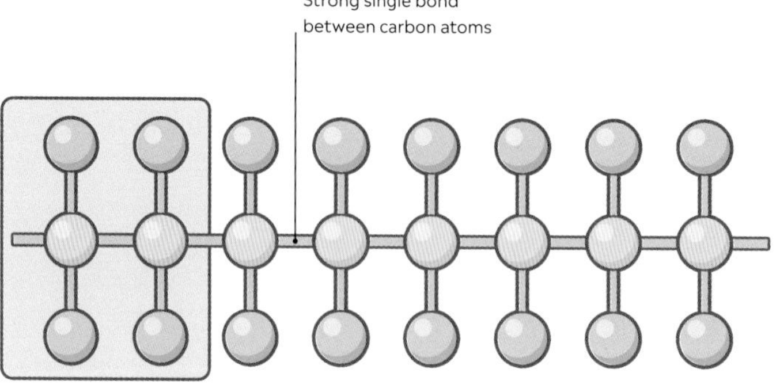

Versatile material
The introduction of plastics in the 20th century provided a versatile alternative to existing materials. Both the telephone and the transparent dome in this 1950s hotel lobby are made of plastic.

1936

THE KREBS CYCLE

German-born biochemist Hans Krebs worked out the series of reactions in living cells in which oxygen is used to break down sugars, fats, and proteins into energy-rich compounds, water, and carbon dioxide. This sequence is a cycle because citric acid is first used and then regenerated in the process.

▽ **Hans Krebs** in his laboratory

▷ **Thylacine** or Tasmanian tiger

1936

THE LAST THYLACINE

The last known living thylacine, *Thylacinus cyanocephalus* (Tasmanian tiger or Tasmanian wolf), a large, carnivorous marsupial, died in captivity in Hobart, Tasmania. Once found in mainland Australia, Tasmania, and New Guinea, it suffered a steep decline through competition with dingos and persecution by sheep farmers.

1936

1936 Earth's solid core is revealed by Danish geophysicist Inge Lehmann's analysis of seismic waves

1936 British physician Leonard Colebrook proves sulfonamide drugs to be effective in treating bacterial infections like streptococcal meningitis

1937 Swedish biochemist Arne Tiselius pioneers electrophoresis, a technique for separating proteins in a suspension using electrical charge

△ **Experimental** jet engine

1937

THE JET ENGINE

After taking out a patent for the jet engine in 1930, Frank Whittle eventually carried out the first tests of his design in 1937. By 1938, the engine was running successfully and producing huge amounts of thrust. However, German engineer Hans von Ohain built the first jet aeroplane, the Heinkel He 178, which had its first flight in 1939.

1937
THE HINDENBURG DISASTER

The LZ 129 Hindenburg was a German passenger airship. Like other airships of its time, it was made buoyant by filling its fuselage with the light but flammable gas hydrogen. As the airship docked at its destination, Lakehurst Naval Air Station in New Jersey, US, the hydrogen burst into flames. The airship was destroyed and 36 people were killed.

◁ **Hindenburg airship** in flames

1937 Theodosius Dobzhansky, a Ukrainian-US biologist, explains the role of mutation in the evolution of species and natural populations

1937 The muon is discovered as a constituent of cosmic-ray particle "showers" by the US physicists Carl D. Anderson and Seth Neddermeyer

1937 British biologist Frederick Charles Bawden discovers that viruses contain nucleic acid (RNA or DNA)

1938 German-US physicist Hans Bethe proposes the mechanism by which elements are created inside stars (nucleosynthesis)

1938

1938
FINDING A LIVING FOSSIL

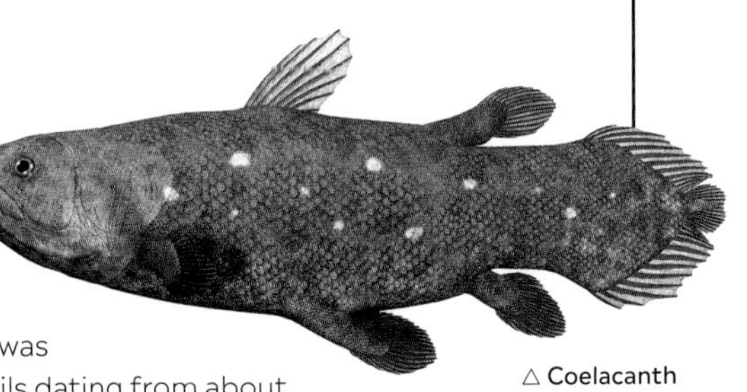

△ Coelacanth

The coelacanth (genus *Latimeria*) is a large fish that was known initially only from fossils dating from about 360 million years to 80 million years ago. A live specimen was caught off the coast of South Africa, prompting it to be dubbed a living fossil. Following several further finds, two species of this remarkable fish have now been recognized.

△ **Hahn and Meitner** in the laboratory

1938
NUCLEAR FISSION

While bombarding uranium with neutrons, German chemists Otto Hahn and Friedrich Wilhelm Strassmann were surprised to find traces of barium. Their former colleague, Austrian-Swedish physicist Lise Meitner, who had moved to Sweden as a refugee from Nazi Germany, proved that the large uranium nucleus had split, or fissioned, releasing energy and producing the stable isotope barium.

1939

NUCLEAR WARNING

Splitting large atomic nuclei releases free neutrons, which can in turn strike other nearby nuclei, causing further fission, and so on. Hungarian-born physicist Leo Szilard realized the consequences of such an uncontrolled nuclear chain reaction. In 1939, soon after fission had been achieved, he wrote an influential letter, also signed by Albert Einstein, to the US president, warning that Nazi Germany might be planning to build a nuclear bomb.

▽ Leo Szilard

△ DDT spraying of barracks

1939

THE INVENTION OF DDT

Swiss chemist Paul Hermann Müller found that the compound dichlorodiphenyltrichloroethane (DDT) was an effective insecticide. Used to fight typhus (spread by lice) and malaria (spread by mosquitos) in World War II, it was later found to be dangerous to other forms of life and its use was prohibited.

1939

1939 Russian-US immunologist Philip Levine recognizes the importance of the rhesus (Rh) factor in human blood

1939 Swiss-US physicist Felix Bloch discovers that the neutron is a composite particle (made of smaller particles)

The ionosphere prevents some radio waves from reaching Earth

IONOSPHERE

Free electrons in the ionosphere reflect radio waves

Path of radio wave

Shortwave radio and the ionosphere
Free electrons in the ionosphere reflect short-wavelength radio; the waves also bounce off the ground, which means that shortwave radio can be used to be broadcast across continents.

1939

INVESTIGATING THE IONOSPHERE

At altitudes above 50 km (30 miles), the atmosphere is increasingly ionized: negatively charged electrons are separated from their atoms, leaving behind positive ions. This mixture of ions and free electrons is called the ionosphere. British physicist Edward Appleton confirmed the existence of the ionosphere in 1927 and continued his research throughout the 1930s. His research on radio signals and the ionosphere proved crucial in the upcoming war.

1939
THE HELICOPTER

Many engineers had tried to build practical rotating-wing aircraft for decades, with some success. In particular, Spanish engineer Juan de la Cierva had invented the autogyro in 1920. However, the first practical machine, and prototype of the modern helicopter, was the brainchild of Russian-US engineer Igor Sikorsky.

◁ **Sikorsky** in his prototype helicopter

1940 The radioactive elements neptunium and plutonium are created by firing neutrons and deuterium respectively at uranium atoms

1940 US geneticists George Beadle and Edward Tatum conclude that the function of a gene is to direct the formation of a particular enzyme – "one gene, one enzyme"

1940 Scottish chemist Alexander Todd examines nucleotides, the constituents of RNA and DNA

1941 The first clinical trials of penicillin prove remarkably successful

1941 British astronomer Harold Spencer Jones accurately calculates the Earth–Sun distance

1941 German-US chemist Fritz Lipmann finds that phosphate molecules with high-energy bonds are important in generating energy for cells

1941

ANTIBIOTICS

Used to treat or prevent some types of bacterial infection, antibiotics work by killing bacteria or stopping them from spreading. They do this by damaging bacterial cells without damaging host (patient) cells. Penicillin was the first antibiotic to be discovered, in 1928, and it forms the basis of a group of treatments used today. These treatments – and many antibiotics – are based on natural products produced by moulds. Antibiotics successfully fight bacterial infections, but they have no effect on viral infections.

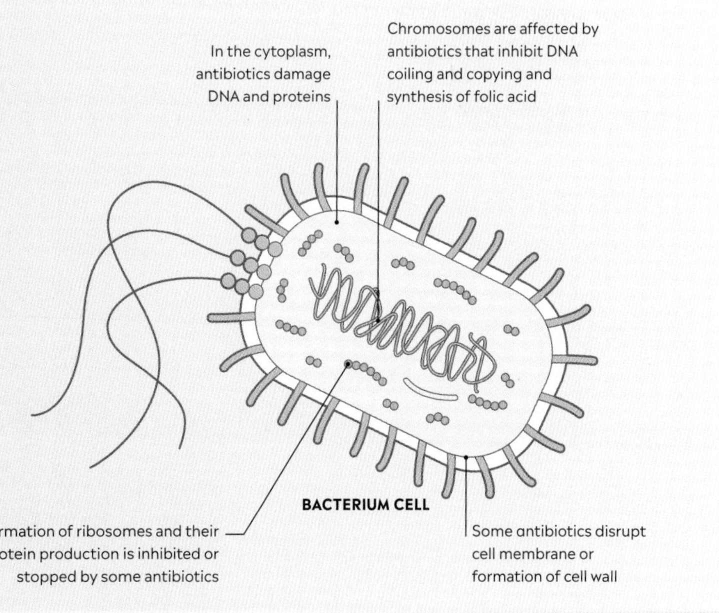
In the cytoplasm, antibiotics damage DNA and proteins
Chromosomes are affected by antibiotics that inhibit DNA coiling and copying and synthesis of folic acid
BACTERIUM CELL
Formation of ribosomes and their protein production is inhibited or stopped by some antibiotics
Some antibiotics disrupt cell membrane or formation of cell wall

How antibiotics disrupt bacteria Antibiotics work by interfering with how bacterial cells function. They either prevent bacteria from reproducing, or they kill bacteria by switching off essential processes.

▽ **The first nuclear** reactor

1942

SUSTAINED NUCLEAR REACTION

As part of the Manhattan Project, a top secret Allied initiative to build a nuclear bomb, a team of scientists led by Italian physicist Enrico Fermi made the world's first nuclear reactor. The reactor contained uranium, uranium oxide, graphite, and wood and was built under a viewing stand in a sports field at the University of Chicago.

1942

RADIO MAPS OF THE UNIVERSE

US astronomer Grote Reber compiled the first survey of radio sources in the sky. Using a steerable, dish-shaped receiver 9 m (30 ft) across, he was able to amplify radio sources and identify their rough direction in the sky, paving the way for later, more sophisticated radio telescopes.

△ **Grote Reber** with his radio telescope

1942

1942 The US government initiates the secret Manhattan Project, to build a nuclear bomb

1942 James Stanley Hey, a British physicist, discovers that large sunspots emit radio waves

1942 US pharmacologists Alfred Gilman and Louis Goodman find that nitrogen mustard reduces lymphoma tumours, a first step in the development of chemotherapy

1912–77

WERNER VON BRAUN

German engineer von Braun was a member of the VfR, a German spaceflight society that flourished in the late 1920s. After the society foundered, von Braun and many of his colleagues worked for the military, developing the V-2 rocket.

1942

TESTING THE V-2 ROCKET

The first test launch of Germany's V-2 rocket-propelled missile reached the edge of space at an altitude of 84.5 km (52.5 miles). Propelled by combustion of ethanol and liquid oxygen, the V-2 was the world's first large liquid-fuelled rocket. However, development problems delayed its first use as a weapon until late 1944, too late to affect the outcome of World War II.

◁ **V-2** launch preparations

Colossus computers deciphered 63 million characters of German code by 1945

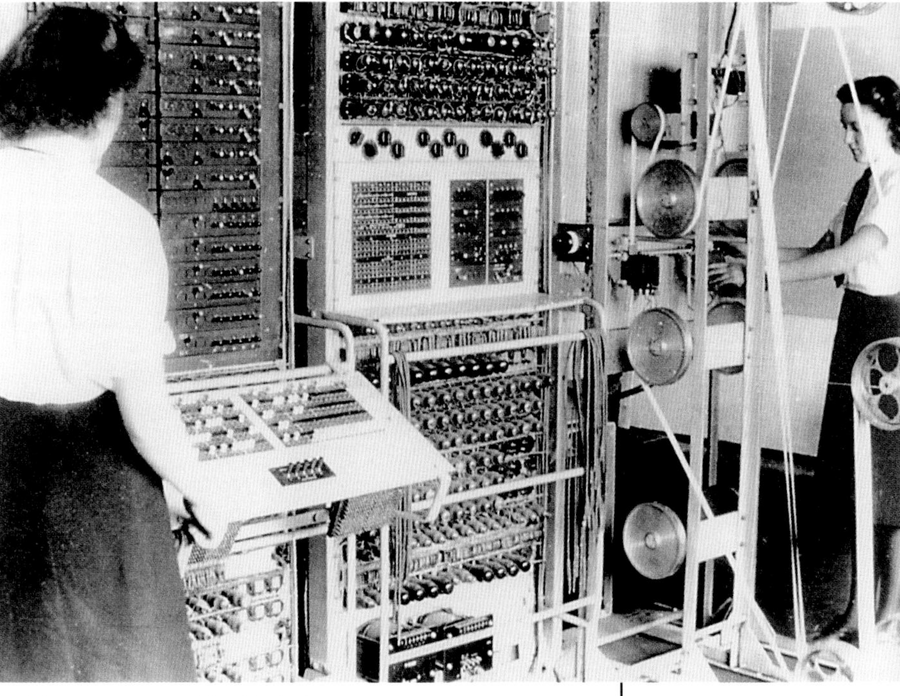

△ **Colossus computer** at Bletchley Park

1943

THE COLOSSUS COMPUTER

The world's first programmable, electronic, digital computer was Colossus Mark I, built at the UK's Government Code and Cypher School at Bletchley Park. It was designed to break a complicated code used by the German army to encrypt their wartime messages.

1943

1942 The Messerschmitt Me 262 V3, the world's first jet-powered fighter aircraft, has its first test flight

1943 French oceanographer Jacques Cousteau and engineer Émile Gagnan invent the Aqua-Lung, the first self-contained underwater breathing apparatus (SCUBA)

1943 Dutch physician Willem Kolff builds the first dialysis machine, an artificial kidney, to treat patients with kidney disease

△ **Parícutin erupts**

1943

LIFE CYCLE OF A VOLCANIC ERUPTION

The Parícutin volcano erupted from a fissure in a Mexican cornfield; over the next nine years it grew into a cone of lava and ash 424 m (1,390 ft) high. The eruption provided scientists with the first opportunity to study the full life cycle of a volcano from birth to extinction.

▽ **Seyfert spiral galaxy** NGC1433

1943

SEYFERT GALAXIES

US astronomer Carl Seyfert identified a class of spiral galaxies that have unusually bright light sources in their nuclei - too bright to be explained by their combined stars alone - and strong emission lines in their spectra. The behaviour of these "Seyfert galaxies" was later linked to supermassive black holes feeding on nearby matter.

NUCLEAR FISSION AND FUSION

Fission reactor
In a fission reactor, the energy released by splitting uranium or plutonium nuclei is used to turn water to steam, which is used to operate a turbine and generate electricity.

Fission and fusion are nuclear reactions – that is, processes in which the nuclei of atoms (or a nucleus and a subatomic particle) collide and transform to produce different nuclei. Usually this causes nuclei to change from one element to another – for example, in nuclear fission, a neutron or other light particle is absorbed by a heavy nucleus, causing the nucleus to split into at least two lighter nuclei. In nuclear fusion, lighter nuclei combine to form a single heavier nucleus.

The nuclei are tightly bound by the strong force (*see pp.288–89*), so nuclear reactions require lots of energy to cause change. Energy can also be released during fission and fusion. During a nuclear reaction, a small amount of mass can seem to disappear. This so-called mass defect is the difference between an atom's mass and the sum of the masses of its constituent particles – for example, a helium atom weighs less than the total mass of two protons and two neutrons. This can be explained by the fact that it takes a certain amount of energy – which is equivalent to mass (*see pp.186–87*) – to break a nucleus into its constituent particles. This quantity of energy is known as the binding energy. Different atoms have different binding energies. When light nuclei fuse or heavy nuclei split, this can result in a dramatic release of excess binding energy. Nuclear power stations use this energy to generate electricity.

As well as being exploited for use in nuclear reactors and weapons, nuclear reactions occur in stars and in interactions between cosmic rays and matter.

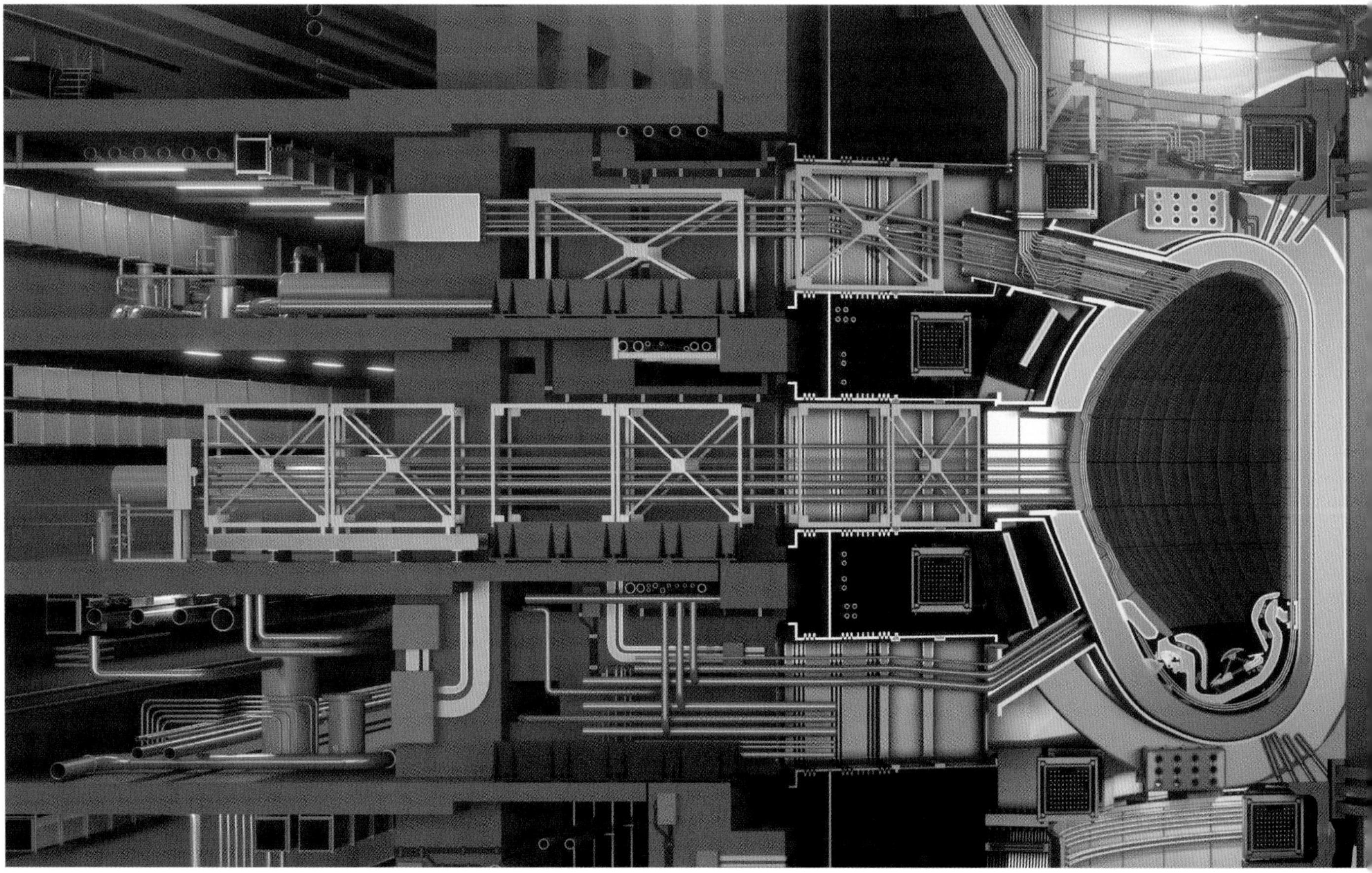

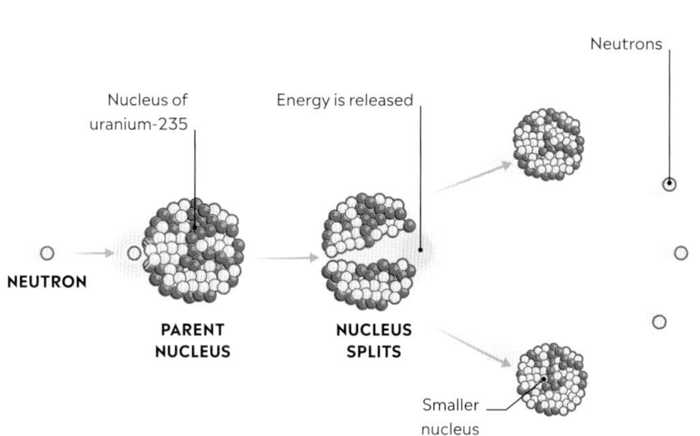

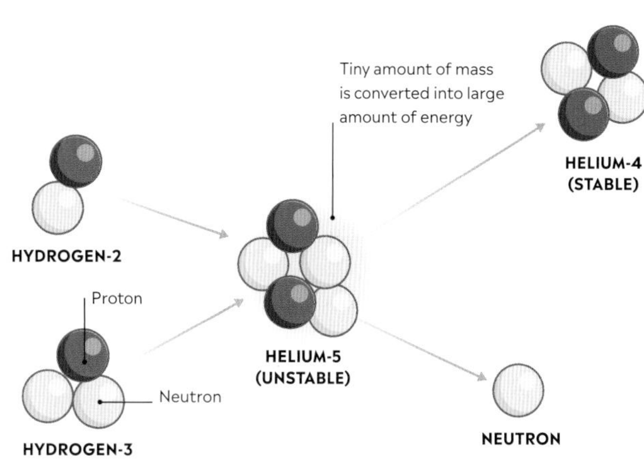

NUCLEAR FISSION

The process of nuclear fission sees a parent nucleus split into two or more smaller daughter nuclei. This often releases a large amount of energy through gamma radiation and fast-moving neutrons that might, under certain conditions, initiate a chain of further fission reactions. Fission can be induced or may happen spontaneously to an unstable nucleus.

NUCLEAR FUSION

In nuclear fusion, nuclei combine to form one or more heavier nuclei - as in, for example, hydrogen nuclei fusing to form helium. When light nuclei undergo fusion, energy is released. Fusion is the process that "powers" the Sun and other stars, thereby producing the diversity of elements we experience on Earth.

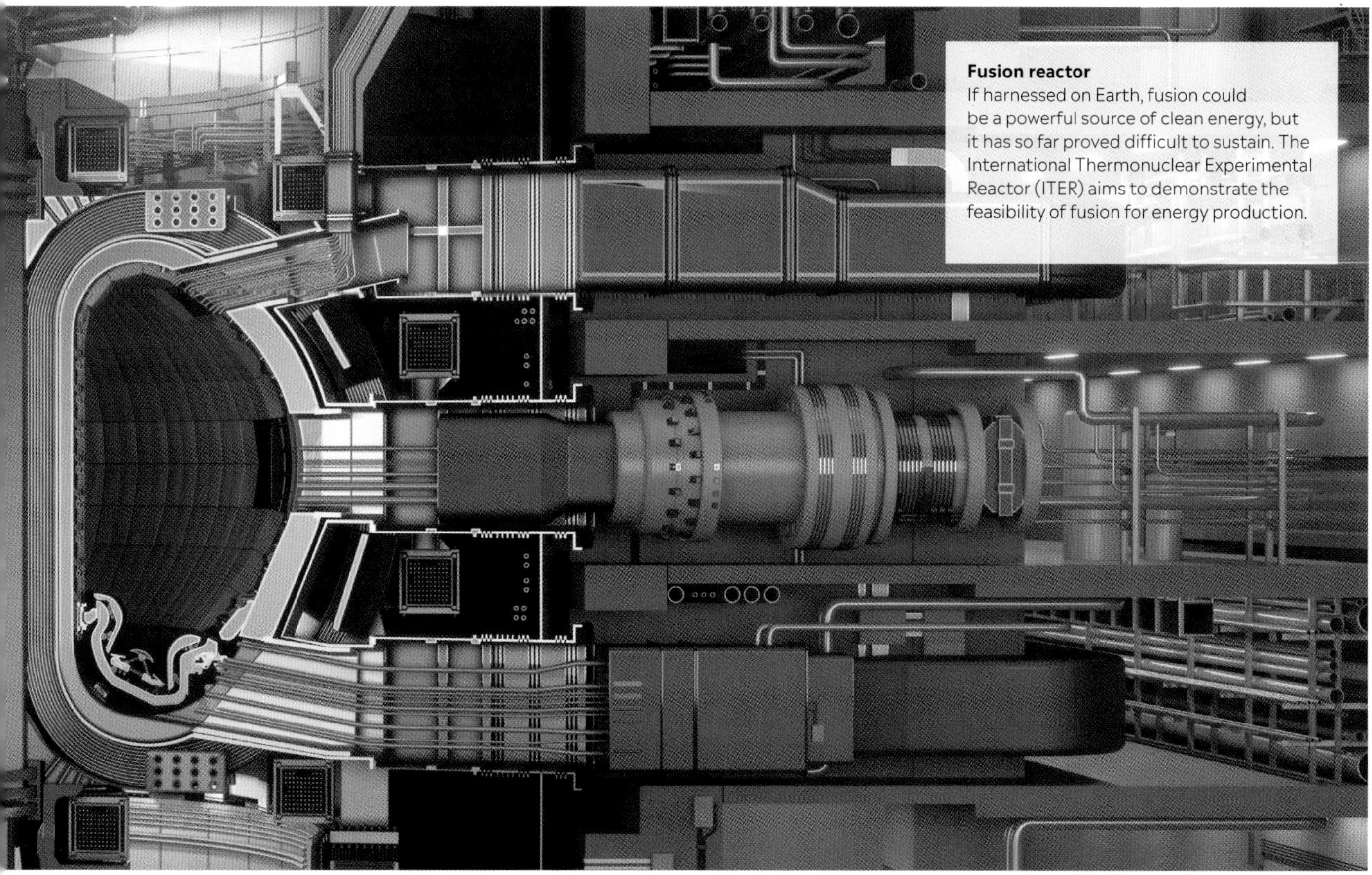

Fusion reactor
If harnessed on Earth, fusion could be a powerful source of clean energy, but it has so far proved difficult to sustain. The International Thermonuclear Experimental Reactor (ITER) aims to demonstrate the feasibility of fusion for energy production.

1944

DNA AND HEREDITY

Canadian-US physicians Oswald Avery and Colin MacLeod, and US geneticist Maclyn McCarty, demonstrated that DNA is the substance that causes bacterial transformation and carries genetic information. They purified strands of a substance from a disease-causing bacterium and showed that this extract was able to transform a harmless form of the same organism into its deadly form. They had effectively isolated DNA and proved that it - rather than proteins - is the agent of heritable change.

◁ **Oswald Avery** at work

1944

1944 German physicist Carl von Weizsäcker models the creation of the planets through the accumulation and merger of smaller bodies called planetesimals

1944 Paper chromatography is developed by British scientists A.J.P. Martin and R.L.M. Synge to separate mixtures of chemicals for study

1945 Nuclear magnetic resonance (NMR) is observed in paraffin wax by physicists in America

1944

ECHOLOCATION

US zoologist Donald Griffin coined the term echolocation to describe how bats navigate in flight, even in the pitch black of caves. Griffin and his colleague Robert Galambos showed that bats emit streams of high-frequency sonic bleeps and that blocking these emissions or the bats' hearing impairs their ability to navigate.

△ **Pipistrelle** bat

1945

RADIO ASTRONOMY AT JODRELL BANK

British physicist Bernard Lovell began conducting radio astronomy tests at Jodrell Bank in England. Initial experiments involved radio signals linked to meteors entering Earth's atmosphere, but from 1947, a new zenith telescope (able to map the strip of the sky that passes directly overhead) enabled cosmic radio sources to be pinned down with new precision. Early discoveries included radio signals from the Andromeda Galaxy.

△ **Jodrell Bank** telescope under construction

DIGITAL COMPUTING

Computers are machines that can automatically carry out a wide range of mathematical operations on data, based on given instructions, known as programs. They typically include devices to input, store, process, and output data. Digital computers work with data in sequences of digits, usually binary. The sequences of 0s and 1s represent the switching of the electric current in tiny components off and on respectively.

Binary numbers

A binary number is a number that is written using only two symbols: 0 and 1. This is the primary language employed in digital computing.

Decimal number	Binary visual 16s	8s	4s	2s	1s	Binary number 16s	8s	4s	2s	1s
1	□	□	□	□	■	0	0	0	0	1
2	□	□	□	■	□	0	0	0	1	0
3	□	□	□	■	■	0	0	0	1	1
4	□	□	■	□	□	0	0	1	0	0
5	□	□	■	□	■	0	0	1	0	1
6	□	□	■	■	□	0	0	1	1	0
7	□	□	■	■	■	0	0	1	1	1
8	□	■	□	□	□	0	1	0	0	0
9	□	■	□	□	■	0	1	0	0	1
10	□	■	□	■	□	0	1	0	1	0

1945

COMPUTER ARCHITECTURE

While helping to develop the ENIAC computer, Hungarian-US mathematician John von Neumann created the blueprint for how most modern computers' internal parts are organized and how instructions and calculations are carried out. At the heart of his scheme was a stored program (set of instructions).

◁ John von Neumann

1945 British author Arthur C. Clarke proposes using satellites in orbit over the equator to provide worldwide communications links

1945 Physicists in the UK and the USA develop high-powered particle accelerators called synchrotrons

1946

1945 The US and USSR capture German rocket scientists and technology for use in future projects

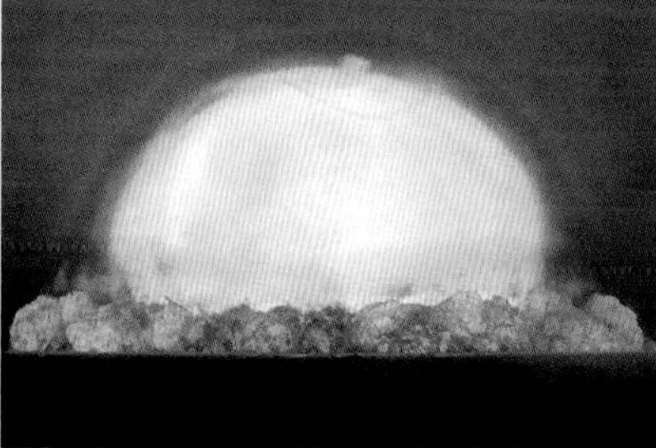

1945

THE FIRST NUCLEAR BOMB

The first explosion of a nuclear weapon was a test carried out in the New Mexico desert. The test was given the codename Trinity. The weapon, a plutonium-based fission bomb nicknamed "The Gadget", was of the same design as the bomb dropped on the Japanese city of Hiroshima a month later. The explosion released the equivalent of 25,000 tonnes of TNT.

◁ **First nuclear detonation**, New Mexico

> "Now I am become Death, the destroyer of worlds."
>
> *ROBERT OPPENHEIMER, DIRECTOR OF THE TRINITY BOMB PROJECT, QUOTING THE* BHAGAVAD GITA, *1945*

"It was as if, suddenly, we had broken into a walled orchard, where protected trees flourished and all kinds of exotic fruits had ripened in great profusion."

CECIL POWELL ON HIS DISCOVERY OF NEW PARTICLES

1947

THE PION

British physicist Cecil Powell used photographic emulsion at high altitude to record the results of collisions by cosmic rays – high-speed particles, mostly protons, arriving from outer space. In 1947, he discovered the pi meson, or pion, the existence of which had been predicted 20 years earlier by Japanese physicist Hideki Yukawa. In this image, a cosmic ray has hit a nucleus (bottom left), producing a star-shaped shower of particles, including a pion that has travelled upwards and to the right, hitting another nucleus and producing a second star-shaped particle shower.

◁ **Pion** emulsion photograph

1947

1947 US test pilot Chuck Yeager becomes the first supersonic pilot, in the X-1 rocket plane

1948 The drug aureomycin is the first of the important family of tetracycline antibiotics discovered by US botanist Benjamin Duggar

1947 The last predicted element, promethium, is discovered by US chemists J.A. Marinsky, L.E. Glendenin, and C.D. Coryell as a product of nuclear fission

▷ **Dead Sea scrolls,** subject to radiocarbon dating

1947

RADIOCARBON DATING

Developed by the US chemist Willard Libby, radiocarbon dating allows scientists to estimate the age of ancient organic material, such as cotton or paper, by measuring the amount of radioactive carbon-14 present. Carbon-14 decays over time, and by comparing its levels with the common but stable isotope carbon-12, the sample's age can be calculated.

△ **Bardeen, Shockley, and Brattain**

c. 1947

THE TRANSISTOR

US physicists John Bardeen, Walter Brattain, and William Shockley Jr invented the transistor while seeking an alternative to vacuum tubes – components used for amplification and switching in electronic circuits. Their transistor was large and crude but sparked a revolution in electronics that touched every area of society.

1948

HALE TELESCOPE

The world's largest reflecting telescope was completed at the Palomar Observatory in California, USA. The 5.1-m (200-in) Hale Telescope utilized new technology for casting large, relatively lightweight mirrors that resist distortion, as well as telescope mounts capable of pointing and steering the huge instrument. It remained the world's most effective telescope until the 1990s.

▷ **The Hale** Telescope dome

1948

SHELL MODEL OF THE NUCLEUS

The shell model of the atomic nucleus provided a new understanding of the organization of protons and neutrons within the atomic nucleus, using similar quantum physics rules as are used to predict energy levels of electrons in atoms. It was developed by German-US physicist Maria Goeppert Mayer, German physicist Johannes Jensen, and Hungarian-US physicist Eugene Wigner.

▷ Maria Goeppert Mayer

1948 US geneticist George Snell publishes his work showing how genetic differences are involved in tissue rejection in experiments with laboratory mice

1948

1948 US physicist Richard Feynman develops quantum electrodynamics, which explains the behaviour of electrically charged particles in terms of quantum physics and relativity

1948 In the USA, physicists Ralph Alpher and George Gamow show how the Big Bang could have created the Universe's lightest elements

TRANSISTORS AND SEMICONDUCTORS

Digital electronic devices process information through an integrated circuit made up of billions of tiny components called transistors. These transistors usually consist of three layers of semiconductor material, each of which has unique electrical properties that can be altered by introducing new elements into its structure (doping). When differently doped semiconductors are sandwiched together, it creates a path through which electric current can only flow a certain way. Semiconductors can be thought of as the brain cells of a device – relaying, switching, and amplifying electrical signals.

INTEGRATED CIRCUIT

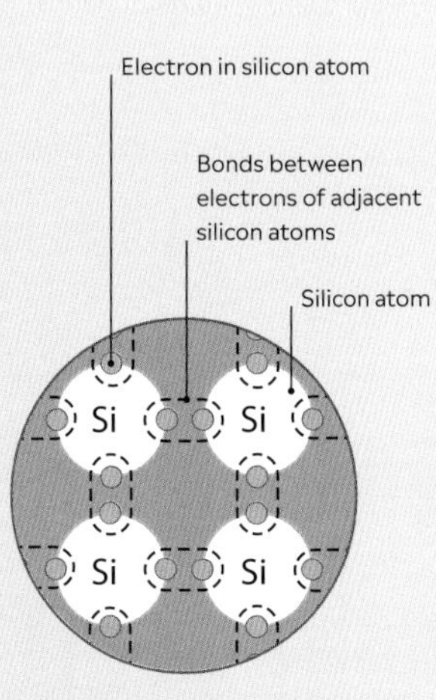

Silicon
Pure silicon conducts electricity when electrons absorb enough energy to break free from atoms.

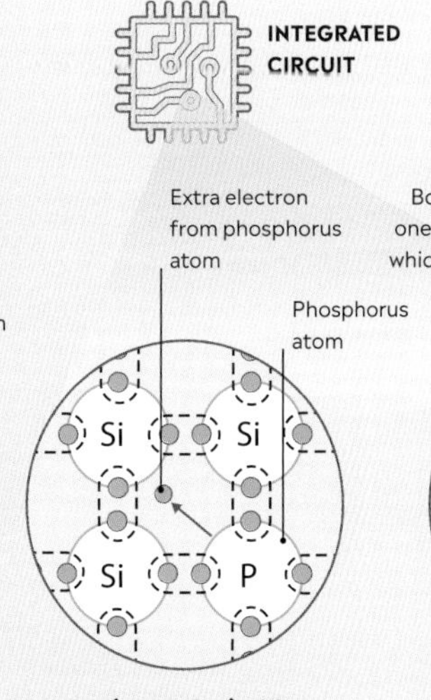

N-type (negative) silicon
Adding phosphorus atoms makes an n-type semiconductor with excess free-moving electrons – an electric current.

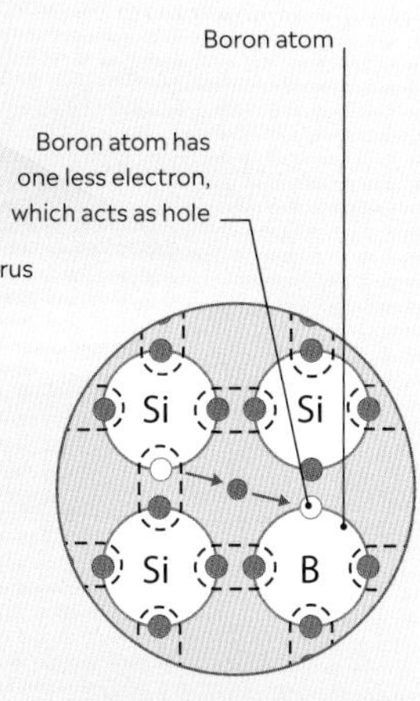

P-type (positive) silicon
This semiconductor has a deficit of electrons, leaving positively charged "holes" that can flow through the silicon.

▽ **Penicillin molecule** model

1949
MOLECULAR MODELLING

To work out the structure of large molecules, British biochemist Dorothy Hodgkin created patterns using X-ray diffraction. She processed the data collected to calculate a model of the penicillin molecule. It was the first time a computer was used directly to solve a biochemical problem.

1949

1949 US astronomer Fred Whipple develops the "dirty snowball" (a mix of rock and ice) model of comets

1949 The USSR secretly conducts its first nuclear weapon test at the Semipalatinsk Test Site in Kazakhstan

1949 US scientist Ralph Baldwin argues that lunar craters were formed by impacts from space, rather than volcanism

1949 The essential amino acids are identified by the US biochemist William Rose, who experiments by carefully controlling the diet of animals

1910–94
DOROTHY HODGKIN

Dorothy Crowfoot Hodgkin was a pioneer in the use of X-ray crystallography to study the structure of large molecules, among them the antibiotic penicillin and vitamin B12, for which she was awarded the Nobel Prize for chemistry in 1964.

1949
ATOMIC CLOCK

The first atomic clock was developed by a team at the US National Bureau of Standards (now called the National Institute of Standards and Technology) led by US physicist Harold Lyons. The clock was based on microwave radiation produced by electrons transitioning between different energy levels in ammonia molecules and heralded a new approach to high-precision timekeeping.

△ **The NIST** atomic clock

> "A man provided with paper, pencil, and rubber, and subject to strict discipline, is in effect a universal machine."
>
> *ALAN TURING,* "INTELLIGENT MACHINERY: A REPORT", *1948*

▷ Alan Turing

1950
THE TURING TEST

British scientist and mathematician Alan Turing proposed a practical test of the (artificial) intelligence of a computer. In the Turing Test, originally called the Imitation Game, a human interrogator asks text-based questions to two players (one human, one computer). The computer is deemed to have human-like intelligence if the interrogator is unable to guess correctly which answers come from which player.

1949 The de Havilland Comet, the world's first jet airliner, has its first test flight in the UK

1950 US biochemist Erwin Chargaff determines that in DNA there are equal amounts of adenine and thymine bases, and of cytosine and guanine, providing clues to the transmission of information on the DNA molecule

1950

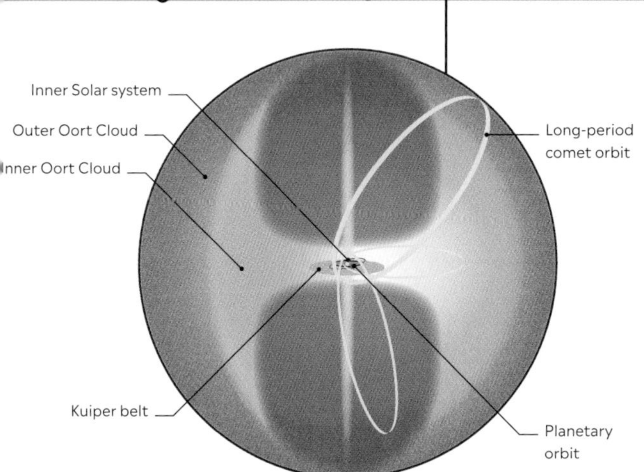

OORT CLOUD STRUCTURE

1950
THE OORT CLOUD

After studying the paths of long-period comets that take thousands of years to orbit the Sun, Dutch astronomer Jan Oort proposed that they originate in a vast spherical shell that surrounds the Solar System at a distance of up to a light-year. This shell soon became known as the Oort Cloud.

1950
GENE REGULATORS

US geneticist Barbara McClintock studied the cell biology and genetics of maize. Each kernel of an ear of maize is an individual embryo, and strains often have kernels that differ in colour. Her major achievement was finding transposable elements (TEs), also known as jumping genes. She discovered that these elements could cause heritable changes, altering how scientists thought about inheritance and inspiring what would become known as epigenetics. These mobile elements are now known as transposons.

△ **Barbara McClintock** studying maize cobs

1951

SHAPE OF THE MILKY WAY

Although evidence that the Milky Way is a rotating disc of stars mounted through the early 20th century, its precise structure remained unclear until 1951. Then US astronomer William Morgan presented his work, which identified concentrations of hot blue supergiant stars and associated areas of star formation tracing a spiral structure across the galactic disc.

▷ **The Milky Way** as it would appear from outside our galaxy (artist's impression)

1951

1951 Designed for business use, the Universal Automatic Computer (UNIVAC) is the first successful commercially produced digital computer

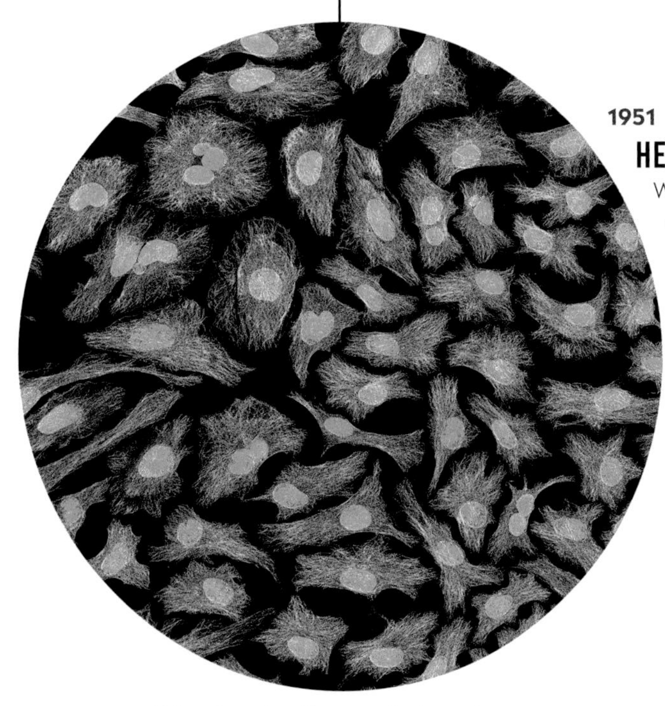

△ **HeLa cells** captured by a scanning laser microscope

1951

HELA CELLS

When American Henrietta Lacks died from cervical cancer, she left a truly remarkable legacy — an immortal lineage of her own cells. Surprisingly, cancer cells from her body continued to replicate prolifically in laboratory culture, unlike most such cells, which divide a few times and then die. These became the first human cell-line and were named HeLa cells by US cell biologist George Otto Gey. HeLa cells are now an essential resource in medicine, instrumental in vaccine development and in many other fields of research.

More than 50 million tonnes of HeLa cells have been grown since 1951

▷ **The Miller-Urey experiment**
To simulate chemical conditions in the early atmosphere of Earth, water vapour and simple gases were hit by electrical discharges (lightning). After a week, this created a "soup" of complex molecules.

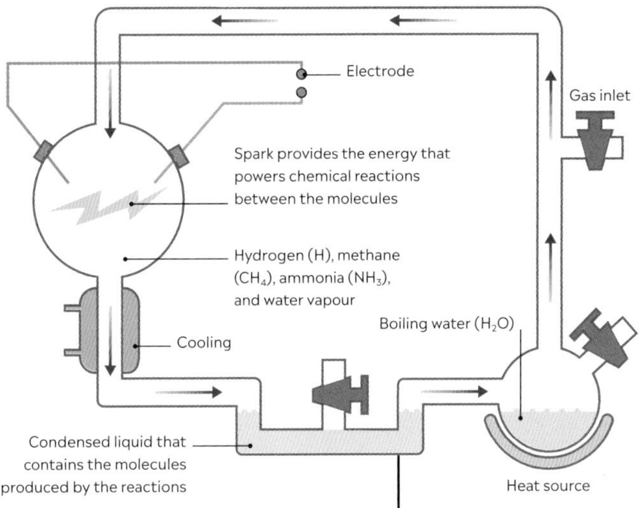

1952
ORIGINS OF LIFE

In an attempt to understand how life on Earth may have originated, US chemists Stanley Miller and Harold Urey conducted an experiment to replicate the conditions on Earth 3.5 billion years ago. Discharging electric sparks through a mix of water, ammonia, hydrogen, and methane, they created amino acids as well as compounds needed to make ribonucleic acid (RNA), which carries genetic information essential for life.

1952

1952 British physiologists Alan Hodgkin and Andrew Huxley create a mathematical model to describe action potential in neurons

1952 US physicist Rosalyn Yalow develops radioimmunoassays for measuring small concentrations of substances in the body

1952 Polish physicists Marian Danysz and Jerzy Pneiwski discover the hyper nucleus – an atomic nucleus that contains protons, neutrons, and one or more hyperons

1952 Dutch astronomer Adrian Blauuw shows that the Zeta Persei star group contains stars only a few million years old

◁ **Bermuda** petrel

1951
PETREL REDISCOVERED

The rare Bermuda petrel or cahow (*Pterodroma cahow*) is an elegant seabird that was assumed to have been extinct for nearly three centuries until a small population of just 18 pairs was found on four tiny rocky islets in Bermuda. After dedicated conservation efforts, its population has slowly recovered, now standing at around 400 individuals.

△ **Detonation** at Elugelab

1952
THERMONUCLEAR EXPLOSION

On the Pacific island of Elugelab, the US conducted the first full-scale test of a thermonuclear weapon, also known as a hydrogen bomb. Most of the energy of a hydrogen bomb is released by nuclear fusion, unlike other nuclear weapons, which rely on fission alone.

1953

DNA STRUCTURE

Four scientists were involved in the discovery of the structure of DNA (deoxyribonucleic acid), the molecule of heredity. British chemist Rosalind Franklin and New Zealand-born biophysicist Maurice Wilkins obtained detailed images of DNA using X-ray crystallography, which helped US zoologist James Watson and British molecular biologist Francis Crick produce molecular models of DNA. Watson and Crick announced their discovery of "the secret of life" in February 1953.

1953

STRANGENESS

The state of any subatomic particle is defined by a set of quantum numbers – values that relate to quantities such as energy and momentum. While studying previously unknown particles in cosmic ray collisions, US physicist Murray Gell-Mann discovered a new quantum number, which he called strangeness.

△ Dr Murray Gell-Mann

1953

1953 US physician John Gibbon invents the heart-lung machine, which maintains the circulation and oxygenation of a patient's blood, allowing surgery on the heart

1953 Catalysts that control branching in polymer molecules are discovered by German chemist Karl Ziegler and Italian chemist Giulio Natta

▷ **DNA structure,** ball-and-stick model

The Global Mid-ocean Ridge marks marks the boundary between several of the tectonic plates that make up Earth's crust

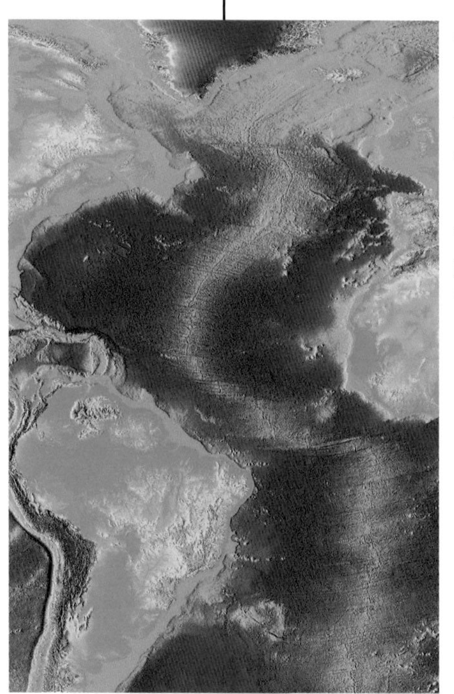

△ **Atlantic** ocean floor topography

1953

MID-OCEAN RIDGE

Using sounding data of ocean floor topography collected at sea by US geologist Bruce Heezen, US cartographer Marie Tharp drew ground-breaking maps of the sea floor of the Atlantic Ocean. The topographic profiles clearly showed for the first time a mountain range called the Mid-ocean Ridge with a distinct and unexpected central rift valley.

1953
REM SLEEP

The two main phases of sleep can be distinguished by movements of the eyes and are known as REM (rapid eye movement) and non-REM sleep. In 1953, US physiologists Nathaniel Kleitman and Eugene Aserinsky first defined REM sleep, linking this phase to dreaming and an increase in brain activity.

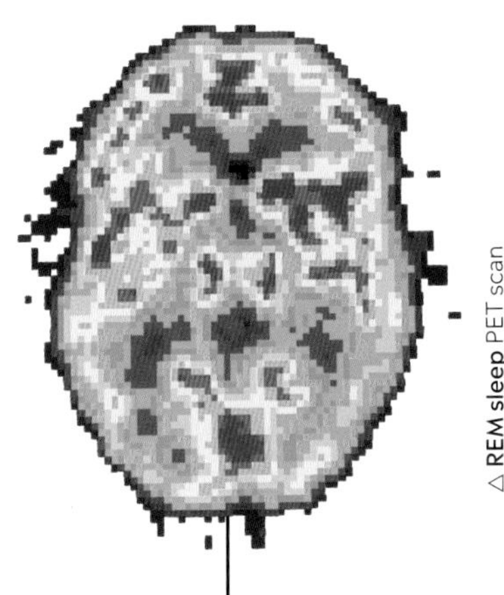

△ **REM sleep** PET scan

▽ **USS *Nautilus*** launch

1954
PRACTICAL NUCLEAR POWER

The world's first nuclear-powered submarine, the USS *Nautilus*, was placed into the water in January 1954, in Connecticut, USA. In June the same year, the world's first power-grid-connected nuclear power station, the Obninsk Nuclear Power Plant, began producing electricity in the USSR.

1954

1954 US physician Jonas Salk creates the first successful vaccine against the crippling disease poliomyelitis (polio)

c. 1954 Dynamo theory shows that fluid motion can generate Earth's geodynamo and its magnetic field

EARTH'S MAGNETISM

The movement of the molten iron in Earth's outer core generates a powerful magnetic field through the whole planet and for thousands of kilometres out into space. The field prevents harmful radiation from the Sun reaching Earth's surface. Like a bar magnet, the field has north and south poles, and this has allowed navigation by compass for many centuries. The magnetized needle of a compass aligns itself with the magnetic field and points north.

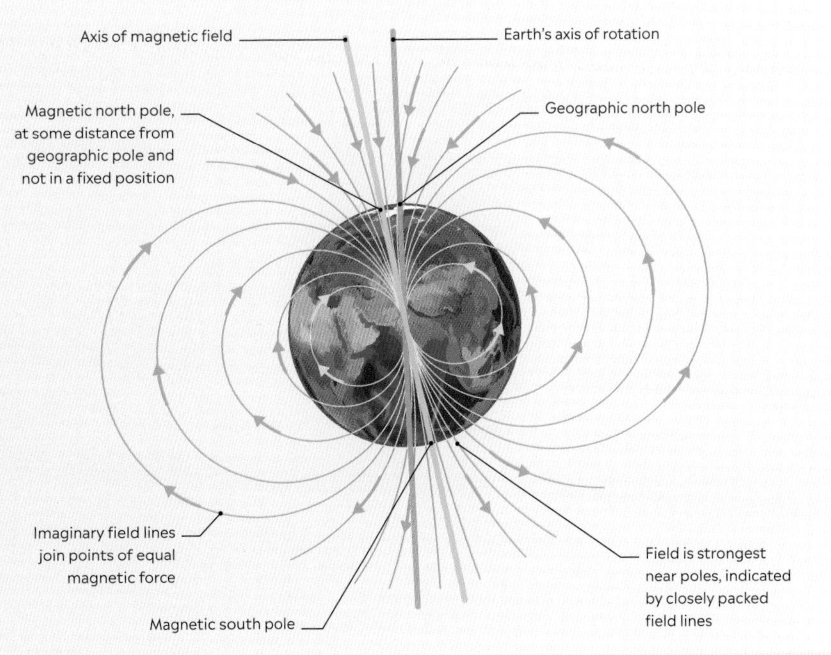

Axis of magnetic field
Earth's axis of rotation
Magnetic north pole, at some distance from geographic pole and not in a fixed position
Geographic north pole
Imaginary field lines join points of equal magnetic force
Field is strongest near poles, indicated by closely packed field lines
Magnetic south pole

Magnetic field and poles
Earth's magnetic field can be represented by imaginary field lines. The positions of the poles and the strength of the field are dynamic and constantly changing.

DNA

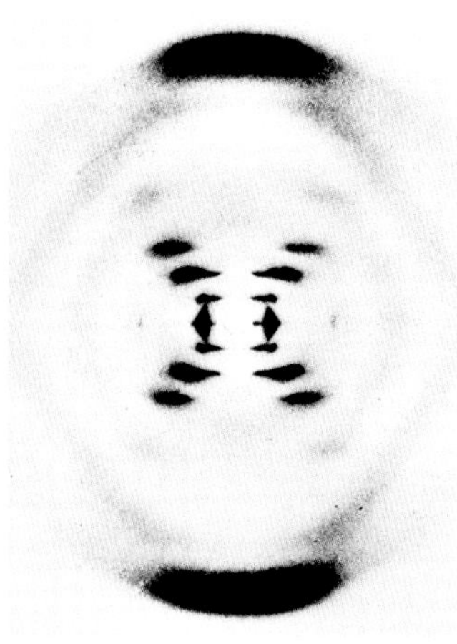

Discovering the helix
This X-ray diffraction image of DNA was obtained by Rosalind Franklin in 1953. The cross of bands indicates the helical nature of DNA.

Deoxyribonucleic acid (DNA) is the molecule that carries all the instructions a living organism needs to grow, reproduce, and function. For years, it was thought of as a long polymer with just four subunits: two so-called purines (adenine and guanine) and two pyrimidines (thymine and cytosine). The double-helix structure of DNA – discovered by Watson, Crick, Franklin, and Maurice Wilkins in the 1950s – explains how the molecule replicates and encodes so much information.

The ladder rungs in the molecule (*see below*) are formed when bases on one strand pair up with those on the other: adenine always pairs with thymine, and cytosine always pairs with guanine. There are 3 billion base pairs in the human genome. Not all DNA encodes the information for making proteins (the genes). In humans, only 3 per cent of the genome is protein-coding. The billions of remaining bases play numerous roles, including switching genes on or off.

Great lengths of DNA (2 m/6 ft in a single human cell) are condensed into coiled strands called chromosomes, which are found in the nucleus. This is true in all organisms with a nucleus. Bacteria (without a nucleus) have a single circular chromosome and smaller circular DNA molecules known as plasmids, which they exchange with other bacteria. Humans have 46 chromosomes in each cell, 23 from each parent.

The genomes from any two people are 99.9 per cent identical. Differences tend to be not in the genes but in the sequences that control the genes. No two individuals, except identical twins, have the same DNA sequence.

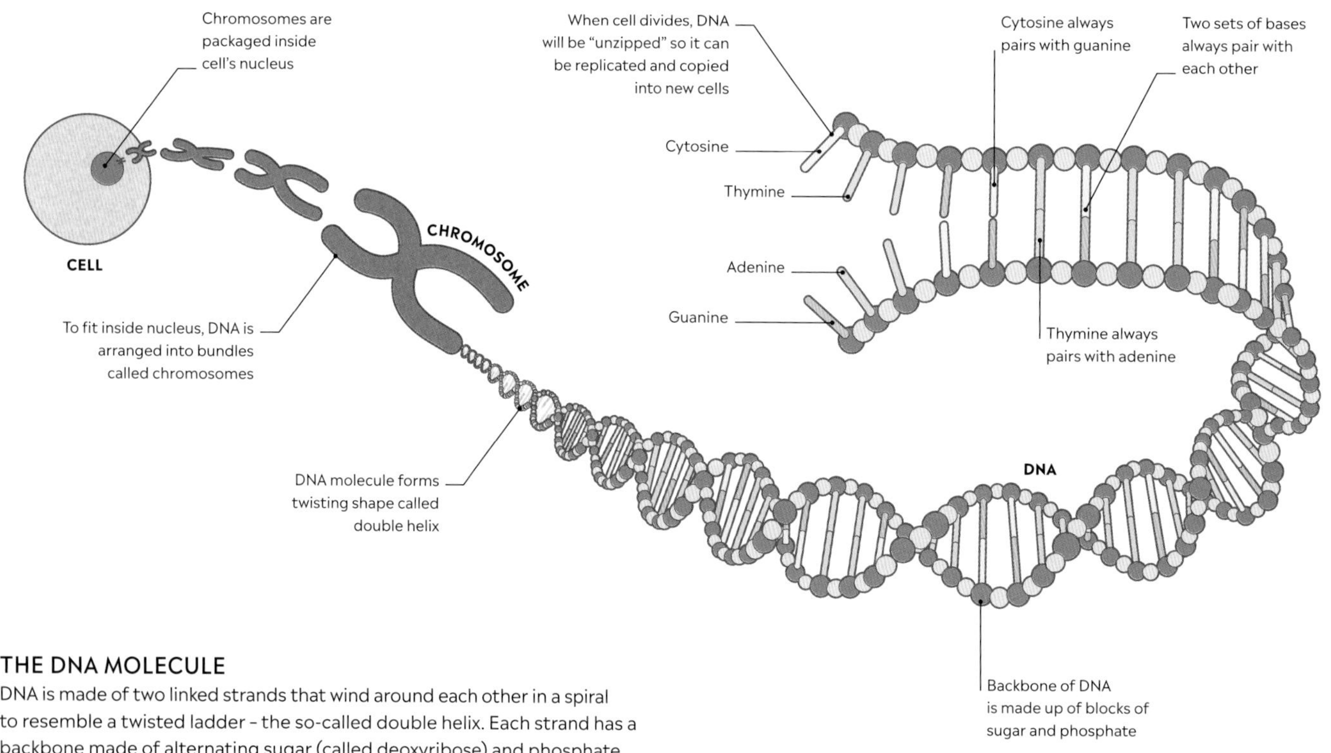
Chromosomes are packaged inside cell's nucleus
CELL
CHROMOSOME
To fit inside nucleus, DNA is arranged into bundles called chromosomes
DNA molecule forms twisting shape called double helix
When cell divides, DNA will be "unzipped" so it can be replicated and copied into new cells
Cytosine
Thymine
Adenine
Guanine
Cytosine always pairs with guanine
Two sets of bases always pair with each other
Thymine always pairs with adenine
DNA
Backbone of DNA is made up of blocks of sugar and phosphate

THE DNA MOLECULE

DNA is made of two linked strands that wind around each other in a spiral to resemble a twisted ladder – the so-called double helix. Each strand has a backbone made of alternating sugar (called deoxyribose) and phosphate groups. Attached to each sugar is one of four bases: adenine (A), cytosine (C), guanine (G), or thymine (T).

Crick and Watson's model
Francis Crick and James Watson used their knowledge of chemical bonds, together with X-ray crystallography results from Rosalind Franklin, to build models of the DNA double helix.

1955

INSULIN SEQUENCED

British biochemist Frederick Sanger worked out the sequence of amino acids in the hormone insulin, which regulates the amount of sugar in the blood. Insulin consists of two amino acid chains. Sanger used a technique that broke up the chains and then identified sections using paper chromatography. He used this information to build the complete sequence of 51 amino acids.

◁ **Insulin molecule,** computer model

1955 German-US biochemist Heinz Ludwig Fraenkel-Conrat discovers that RNA controls the reproduction of viruses and is located in the core of the virus

1955

1955 Spanish biochemist Severo Ochoa discovers an enzyme that synthesizes RNA; US biochemist Arthur Kornberg later isolates the enzyme that assembles DNA – DNA polymerase

1955 Fred Hoyle and Martin Schwarzschild, a British and German-US team, model the development of stars near the end of their lives into red giants

The largest known red giants are more than 1,000 times wider than the Sun

1955

ATOMS VISUALIZED

In 1951, German physicist Erwin Müller invented the field ion microscope. This device uses a powerful electric field to fling ions off a sharp metal tip; the ions hit a phosphor-coated screen in patterns that correspond directly to the arrangement of atoms at the tip. In 1955, Müller's device produced the first clear, faithful atomic-scale images.

△ **Iridium atoms** imaged by a field ion micrograph

△ Bevatron accelerator

1955
ANTIPROTONS DETECTED

The antiproton is the antiparticle of the proton: identical in mass, but with opposite electric charge. Its existence had been predicted by Paul Dirac in 1933, but it was not detected until 1955. Then, Italian-US physicist Emilio Segrè and US physicist Owen Chamberlain created antiprotons in energetic collisions at the Bevatron, a particle accelerator at Lawrence Berkeley National Laboratory, USA.

1956
RIBOSOMES

While investigating the components of cells using an electron microscope, Romanian-US cell biologist George Palade identified ribosomes – previously unknown organelles that are the site of protein synthesis. These are sometimes referred to as Palade granules in his honour.

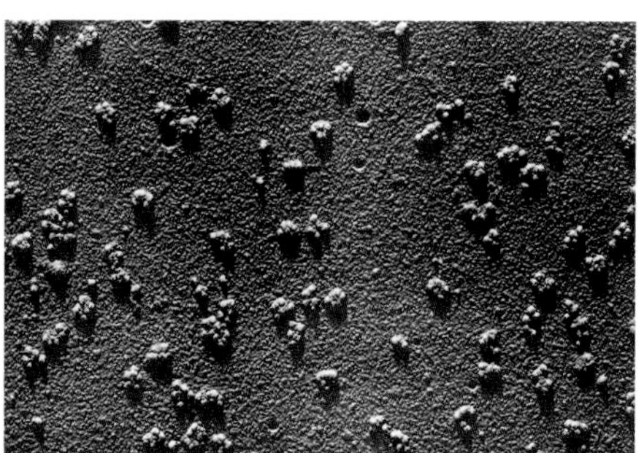

△ **Clusters** of rabbit ribosomes

1956 The first transatlantic telephone cable, TAT-1, begins operation

1956

1955 Indian-US physicist Narinder Singh Kapany makes important breakthroughs in optical fibre technology

1956 By measuring microwave emissions from Venus, US astronomer Cornell Mayer shows the planet's surface must be searing hot

1956 US physicists Clyde Cowan and Frederick Reines prove the existence of neutrinos, proposed by Wolfgang Pauli in 1930

1955
RADIO INTERFEROMETER

British astronomer Martin Ryle showed how interferometry (the comparison of radio signals received simultaneously at two or more separated antennae) can be used to pin down the location of cosmic radio sources. As interferometry developed, Ryle constructed antennae networks up to 5 km (3 miles) long. Using his techniques, astronomers were able to resolve detail equivalent to optical instruments for the first time.

▷ **Ryle** with his early interferometer

1918–2013
FREDERICK SANGER

British biochemist Sanger is one of only five people to have won two Nobel Prizes. He was awarded his first in 1958 for sequencing insulin and his second in 1980 for finding the first complete sequence of nucleotide bases of a small virus's DNA.

▷ **Sputnik 1** with its antennae extended

1957
SPUTNIK 1

On 4 October, the Soviet Union launched the world's first artificial satellite, Sputnik 1, using a modified R-7 Semyorka missile. The surprise launch of this 84-kg (184-lb) metal sphere fitted with a simple radio transmitter shocked the world and upstaged America's widely publicized plans for a satellite launch as part of 1957's International Geophysical Year. Sputnik continued to send signals for 21 days, until its batteries ran out.

◁ **Sputnik** on the launchpad

1957

1957 A giant 76-m (250-ft) steerable radio telescope is completed at Jodrell Bank, UK

1957 British ecologist George Evelyn Hutchinson defines the ecological niche as "a hypervolume shaped by the environmental conditions under which a species can exist indefinitely"

1957
THE CHEMISTRY OF PHOTOSYNTHESIS

The US biochemist Melvin Calvin used radioactive carbon-14 isotopes to mark chemicals created in the process of photosynthesis and identified them using chromatography. As the chemical reactions involved were extremely fast, it was painstaking work, but Calvin slowly discovered all the chemicals and processes involved.

△ **Melvin Calvin**

△ **Effect of gibberellic acid** on cabbage

1957
PLANT HORMONES

Discovered in the 1920s, gibberellins and auxins are plant hormones that regulate developmental processes such as germination, dormancy, flowering, and growth. In 1957, gibberellic acid (gibberellin A_3) was isolated, unlocking its use in agriculture and horticulture to promote flowering and increase fruit size.

▽ **Explorer 1** being launched into space

1958

VAN ALLEN BELTS

Following the failure of its Vanguard rocket system, the USA used Juno I, a modified missile designed by Wernher von Braun, to launch its first successful satellite – Explorer I. Fitted with scientific instruments alongside its radio transmitter, Explorer I discovered belts of high-energy particles trapped by Earth's magnetic field. These were later named the Van Allen belts, after mission scientist James Van Allen.

ПРАВДА

СОВЕТСКАЯ НАУКА ОДЕРЖАЛА НОВУЮ БЛЕСТЯЩУЮ ПОБЕДУ

БЕСПРИМЕРНЫЙ НАУЧНЫЙ ПОДВИГ

СООБЩЕНИЕ ТАСС

О движении третьей советской космической ракеты

△ **Reports in Soviet press**

1959

EXPLORING THE FAR SIDE OF THE MOON

The Soviet Luna 3 mission flew past the Moon to send back the first images of the lunar far side (the hemisphere that is hidden from Earth). With electronic imaging still in its infancy, its pictures were captured on photographic film, which was then scanned for conversion into radio signals.

1958 US physicist Eugene Parker models the structure of the solar wind – particles streaming across space from the surface of the Sun

1957 British astronomer Geoffrey Burbidge and others demonstrate that elements heavier than iron are created by the death of massive stars

1958 British doctor Ian Donald pioneers the diagnostic use of ultrasound

1959 In Cambridge, UK, biochemists Max Perutz and John Kendrew work out the 3D structure of proteins

1958

THE FIRST MICROCHIP

US electrical engineer Jack Kilby built a working circuit, made of several transistors, resistors, and capacitors, on a small piece of germanium (a semiconductor). He had built the first integrated circuit, or microchip.

▷ **Prototype microchip** from 1958 encased in plastic

Modern microchips can contain billions of transistors measuring a few nanometers

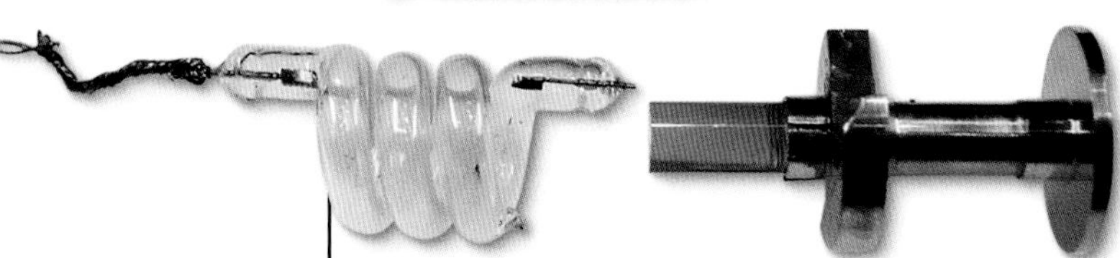

▷ **Components** of Maiman's ruby laser

1960

THE LASER

Several people were working in the 1950s to develop a device that would produce coherent light (a beam of photons that have the same frequency and whose wavelengths are in phase). In 1960, US engineer Theodore Maiman was successful, building the first laser (an acronym for "light amplification by stimulated emission of radiation"). He used a ruby crystal as the "lasing medium", the part that produces the bright, coherent light.

1960

WEATHER SATELLITE

The US space agency NASA launched TIROS-1 (Television Infrared Observation Satellite), the first full-scale satellite designed to monitor Earth's weather. From its position in low Earth orbit, TIROS returned visible light pictures as well as infrared images that could highlight cloud cover. The TIROS satellites demonstrated the potential of space for so-called "remote sensing" of Earth's environment.

△ **TIROS** images of Earth

1960

1960 US biologists Kenneth Norris and John Prescott demonstrate dolphin echolocation using a temporarily blinded dolphin

1960 US physicist Luis Alvarez discovers several short-lived subatomic particles via peaks in energy curves in particle accelerators

1960

SPREADING SEAFLOOR

Harry Hess, a US geologist, proposed that, over time, the ocean floor crust moves away on either side of volcanically active mid-ocean ridges as new ocean floor lavas are erupted. This process, known as seafloor spreading, explains why the age of ocean floor rocks increases with their distance from the ridge.

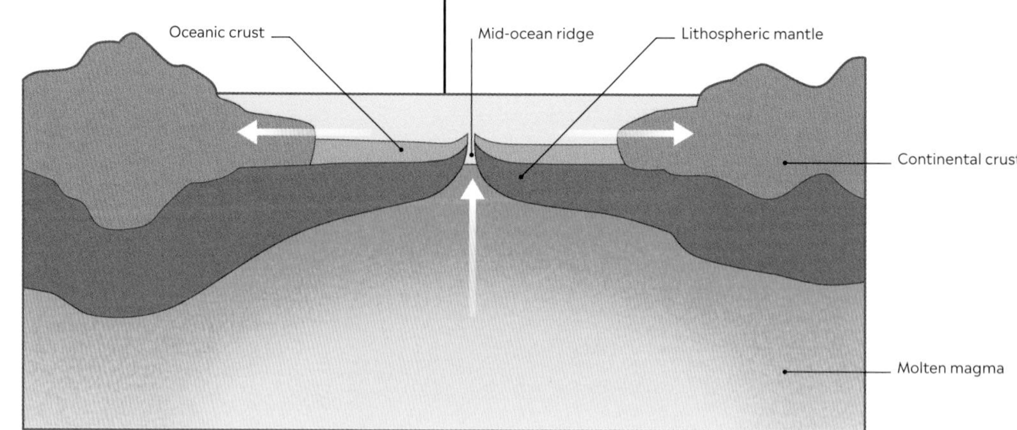

△ **Magma rising below mid-ocean ridges** erupts as new ocean floor lavas. As ocean floor rocks spread away on either side of the ridge, the ocean widens, pushing the continents further apart.

“I see Earth! It is so beautiful!”

YURI GAGARIN, 1961

1961

MAN IN SPACE

Russian pilot Yuri Gagarin became the first human to enter space, aboard the Soviet Vostok 1 spacecraft. His 108-minute, mostly automated flight made a single orbit around Earth. Following re-entry, he ejected from the spacecraft, parachuting to Earth from an altitude of 7 km (4.3 miles).

▷ **Yuri Gagarin** aboard Vostok 1

1961 US president John F. Kennedy commits NASA to landing people on the Moon by the end of the 1960s

1961 French biochemists Jacques Monod and François Jacob reveal how an inducer (sugar) attaches to a repressor, obstructing it, so that genes are activated to make enzymes

1961

1961

CRACKING THE CODE

In a famous experiment, US biochemist Marshall Nirenberg and German biochemist Heinrich Matthaei deciphered the first three-letter (triplet) codon in the genetic code. They found that an extract from bacterial cells was able to make protein outside of living cells. When a form of RNA was added to the extract, a protein formed of phenylalanine was produced, showing that RNA controls the production of a specific protein.

△ Marshall Warren Nirenberg

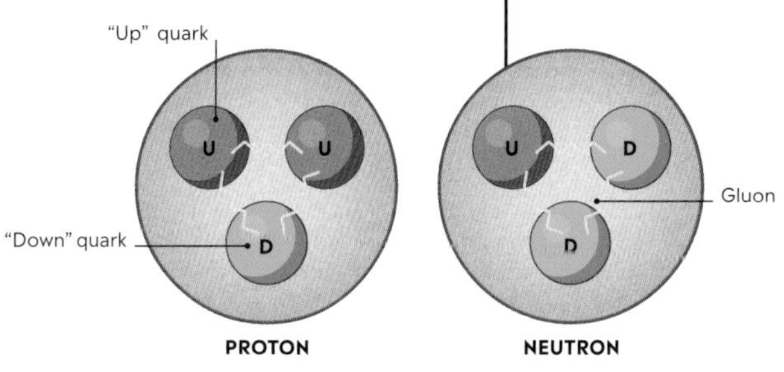

△ **Hadrons:** protons and neutrons are hadrons – each has a different combination of "up" and "down" quarks.

1961

QUARK THEORY

Attempting to classify the growing number of subatomic particles being discovered, US physicist Murray Gell-Mann and Israeli physicist Yuval Ne'eman independently suggested a scheme that included a family of particles now known as hadrons. Gell-Mann later proposed a theory that explained what all hadrons had in common: they are composed of smaller particles, which he called quarks, bound together by gluons.

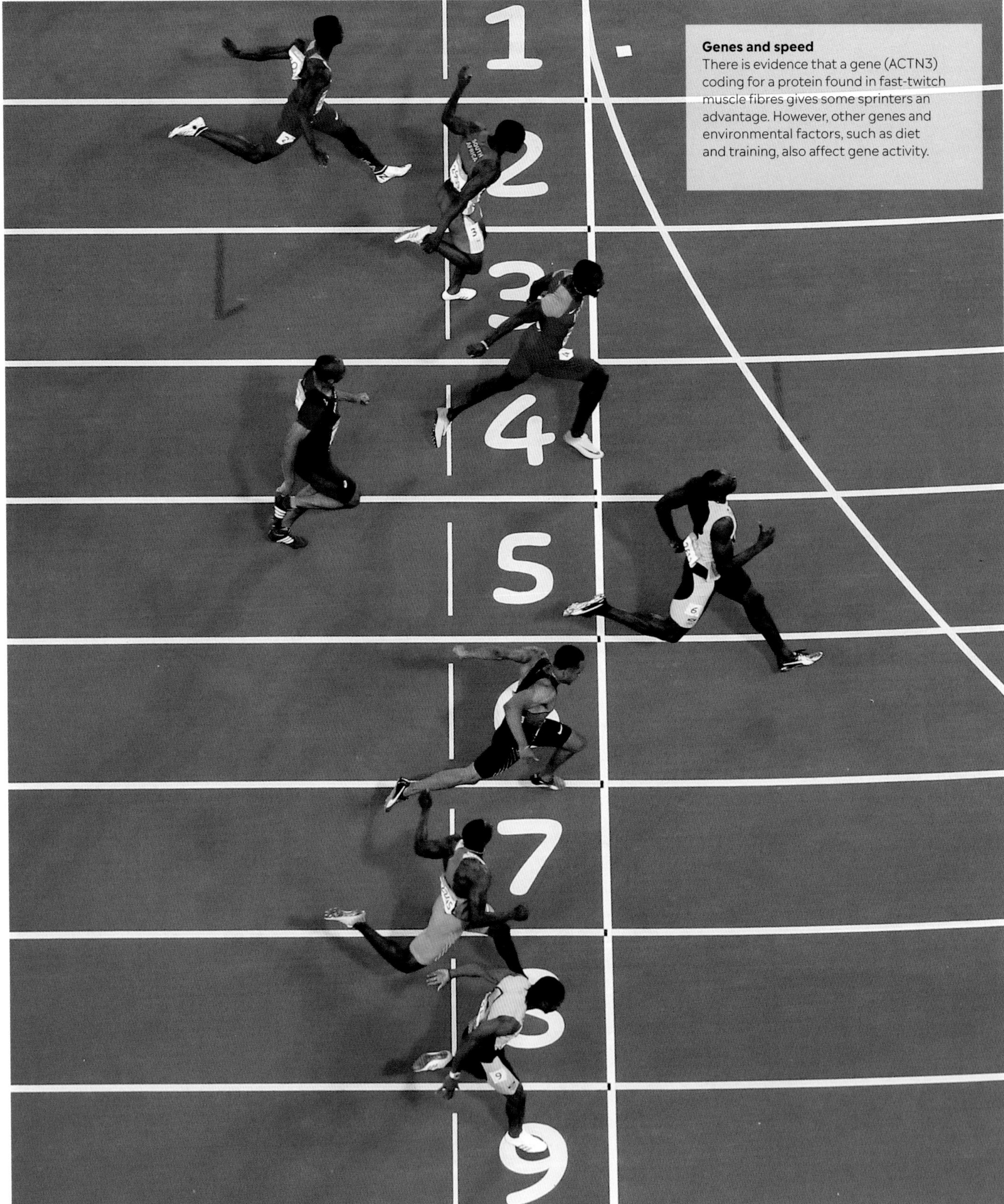

Genes and speed
There is evidence that a gene (ACTN3) coding for a protein found in fast-twitch muscle fibres gives some sprinters an advantage. However, other genes and environmental factors, such as diet and training, also affect gene activity.

GENE EXPRESSION

Gene expression is the process by which structural genes are activated to produce proteins. The process involves transcription of DNA into a related molecule called messenger RNA (mRNA) and translation of mRNA into protein.

There are three types of RNA involved in gene expression: mRNA, which carries a copy of the gene to be expressed from the DNA in the nucleus; transfer RNA (tRNA), which carries amino acids, which are the building blocks of proteins; and ribosomal RNA (rRNA), which joins with protein to form a structure called the ribosome to regulate protein synthesis.

Every cell in an organism contains every structural gene in the genome, but only a small fraction of these genes is expressed in any cell. Cells have different functions – liver cells or nerve cells, for example – and each cell needs different proteins to be able to perform its role. This means gene expression is tightly regulated.

Transcription begins when an enzyme called RNA polymerase binds to a so-called promoter gene on the DNA strand. Normally, the promoter gene is just ahead of where the transcription will start. In addition to promoter genes, transcription is regulated by operator genes, which provide the binding site for repressor proteins. Repressor proteins halt transcription and are coded for by regulator genes.

Operons, which were first discovered in prokaryotes (organisms with nuclei in their cells), are clusters of genes under the control of a single promoter. Genes contained in an operon are either expressed together or not at all.

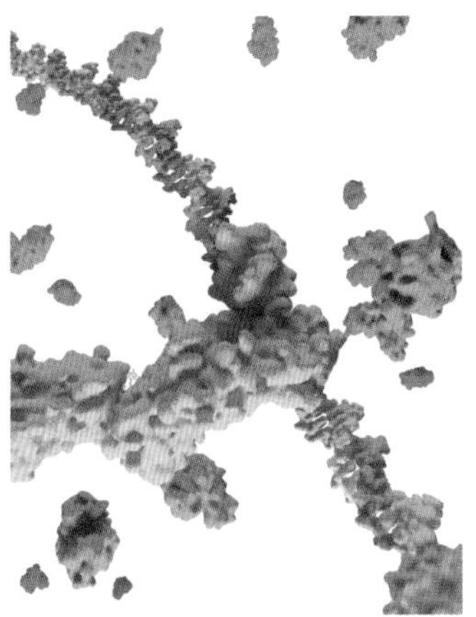

The lac operon
Required for the transport and metabolism of lactose in bacteria such as *E. coli*, the lactose operon (lac operon) was the first genetic regulatory mechanism to be understood.

REGULATION

Regulatory proteins help control the synthesis of protein in cells. Transcription of a required gene is controlled by a series of genes that sit in front of it. These include regulator, promoter, and operator genes. The gene will only be transcribed if the conditions are right.

REPRESSION

Repressor proteins inhibit the expression of one of more genes by binding to the operator gene. If a repressor protein is blocking the gene, transcription cannot take place. The gene can only be turned on when a change in the environment removes the repressor protein.

ACTIVATION

Transcription can start when an activator binds to the regulator, and no repressors are blocking the gene. Activators have a sequence-specific DNA-binding domain and an activation domain that increases gene transcription. The sequence-specific binding domain means that only certain genes can be activated.

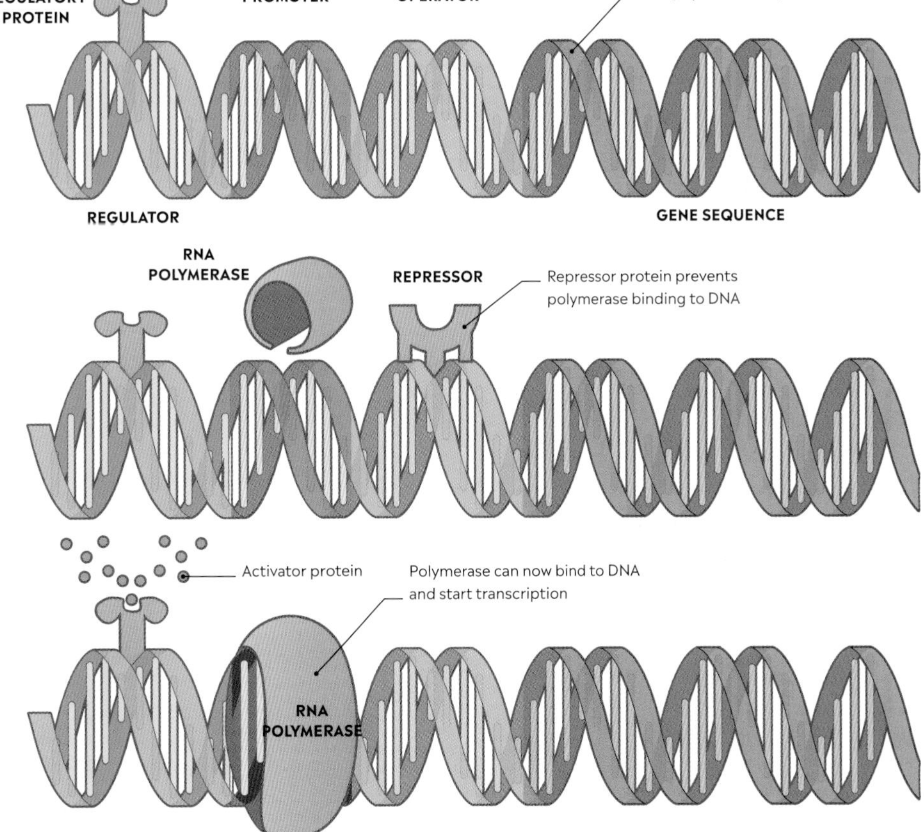

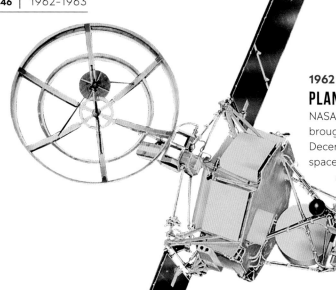

1962

PLANETARY PROBES

NASA launched the Mariner 2 spacecraft into an orbit that brought it within 35,000 km (22,000 miles) of Venus on 14 December. It was the first successful encounter between a spacecraft and another planet. Mariner 2's instruments revealed Venus's dense atmosphere and searing surface temperatures, and confirmed the presence of a solar wind of particles streaming off the Sun's surface and filling interplanetary space.

△ Mariner 2

1962

1962 John Glenn is the first US astronaut to orbit Earth

1962 US-based scientists Linus Pauling and Emile Zuckerkandl introduce the molecular clock – the idea that the relatively constant evolution of DNA and protein sequences can be used to estimate species divergence

1962 Telstar 1, the first television communications satellite, is launched

△ Rachel Carson

1962

ENVIRONMENTAL AWAKENING

US biologist Rachel Carson published *Silent Spring*, a book that inspired the global environmental movement. It addressed the role of pesticides in harming nature, and eventually led to bans of certain insecticides, notably the highly toxic compound DDT (dichlorodiphenyltrichloroethane).

> "The real wealth of the Nation lies in the resources of the Earth – soil, water, forests, minerals, and wildlife."
>
> *RACHEL CARSON, IN A LETTER TO* THE WASHINGTON POST, *1953*

1963

ARECIBO OBSERVATORY

With a 305-m (1,000-ft) span, the Arecibo radio telescope in Puerto Rico was completed in 1963. The world's largest telescope for five decades, it was used in the discovery of Mercury's rotation and of several pulsars (fast-spinning stellar remnants), and to search for extraterrestrial radio signals. In 1974, it sent humanity's first deliberate message to the stars.

△ **Arecibo telescope's** reflector dish

1963
POLE REVERSALS

Earth's magnetic field is generated by the flow of the molten outer core. At present, the field has North and South poles which are close to Earth's axis of rotation. Investigation of the magnetization of certain minerals in the rock record has revealed that, over geological time, this orientation has reversed many times. The 1963 Vine–Matthews–Morley hypothesis found that the Earth's oceanic crust acts as a recorder of these reversals.

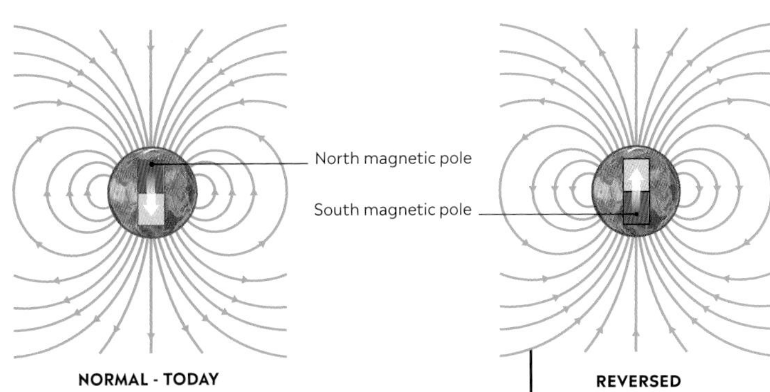

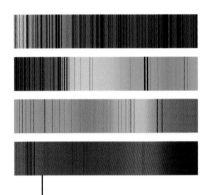

1963
DISTANT GALAXIES

Astronomers showed that a mysterious class of rapidly varying radio signals appeared to come from star-like objects, which they named quasars (quasi-stellar radio sources). When Dutch astronomer Maarten Schmidt analyzed the spectrum of quasar 3C 273, he realized that its light had been red shifted, indicating that it was embedded in a distant galaxy moving rapidly away from Earth.

◁ **Redshifted** absorption lines in quasar spectra

1963

1963 US meteorologist Edward Lorenz coins the term "butterfly effect", describing a property of complex systems (such as the atmosphere) by which small changes can lead to large-scale, unpredictable variations

1963 Russian cosmonaut Valentina Tereshkova becomes the first woman in space on the Soviet Vostok 6 mission

QUASARS

A galaxy that emits an exceptional amount of energy is known as an active galaxy. Among the most violent types of active galaxies are quasars. At the centre of all active galaxies, a supermassive black hole feeds on matter from its surroundings. As material falls into its powerful gravitational field, it is heated to extreme temperatures, emitting brilliant radiation and jets of high-speed particles. The quasars that we observe are in the process of formation or rapid growth. However, we see these distant objects as they were far in the past, because radiation from them has taken a long time to reach us. This phase of galaxy evolution actually peaked billions of years ago.

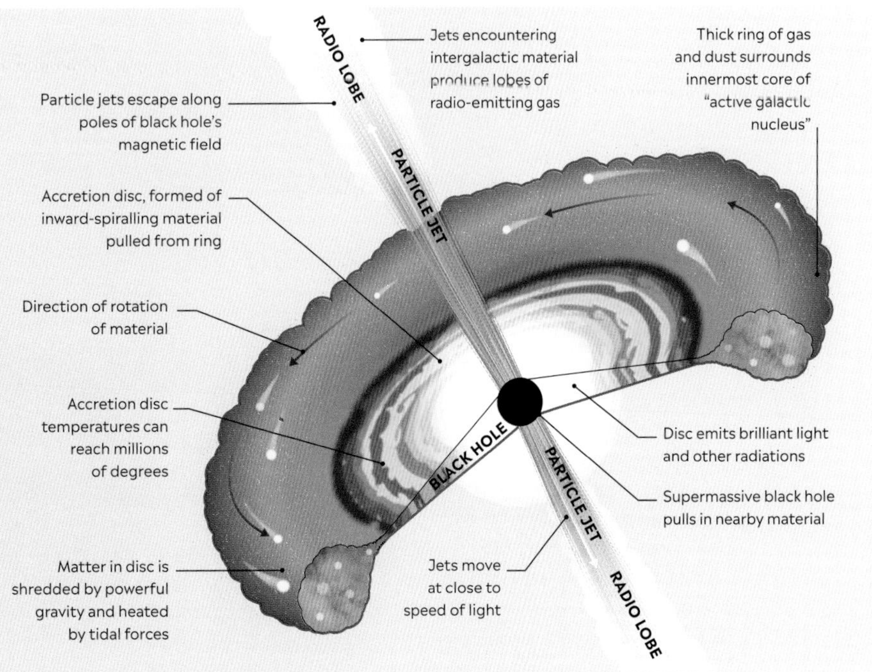

1964

COSMIC MICROWAVE BACKGROUND

One of the most important breakthroughs in cosmology – and in physics more generally – was the discovery of the radiation left over from the very early history of the Universe. This "cosmic microwave background" comes from all directions and provides compelling support for the Big Bang theory. The radiation was discovered by US physicists Arno Penzias and Robert Wilson using a communications radio antenna in Holmdel, New Jersey, US.

◁ **Holmdel microwave** antenna

1964

1964 The USSR's Voskhod 1 mission puts three cosmonauts in orbit on the first multi-crewed spaceflight

1964 British physicist Peter Higgs is among six physicists who predict the existence of a field that gives elementary particles mass – later called the Higgs field

1964

KIN SELECTION

British biologist William Donald Hamilton published the principle of kin selection, which accounted for many examples of altruistic behaviour that standard evolutionary theory could not adequately explain. Central is the concept of inclusive fitness, whereby individuals may increase the propagation of their shared genes by helping to raise the young of close relatives.

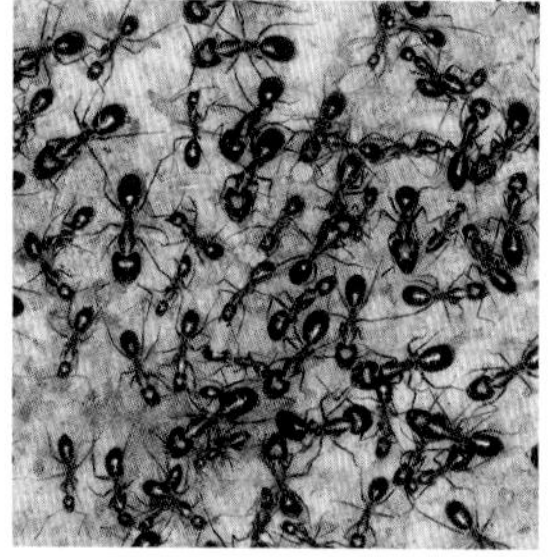

△ **Altruistic** carpenter ants

"Cosmology is a science which has only a few observable facts to work with."

ROBERT WOODROW WILSON,
NOBEL LECTURE, *1978*

1965

VENUS PROBE

The USSR launched its Venera 3 mission towards Venus. It was made up of a flyby vehicle and a cone-shaped lander designed to parachute into the Venusian atmosphere. The craft returned much data about interplanetary space, but communications broke down shortly before its encounter with Venus. The lander was released successfully, becoming the first artificial object to reach the surface of another planet.

△ **Venera 3 medallion** carried into space

1965

NEUTRINO ASTRONOMY

Astronomers in South Africa and India independently detected solar neutrinos – near-massless particles emitted by nuclear fusion reactions in the core of the Sun. Both groups used detectors installed in deep mines. Neutrinos passed through the overlying rock to enter the detector tanks, while other particles were blocked out. Construction also began on the first large-scale neutrino detector at Homestake Mine in South Dakota, US.

△ **Detector tank** at Homestake Mine

1965

TRANSFER RNA

Francis Crick had earlier suggested that there must be a molecule involved in translating the genetic code of DNA and RNA into the chains of amino acids that make up proteins. In 1965, US biochemist Robert W. Holley found the sequence of 77 nucleotides of just such a compound – transfer RNA (tRNA) – in yeast. It was soon shown that a tRNA molecule codes for a particular amino acid: Messenger RNA (mRNA) carries genetic information from DNA, and tRNA is the physical link between mRNA and protein synthesis.

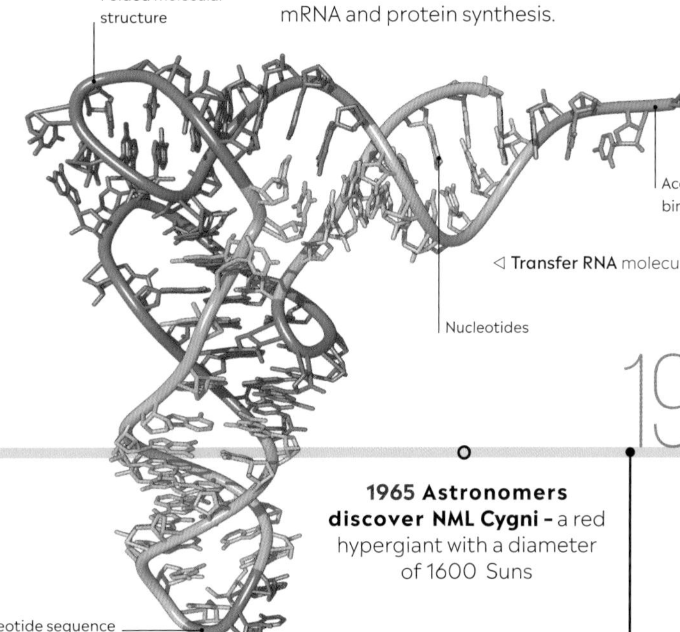

◁ **Transfer RNA** molecule model

1965 A hominid fossil found at Olduvai Gorge, Tanzania is named *Homo habilis* and dated to 2.4–1.4 MYA

1965 The US Mariner 4 spacecraft flies past Mars and returns the first close-up images of the planet

1965

1965 Astronomers discover NML Cygni – a red hypergiant with a diameter of 1600 Suns

1965

HOLOGRAPHY

The notion that a two-dimensional photographic plate could hold a three-dimensional image had been conceived by Hungarian-British physicist Dennis Gabor in 1948. Gabor's invention, for which he coined the term hologram, was aimed at improving the images from electron microscopes, and used electrons rather than light to produce images. Laser light made it possible to use Gabor's idea to create holographic images of everyday objects. US electrical engineer Emmett Leith and Latvian-US physicist and inventor Juris Upatnieks were among the first to make holograms in 1965.

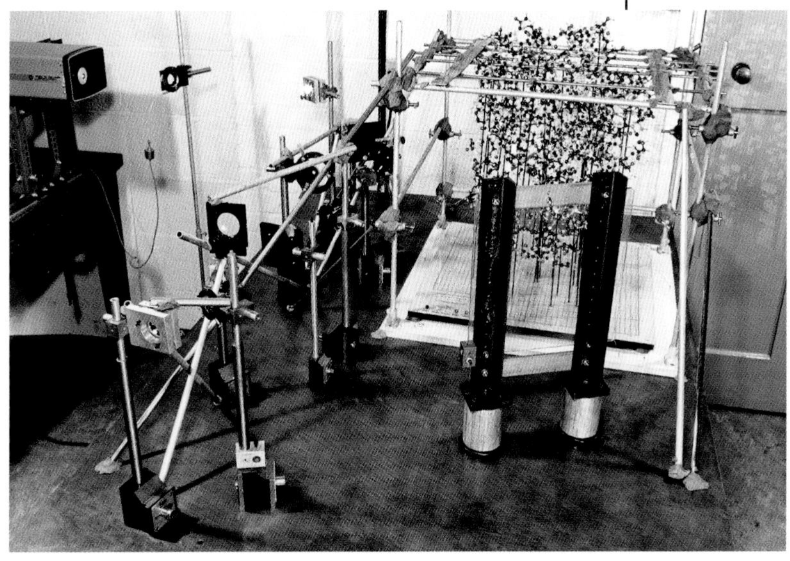

▷ **Creating** a hologram

1966
MOON MISSION

In the next stage of the space race between the USSR and US, both countries landed probes on the surface of the Moon and beamed back pictures and other data. The Soviet Luna 9 mission landed a spherical instrument canister on 3 February 1966. On 2 June, NASA's Surveyor 1 landed a more sophisticated, spider-like landing platform to test procedures later used in the Apollo lunar lander. The landings proved the lunar soil could support spacecraft.

▷ **Luna 9** in the Moon's Ocean of Storms

1966

1966 US surgeon Michael E. DeBakey attaches the first artificial heart to a patient

1966 US biochemist Marshall Nirenberg cracks the DNA code by completing the deciphering of the 64 RNA three-letter code words (codons) for all 20 amino acids

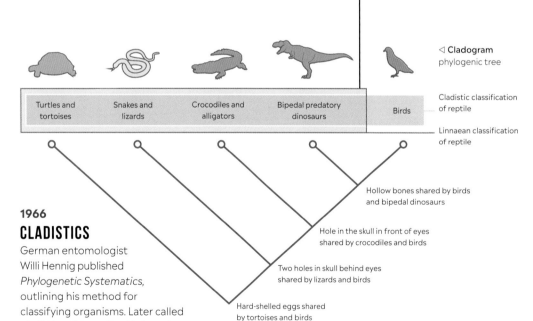

◁ **Cladogram** phylogenic tree

Turtles and tortoises | Snakes and lizards | Crocodiles and alligators | Bipedal predatory dinosaurs | Birds

Cladistic classification of reptile

Linnaean classification of reptile

Hollow bones shared by birds and bipedal dinosaurs

Hole in the skull in front of eyes shared by crocodiles and birds

Two holes in skull behind eyes shared by lizards and birds

Hard-shelled eggs shared by tortoises and birds

1966
CLADISTICS

German entomologist Willi Hennig published *Phylogenetic Systematics,* outlining his method for classifying organisms. Later called cladistics, this method is based on determining characteristics that are deemed to be inherited from a common ancestor. It results in the construction of diagrams (cladograms) that indicate evolutionary relationships.

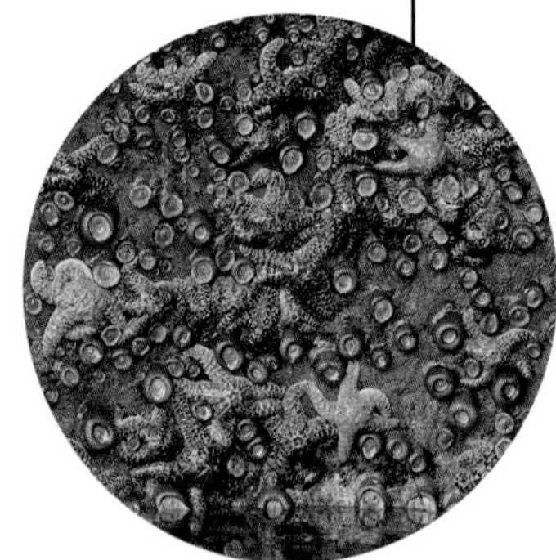

△ **Starfish** *Piaster chraceus*

1966
KEYSTONE SPECIES

US ecologist Robert T. Paine coined the term "keystone species" to describe those species that exert influence on others lower down the food chain in an ecosystem. His classic study was of a rocky marine habitat where the top predator – the starfish *Pisaster ochraceus* – was removed. This resulted in a gradual reduction in the number of species present.

1967

THE FIRST PULSAR

While studying quasars at a radio telescope in Cambridge, UK, British astronomer Jocelyn Bell recorded a signal unlike anything detected before. She later found that the radio signal was pulsing extremely regularly and nicknamed it LGM-1 - a reference to "little green men", as it sounded as if it might emanate from an alien civilization. However, the following year, the source of the unusual signal was identified as a rotating neutron star - the first known pulsar.

△ Jocelyn Bell

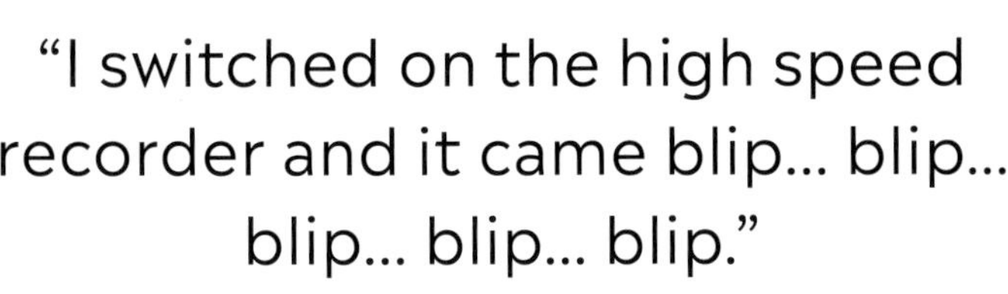

"I switched on the high speed recorder and it came blip... blip... blip... blip... blip."

JOCELYN BELL, BEAUTIFUL MINDS, *2010*

1966 Low-temperature superconducting magnets make NMR spectroscopy an accurate tool for determining the structure of complex organic molecules

1966 Scottish scientist June Almeida identifies a new group of viruses, which have a corona of spikes, naming them coronaviruses

1967

1967 The US Moon program is delayed after rehearsals for the first crewed Apollo mission end in a fire that kills three astronauts

1967

HEART TRANSPLANT

South African surgeon Christiaan Barnard carried out the first human heart transplant in Cape Town. His feat received worldwide publicity and inspired further progress in the technique. While the first patient, Louis Washansky, died after only eighteen days, Barnard's fifth and sixth patients lived for over 13 and 24 years after their transplants.

△ Dr Christiaan Barnard

1967

THEORY OF ISLAND BIOGEOGRAPHY

Based on research into ant populations on Melanesian islands, US biologists Edward O. Wilson and Robert MacArthur proposed that the number of species inhabiting an island is a dynamic balance between the immigration of new species and extinction of resident species.

△ Melanesian islands

Star rotates about central axis many times each second

Neutron star

Powerful magnetic field

Magnetic poles sweep across sky as star rotates

Beams of radiation escape at magnetic poles

N

S

1968
SPINNING NEUTRON STARS

US astronomer Richard Lovelace used the giant Arecibo radio telescope to measure the precise period of signals from the newly discovered LGM-1 "pulsar". He and others demonstrated that the star's characteristics matched the properties of a collapsed, fast-spinning stellar remnant called a neutron star, channelling beams of radiation out of an intense magnetic field. Pulsars provided the first evidence that neutron stars really exist.

Collapsing stars

When a stellar core collapses into a city-sized neutron star at the end of its life, it spins much more rapidly, and its magnetic field is intensified. This produces narrow, lighthouse-like beams of escaping radiation.

1968

1968 US physician Robert A. Good performs the first successful human bone marrow transplant between people who are not identical twins

1969 Swiss geneticist Werner Arber discovers restriction enzymes, leading to investigations that explain the order of genes on the chromosome

△ **Glashow, Salam, and Weinberg (l-r)**

1968
ELECTROWEAK INTERACTION

The weak interaction, involved in radioactive decay, is one of four fundamental interactions (forces) acting between particles. In order to account for certain disparities between the weak interaction and the other three fundamental forces (namely gravitation, electromagnetism, and the strong force), South Asian physicist Mohammad Abdus Salam, with US physicists Steven Weinberg and Sheldon Glashow, arrived at a theory that united the weak and electromagnetic interactions into a single force: electroweak interaction.

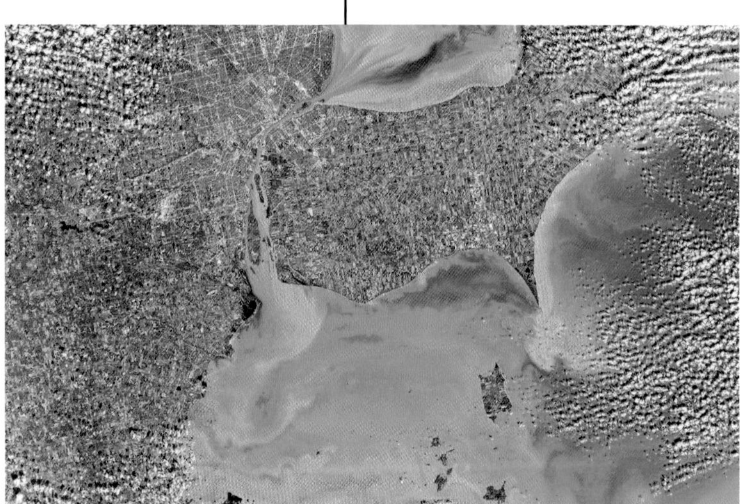

△ **Lake Eyrie** and environs

1968
ENVIRONMENTAL WARNINGS

The US Federal Water Pollution Control declared that "man is destroying Lake Eyrie" by dumping so much untreated municipal waste into it that it would take 500 years to recover. At the same time, presidential science advisor Donald F. Hornig warned utility companies that burning fossil fuels and releasing CO_2 in the atmosphere might "produce major consequences on the climate – possibly even triggering catastrophic effects such as have occurred... in the past".

1969

DISCOVERY OF LYSTROSAURUS

US palaeontologist Edwin H. Colbert and his team found *Lystrosaurus* fossils at Coalsack Bluff in the Transantarctic Mountains of Antarctica. Since fossils of this dog-like herbivore were already known from the lower Triassic period in southern Africa, the find helped support the hypothesis of continental drift through plate tectonics.

▽ ***Lystrosaurus*** artist's impression

1969

FIVE KINGDOMS

US ecologist Robert H. Whittaker proposed a system of classification in which organisms were categorized by cell structure, mode of nutrition, mode of reproduction, and body organization into five kingdoms: Monera (bacteria); Protista (mainly protozoa and algae); Fungi (moulds, yeasts, and mushrooms); Plantae (plants); and Animalia (animals).

▷ **Parrot toadstool,** a fungus

1969 US physicist Joseph Weber makes the first serious attempt to detect gravitational waves but does not succeed

1969

1969 ARPANET (the Advanced Research Projects Agency Network), the precursor to the internet, opens across the USA

"For one priceless moment in the whole history of man, all the people on this Earth are truly one..."

US PRESIDENT RICHARD NIXON IN CONVERSATION WITH ALDRIN AND ARMSTRONG, 1969

▽ **Buzz Aldrin** steps on to the Moon

1969

MOON LANDING

Neil Armstrong and Buzz Aldrin became the first people to walk on the surface of the Moon after their lunar module Eagle touched down in the Sea of Tranquility on 20 July, 1969. The Apollo 11 mission followed four crewed rehearsals and was the first of six successful lunar landings that continued until 1972.

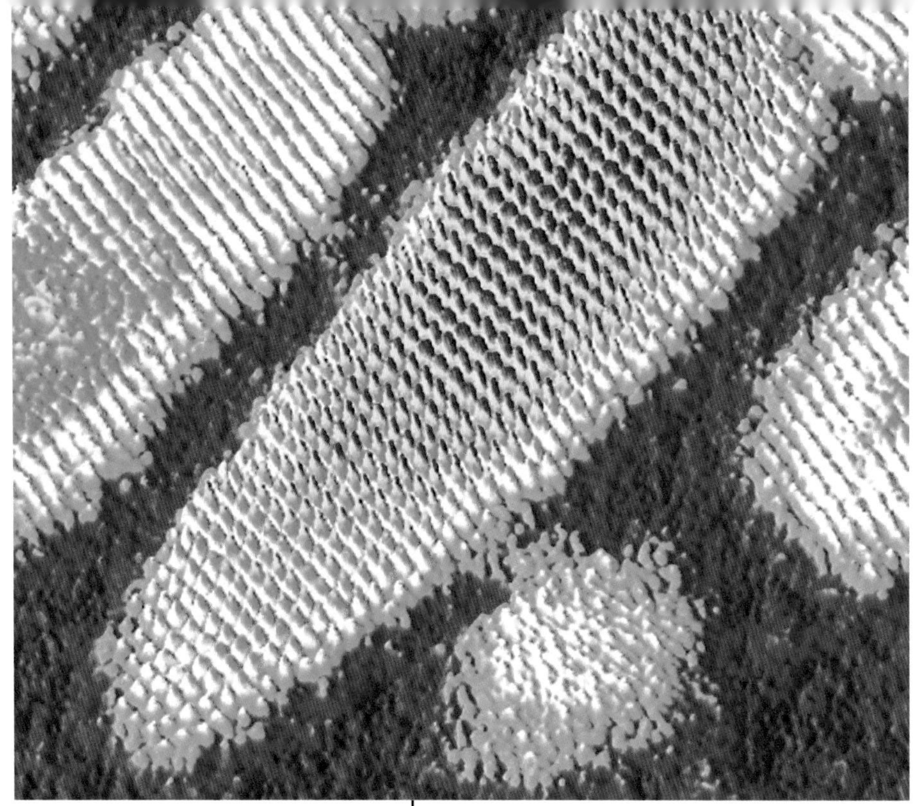

1970
SCANNING TRANSMISSION ELECTRON MICROSCOPE

A scanning transmission electron microscope (STEM) is a very high resolution instrument in which an electron beam passes through a thin sample, scanning back-and-forth as it does so. This allows it to produce images so detailed that individual atoms can be imaged. It was invented by British-US physicist Albert Crewe utilizing a high-precision electron gun he had invented in 1964.

◁ STEM image of a rod just 13 nm in length

1970

1970 Indian chemist Har Gobind Khorana completes the synthesis of a gene outside a living organism, showing that it can function in a bacterium

1971 A radio telescope 100 m (328 ft) across is completed at Effelsberg, Germany. It remains the world's largest steerable telescope for 29 years

1970 Uhuru, the first X-ray satellite telescope, is launched by an international team off the Kenyan coast

▽ Andromeda galaxy's spiral arms

1970
MISSING MASS

In their pioneering studies of galaxy rotation in the Andromeda Galaxy, US astronomers Vera Rubin and Kent Ford found that its outer reaches orbit faster than expected. This suggested that spiral galaxies contain several times more mass than the light from their visible parts would suggest, providing crucial evidence that much of the Universe is made of dark matter only detectable by its gravitational effect.

1928–2016
VERA COOPER RUBIN

Rubin was a pioneering American astronomer who found early evidence for the existence of galaxy clusters and superclusters. Known for her research into galaxy rotation, she was the second woman to be elected to the US National Academy of Science.

1971
MARTIAN SURFACE

The US Mariner 9 mission became the first space probe to orbit Mars. The probe arrived during a global dust storm but as the atmosphere cleared, it sent back images that revealed the Red Planet's dried-up riverbeds, canyons, and towering volcanoes for the first time.

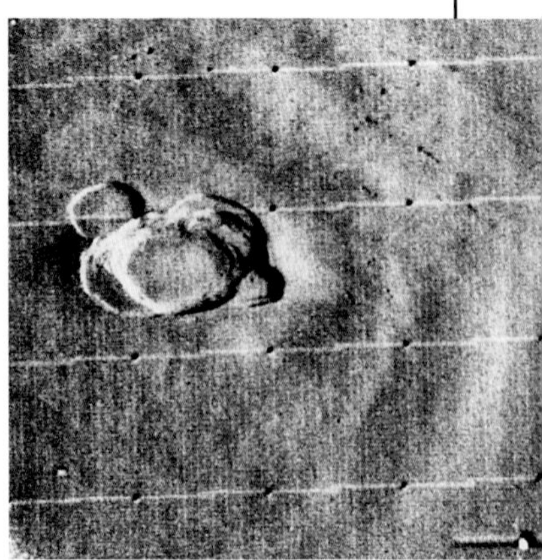

▷ **Mariner 9** view of Olympus Mons volcano on Mars

1972
HAWAII'S ORIGINS

US geologist Jason Morgan suggested that the Hawaii-Emperor chain of volcanic islands was created over time by the Pacific tectonic plate crossing a plume of heat deep in Earth's mantle. Lava repeatedly erupts on the seabed and builds into a volcano above sea level. The volcano then sinks as it cools, forming an island.

△ **Hawaiian islands** satellite image

▽ Land snail fossil

1972
PUNCTUATED EQUILIBRIUM

After conducting studies of species such as land snails, US palaeontologists Stephen Jay Gould and Niles Eldredge challenged the idea of gradual evolution. They suggested that most species change little for millions of years, but that this equilibrium can be "punctuated" by periods of rapid change in which new species emerge but leave little evidence in the fossil record.

1971 Astronomers discover Cygnus X-1, a binary star system containing a possible black hole

1972 US mathematician Edward Lorenz publishes a landmark paper in chaos theory that establishes the phrase "the butterfly effect"

1972

"Chaos: when the present determines the future, but the approximate present does not approximately determine the future."

SUMMARY OF CHAOS THEORY BY EDWARD LORENZ

CHAOS THEORY

The idea behind chaos theory is that systems that are very sensitive to initial conditions display seemingly (though not truly) random behaviour. Although it is difficult to predict how such systems will evolve, it is not impossible. Scientists draw on chaos theory in a bid to understand the patterns that underlie such behaviour. Chaos is seen in many natural systems, such as weather patterns and ocean turbulence, as well as in human systems like road traffic and stock markets.

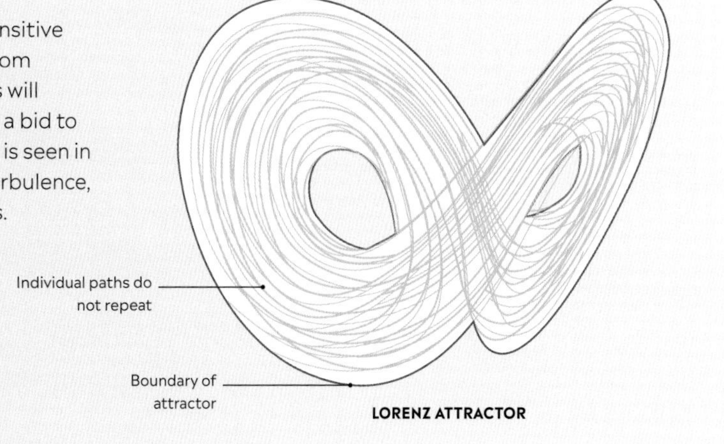

LORENZ ATTRACTOR

Interaction between variables

In modelling Earth's climate, US mathematician Edward Lorenz considered how three simple climate variables might interact. His Lorenz attractor demonstrates how a system with these variables keeps changing, never retracing its path.

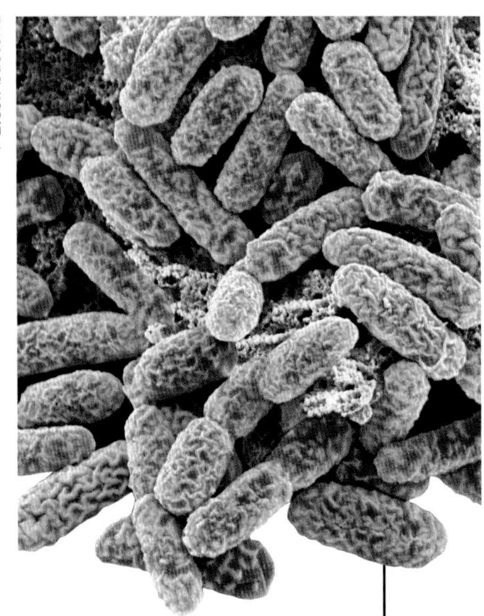
▽ *E. coli* bacteria

1973
GENE TRANSFER
US scientists Herbert W. Boyer and Stanley N. Cohen successfully transferred a plasmid (a strand of DNA in a cell that can replicate independently of the chromosomal DNA) to a bacterium, *Escherichia coli*, making it resistant to the antibiotic tetracycline. Their work showed that genetic material could be transferred between species and inspired further investigations, including research into the use of bacteria to transfer human genes.

1973
GRAND UNIFICATION THEORY
Developed by South Asian physicist Abdus Salam and Indian-US theoretical physicist Jogesh Pati, grand unification theory (GUT) is an attempt to simplify the interactions of subatomic particles by combining them into one "superforce". This force would act at very high energies like those that would have existed just after the Big Bang.

◁ Abdus Salam

1973

1973 NASA's Pioneer 10 becomes the first spacecraft to fly past Jupiter, and the first to chart a course out of our Solar System

1973 John Ostrom, a US palaeontologist, argues that birds are direct descendants of dinosaurs

1973 The Soviet Mars 5 probe is launched. After entering orbit, it measures the chemical content and temperature of the Martian surface

1973 Endorphins are discovered by British-German team John Hughes and Hans Kosterlitz, who call them enkephalins

GENETIC ENGINEERING
The process of genetic engineering (or genetic modification, GM) involves introducing a gene from the genome of one organism into the genome of another. In this way, crops can be altered to carry genes that help produce, for example, improved nutrients. GM can be controversial, with some people worried that changing genomes presents unknown risks. Scientists are developing ways of treating inherited disease with a type of genetic engineering called gene therapy.

Making insulin
Bacterial cells have been engineered to produce the insulin needed to treat patients with diabetes. The human insulin gene is inserted into the DNA of a bacterium.

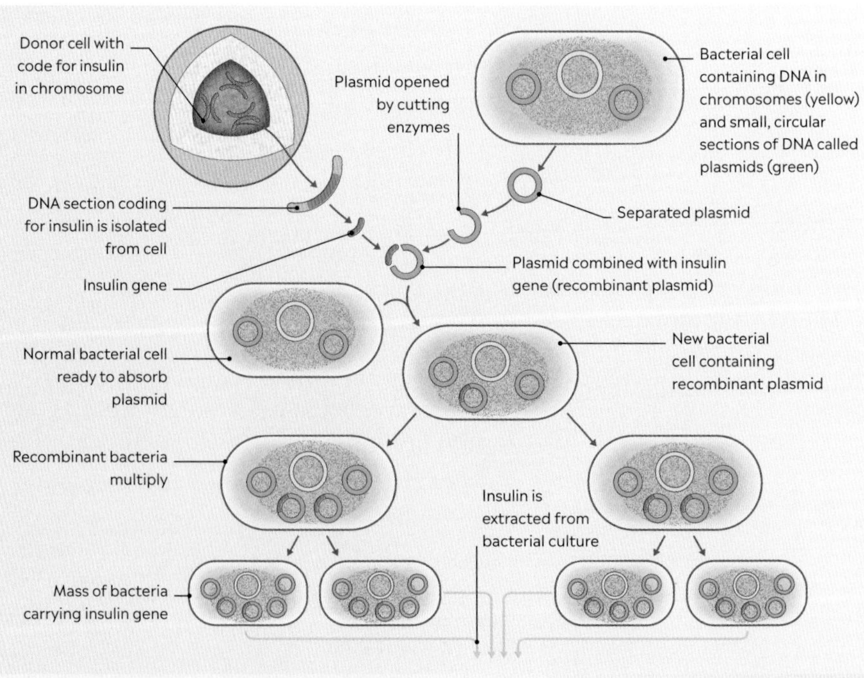

1973
SKYLAB

NASA's first space station, Skylab, was launched from the Kennedy Space Centre in Florida. Skylab was damaged during launch, but astronauts from its first crewed mission were able to make repairs. Although the USSR had launched its first short-lived space station, Salyut 1, in 1971, it did not successfully launch another until 1974, by which time the US had sent three crews to Skylab.

◁ **Skylab** in Earth orbit

Skylab was occupied by nine astronauts for 171 days

1973 US physicist Edward Tryon suggests that the Universe emerged from a quantum fluctuation (a temporary random change in the amount of energy in a point in space)

1974 US astronomers William K. Hartmann and Don Davies propose that the Moon formed after a giant impact smashed into Earth shortly after its formation

1974 US physicist Martin Perl discovers the tau lepton, an elementary particle

1974

1974
THE OZONE HOLE

US chemists Mario Molina and F. Sherwood Rowland were the first to calculate the risk posed to Earth's atmosphere by the release of industrial chlorofluorocarbons (CFCs), commonly used as refrigerants. CFCs allow chlorine to build up in the stratosphere, where it promotes the destruction of the ozone layer, the protective layer that prevents harmful ultraviolet radiation from sunlight damaging life on Earth's surface.

▷ **Polar ozone hole,** satellite image

◁ **The Mariner 10** spacecraft

1974
MISSION TO MERCURY

NASA's Mariner 10 probe returned close-up images of Mercury during its first fly-by encounter. In order to reach the fast-moving innermost planet, the spacecraft used Venus's gravity to make a "slingshot" manoeuvre, entering an orbit that intercepted Mercury every 177 days and enabled three separate encounters.

1975

MONOCLONAL ANTIBODIES

Argentinian biologist Cesar Milstein and German colleague Georges Koehler fused cells from an immortal myeloma cell-line with B-cells that produced a specific antibody. These hybrid cells (hybridomas) were then able to produce large quantities of identical antibodies, known as monoclonal antibodies.

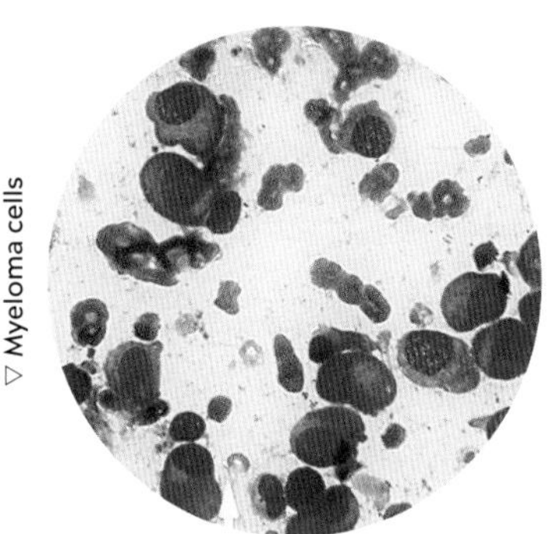

▷ Myeloma cells

Hydrothermal vent chimneys can grow up to 9 m (30 ft) in 18 months and reach heights of 60 m (200 ft)

1977

HYDROTHERMAL VENTS

US oceanographer Tjeerd van Andel was the first to observe a deep-sea hydrothermal vent rising from the Galapagos mid-ocean ridge in the Pacific. These vents are created when magma underlying the ridge heats metal-enriched seawater to 400 °C (750 °F) and vents it into near-freezing ocean water. The vents create a unique ecosystem, which is independent of sunlight.

▷ **Black smoker** hydrothermal vent

1975

1975 US biologist E.O. Wilson describes the biological basis of social behaviour, explaining how social and ecological conditions drive behavioural evolution

1975 US and Soviet astronauts meet in orbit during the Apollo-Soyuz mission

▷ **Viking 1** Mars lander

1976

MARS LANDINGS

NASA's Viking 1 and 2 missions arrived at Mars. Each comprised an orbiter that returned colour views of the planet from space, and a lander that descended to the surface and sent back images and data. The landers carried robot arms to take rock samples and test them for signs of life, as well as instruments for studying atmospheric conditions.

△ **Viking 2's** landing site in Utopia Planitia

THE STRUCTURE OF THE OCEANS

Detailed mapping of seabed topography reveals major features that result from the movement of tectonic plates (see p.191). At the margins of continents are shallow submerged shelves that slope steeply down to the deep ocean floor, which is covered with layers of sediment. A mid-ocean ridge marks the original joint between continents; here, volcanic rocks erupt on the seafloor and spread apart on either side of the ridge to form new oceanic crust.

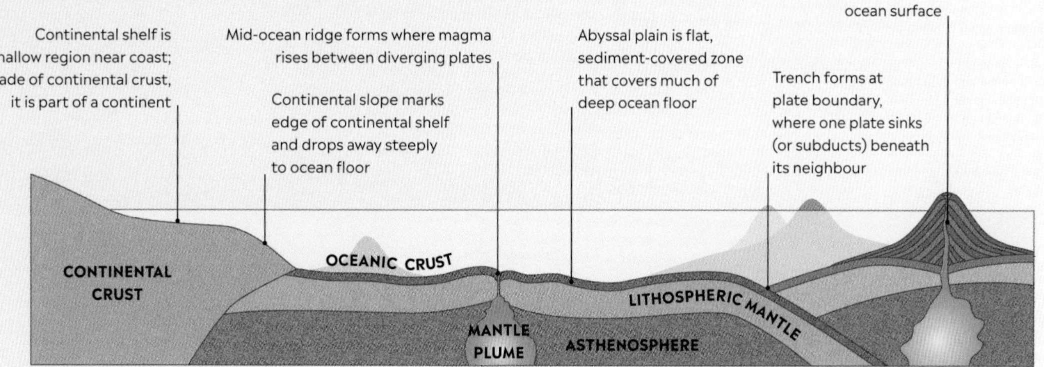
Volcanic island forms where undersea volcano grows tall enough to rise above ocean surface
Continental shelf is shallow region near coast; made of continental crust, it is part of a continent
Mid-ocean ridge forms where magma rises between diverging plates
Abyssal plain is flat, sediment-covered zone that covers much of deep ocean floor
Trench forms at plate boundary, where one plate sinks (or subducts) beneath its neighbour
Continental slope marks edge of continental shelf and drops away steeply to ocean floor
CONTINENTAL CRUST
OCEANIC CRUST
LITHOSPHERIC MANTLE
MANTLE PLUME
ASTHENOSPHERE

Tectonics and the seabed
The ocean floor is made up of volcanic crust. Its formation is the result of millions of years of tectonic movement, spreading out from a mid-ocean ridge.

1977 British biochemist Richard Roberts and US geneticist Phillip Sharp independently discover that genes are broken up into segments and scattered along chromosomes

1977 The disease smallpox is eradicated following a global campaign of immunization that began in 1967

1977

1977 British and US teams develop different methods of sequencing the nucleotides in strands of DNA

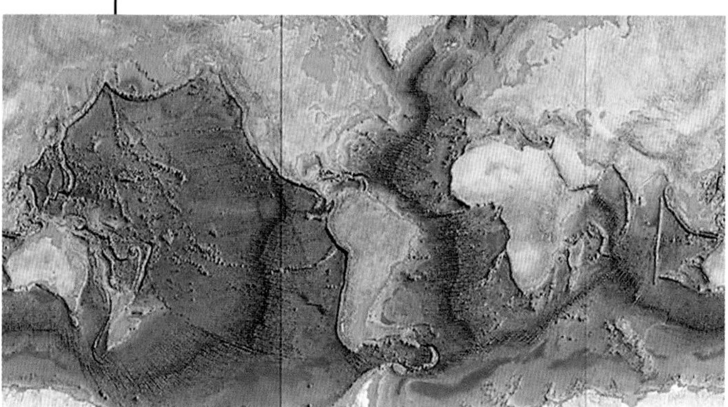

△ Heezen-Tharp World ocean floor map

1977

MAPPING THE OCEAN FLOOR

US geologist Marie Tharp used data from US geoscientist Bruce Heezen to create the first map of the world's ocean floor. It revealed the main features associated with plate tectonics, such as the mid-ocean spreading ridge, submarine volcanoes, and subduction zones.

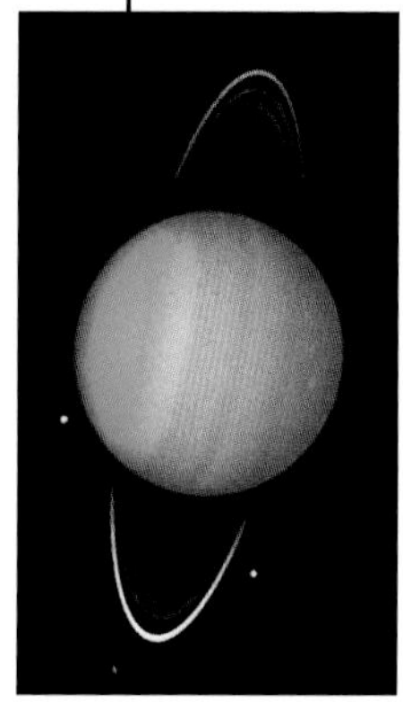
△ **Uranus** and its rings

1977

RINGS OF URANUS

US astronomer James Elliot and colleagues discovered rings around the planet Uranus. Intending to measure changes to light from a distant star as it passed behind the planet itself, they noticed that the star disappeared briefly several times on either side of Uranus. Eventually they identified a system of nine rings, much darker and narrower than those of Saturn; four more have since been discovered.

△ **Princeton** large Torus reactor

1978

1978
FUSION REACTOR
The Princeton Large Torus (PLT) was an early tokamak - a doughnut-shaped chamber able to contain plasma (a gas made of ions and free electrons) at the extremely high temperatures necessary for nuclear fusion reactions to occur. In July, the PLT held plasma at above 60 million °C (108 million °F) - a major milestone in fusion research.

1978
DINOSAUR SOCIAL BEHAVIOUR
The discovery of a nest of 15 hadrosaurian ("duck-billed") skeletons in Montana suggested to US paleontologists John Horner and Robert Makela that young dinosaurs were cared for by their parents. Like birds today, the *Maiasaura* dinosaurs they studied nested communally, and probably returned to the same nesting sites year after year.

▽ **Hatching dinosaur egg** fossil

1978 US astronomer James Christy discovers Charon, the largest satellite of the dwarf planet Pluto

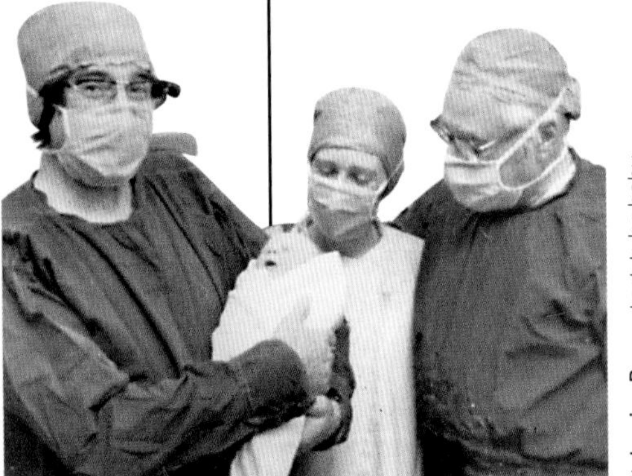

△ **Louise Brown** test-tube baby

1978
TEST-TUBE BABY
Louise Joy Brown, the first human to be conceived using the technique of in vitro fertilization (IVF), was born. In IVF, a method developed by British scientists, egg and sperm cells are united outside the body and the early embryo is then inserted into the mother's womb to continue development to term.

1978
PIONEERS AT VENUS
Two NASA Pioneer missions arrived at Venus. The Pioneer Venus Orbiter carried an array of instruments, including a radar to map the surface beneath the planet's thick clouds. The Pioneer Venus Multiprobe released four separate probes that descended into the atmosphere to send back data on the searing, high-pressure conditions.

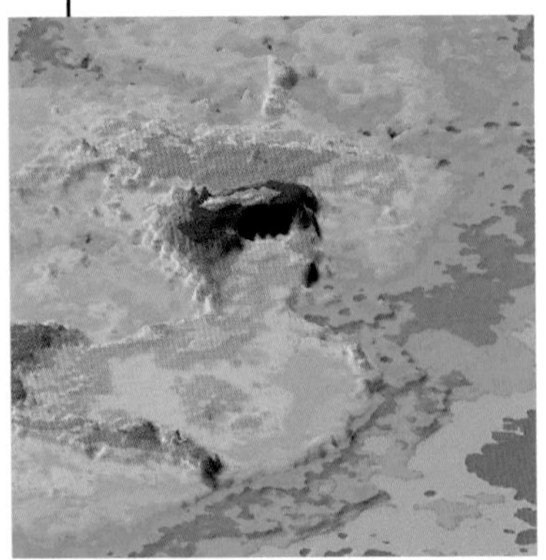

△ **A radar map** of Venus's Ishtar Terra plateau

1979
METEOR IMPACT

While analysing seabed clay from the end of the Cretaceous period (when dinosaurs died out), US scientist Louis Alvarez and his son Walter discovered a spike in the concentration of iridium. As iridium is normally introduced to Earth from space in very small amounts, they theorized that Earth must have been hit by a large extra-terrestrial asteroid, and that this was responsible for the extinction of the dinosaurs. The vast Chicxulub crater was later identified as the impact site of that asteroid.

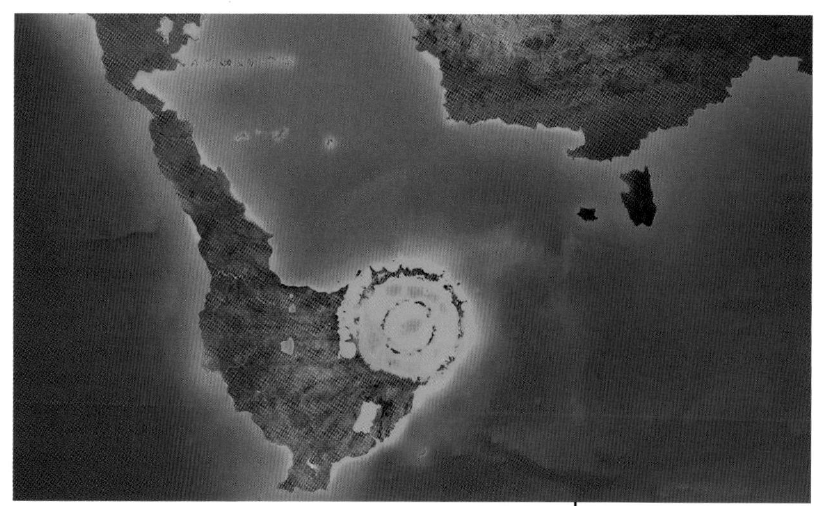

▷ **Chicxulub crater** soon after impact, artist's impression

The Chicxulub impact crater was discovered in the Yucatan Peninsula, Mexico in 1990

1979

1979 NASA's Voyager 1 and 2 spacecraft fly past Jupiter, imaging the giant planet and its moons

1979 Evidence for gluons (the particles that hold together quarks) is found in experiments at the DESY laboratory in Germany

1979 The partial meltdown of one of the reactors at the Three Mile Island nuclear plant raises concerns about nuclear power

MASS EXTINCTIONS

Although species continually appear and disappear, sometimes a large number is lost suddenly in a mass extinction. The disappearance of the dinosaurs 65 million years ago is one example, but there have been five mass extinctions over the past 500 million years, due to asteroid impact, volcanic eruption, or change in sea level. A sharp recent rise in extinction rates suggests that a sixth mass extinction is currently underway.

KEY TO GEOLOGICAL PERIODS

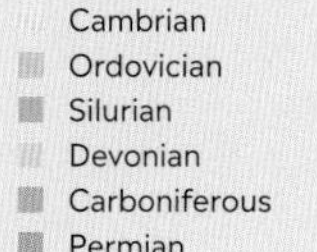

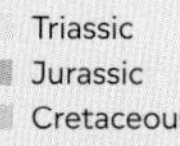

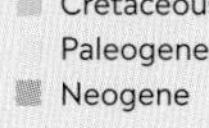

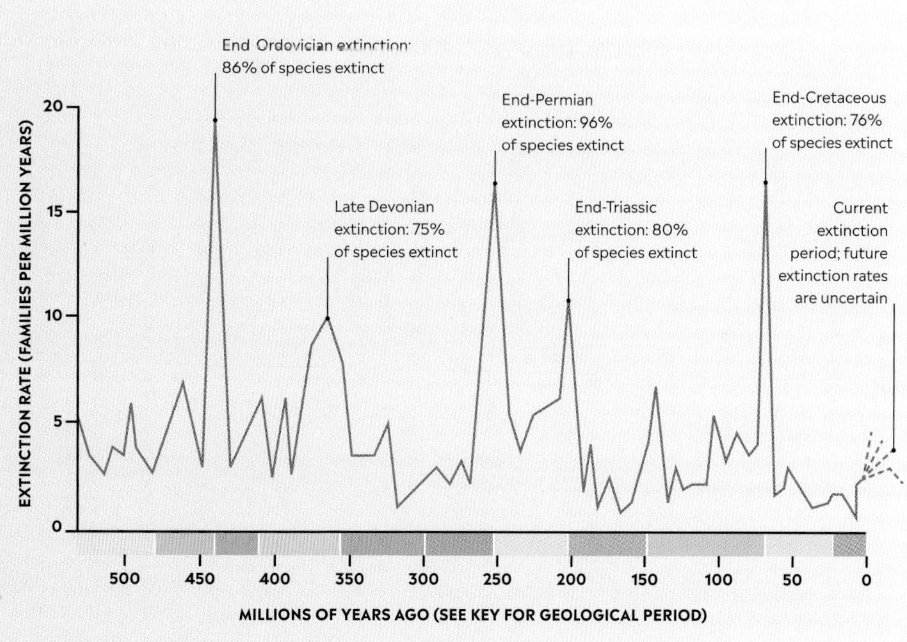

▷ **Titan**, a moon of Saturn

1980
SATURN'S MOONS PHOTOGRAPHED

Voyager 1 returned close-up pictures of Saturn's moons a year after Pioneer 11 first encountered the ringed planet itself. The probe imaged several icy moons and Titan's thick atmosphere before deflecting to a course towards interstellar space. Voyager 2 followed on a different trajectory nine months later.

1980
QUANTUM BEHAVIOUR

The Hall effect is a phenomenon in which a magnetic field affects the trajectory of electrons through a conductor. While experimenting with field-effect transistors at very low temperatures, German physicist Klaus von Klitzing discovered a quantum version of the effect that could be used to store data in quantum computers.

▷ **Klaus von Klitzing**

1980

1980 Tim Berners-Lee, an English computer scientist, begins work on ENQUIRE, a precursor to the World Wide Web

1981 US company IBM releases the IBM Personal Computer – the blueprint for most personal computers for many years

△ **VLA dishes** in the Socorro desert

1980
VERY LARGE ARRAY

The Very Large Array (VLA) radio telescope entered operation near Socorro, New Mexico, in the US. The instrument consists of 28 steerable radio dishes, each 25 m (82 ft) in diameter, mounted on Y-shaped rail tracks, each arm of which is 21 km (13 miles) long. Signals collected by the dishes can be combined by a technique called interferometry, enabling the VLA to see the radio Universe at a level of detail equivalent to powerful visible-light telescopes.

"By means of the shuttle, we will be able to build space stations and power stations, laboratory facilities, and habitations, and everything else in space."

ISAAC ASIMOV IN AN INTERVIEW, 1979

1981

SPACE SHUTTLE

NASA's space shuttle blasted off on its first mission. The revolutionary vehicle consisted of a plane-like orbiter with a large external fuel tank, assisted at launch by a pair of booster rockets. The shuttle could be used for satellite launches and repairs as well as acting as a temporary space station.

◁ **Space shuttle *Columbia*** lifts off

1981

1981 US theoretical physicist Andrei Guth publishes his theory of cosmic inflation – the idea that the early Universe expanded exponentially fast for a fraction of a second after the Big Bang

1981 A new disease with a range of symptoms is reported in California, USA. It is later named acquired immune deficiency syndrome (AIDS)

1981 US molecular biologist George Streisinger successfully clones a vertebrate animal – a zebrafish

1981

MAGNETIC RESONANCE IMAGING

British scientists built a full-body magnetic resonance imaging (MRI) scanner and used it on a patient. MRI uses a strong magnetic field to align the hydrogen atoms in the body. Radio waves at different frequencies are used to dislodge these protons, and the vibrations they give can be measured and mapped. 3D images of organs inside the body, especially the brain, are created and can be used to detect changes over time.

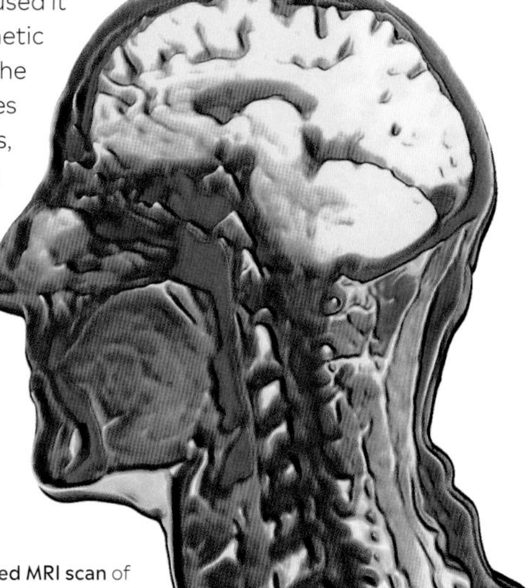

▷ **Coloured MRI scan** of the human head

△ **Foraminiferan shell** SEM

1981

PAST CLIMATES

The CLIMAP project analysed seafloor sediments that contained the chalky remains of organisms such as foraminiferans. CLIMAP scientists could deduce past ocean conditions and map water temperatures that prevailed during the last glacial maximum, between 31,000 and 16,000 years ago.

1982
VENUS PROBES

The Soviet Venera 13 and 14 missions arrived at Venus. Each consisted of a carrier craft that released a lander, which parachuted to the surface through the hostile Venusian atmosphere. The landers survived for only 1–2 hours, in which time they returned sounds and the first colour images from the surface. They also analysed the rocks at their landing sites, confirming that they are igneous (volcanic in nature).

◁ **Venera 13 lander,** model

1982

1982 US biochemists transfer a rat growth gene into mouse eggs in the first successful gene transfer between mammal species

1982 Spanish physicist Blas Cabrera detects a magnet with just one pole – a phenomenon predicted by the grand unified theory

1982
PRIONS

US biochemist Stanley Prusiner isolated a protein from brain cells of people suffering from the degenerative Creutzfeldt-Jakob disease. He suspected the protein to be the agent of infection and named it a prion. The protein was also present in healthy people but was folded differently; Prusiner found that prions transmit their misfolded shape onto normal types of the same protein.

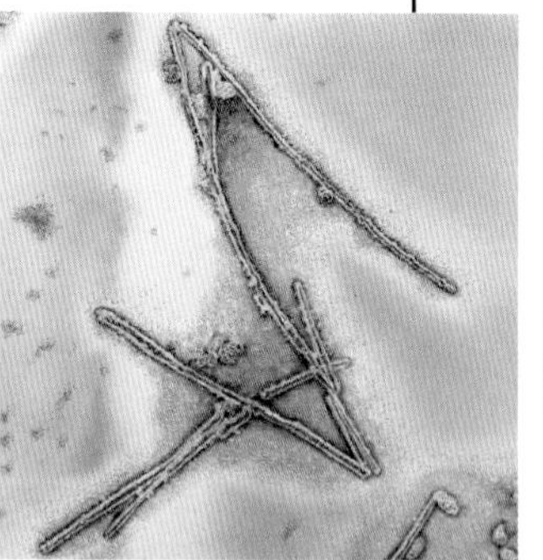

△ **Fibrils composed of prions**

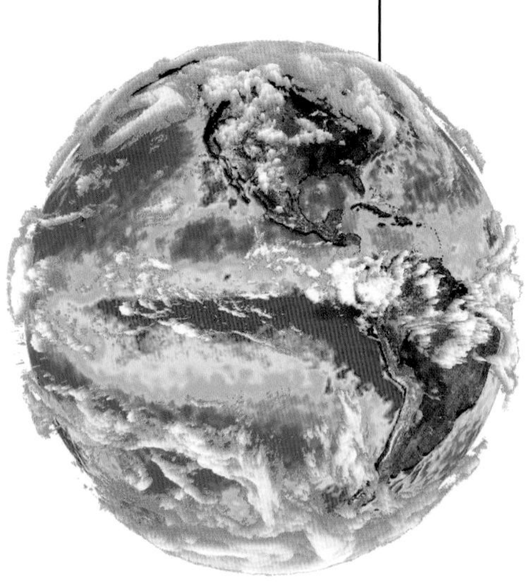

△ **Earth's surface temperature** in El Niño year

1982
EL NIÑO EVENT

From May 1982 to June 1983, the planet experienced one of the most extreme El Niño events on record. These climatic events (named for "the boy" in Spanish, as they often peak around Christmas, the traditional birth date of Jesus), occur every few years and involve the warming of seas in the tropical eastern Pacific. The prolonged 1982–83 event caused droughts and fires in Australia, Africa, and Indonesia, while Peru suffered torrential rainfall.

1983

INFRARED ASTRONOMY

The launch of the US/Dutch/British Infrared Astronomy Satellite (IRAS) heralded a new type of astronomy. The mission carried a telescope chilled to a temperature of -271°C (-456°F), allowing it to capture weak infrared radiation. During 300 days of observation, IRAS detected objects and materials too faint to emit visible light, such as cool dust within galaxies and star-forming nebulae, and planet-forming discs around stars.

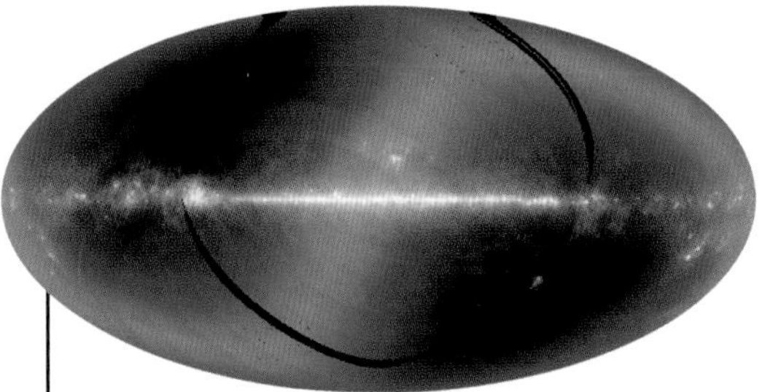

△ **IRAS image** of whole sky, centred on the Milky Way

1983

CAUSE OF AIDS

French virologist Luc Montagnier and colleagues examined samples from a patient suffering from acquired immune deficiency syndrome (AIDS) and found a virus in the lymph nodes, naming it lymphadenopathy-associated virus (LAV). They were not sure that the virus was responsible for causing AIDS; this was confirmed by US biochemist Robert Gallo, and the virus became known as human immunodeficiency virus (HIV).

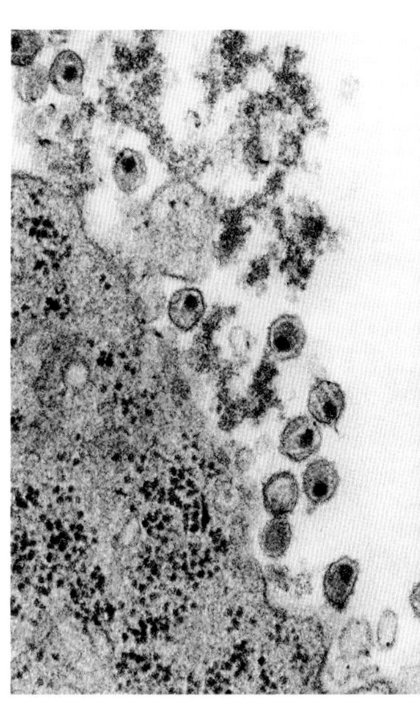

▷ **HIV,** electron micrograph

1983 The Motorola DynaTAC 8000X, released in Chicago, US, is the first commercially available cellphone

1983 US geneticist Kary Mullins discovers the polymerase chain reaction, a method of creating millions of copies of a tiny sample of DNA, revolutionizing biotechnology

1983

1983

NUCLEAR WINTER

US scientist Carl Sagan calculated the effects of a nuclear war on the global atmosphere. He projected that the resulting firestorms would send hundreds of millions of tonnes of smoke and soot high into the atmosphere where they would block sunlight. His prediction was that the resulting low light levels and freezing temperatures would destroy plant life, leading to global famine and mass casualties.

△ **Missile alert system facility,** Wyoming, USA

"We have placed our civilization and our species in jeopardy."

CARL SAGAN, PARADE *MAGAZINE,* 1983

THE STANDARD MODEL OF COSMOLOGY

Measuring the balance
The Planck satellite measured the energy content of today's Universe as 68.5 per cent dark energy, 26.6 per cent dark matter, and just 4.9 per cent visible matter.

Most cosmologists (experts in the formation and evolution of the cosmos) today believe that the Universe can be described by a standard model they call lambda-CDM. This is a small set of assumptions added to the Big Bang theory to create a modern-day Universe with the properties astronomers have observed.

Lambda (the Greek letter Λ) is a cosmological constant that makes the expansion of space accelerate over time – the mysterious phenomenon known as dark energy. CDM (cold dark matter), meanwhile, is the "missing mass" of the cosmos, which influences visible objects through its gravity but is itself both completely dark and transparent to light and other radiations.

Dark energy and cold dark matter seem to explain a number of observations that cannot otherwise be accounted for by the Big Bang theory alone. For example, dark matter of some kind is needed to explain how knots of increased density had already developed even before the Universe became transparent, as shown by ripples in the cosmic microwave background radiation. In addition, the coldness of dark matter explains observations that galaxies started out small and grew through collisions and mergers. Hot dark matter, by contrast, would initially produce larger objects on the scale of galaxy superclusters that would then gradually separate into smaller fragments.

The cosmological constant lambda explains why distant exploding stars appear fainter and more remote than would be expected if cosmic expansion was steady or slowing down (*see below*).

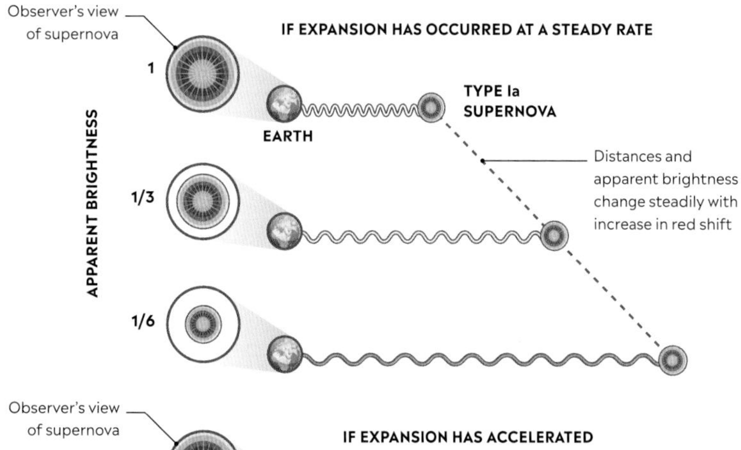
Observer's view of supernova
IF EXPANSION HAS OCCURRED AT A STEADY RATE
APPARENT BRIGHTNESS
1
1/3
1/6
EARTH
TYPE Ia SUPERNOVA
Distances and apparent brightness change steadily with increase in red shift

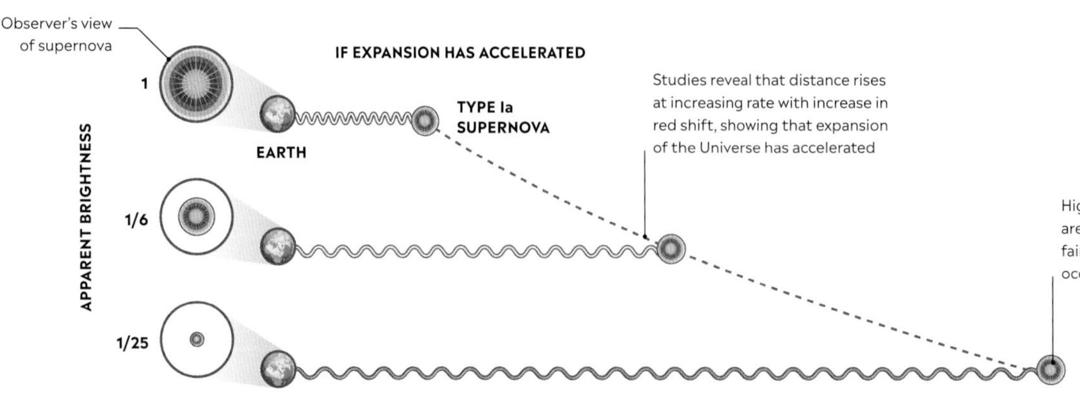
Observer's view of supernova
IF EXPANSION HAS ACCELERATED
APPARENT BRIGHTNESS
1
1/6
1/25
EARTH
TYPE Ia SUPERNOVA
Studies reveal that distance rises at increasing rate with increase in red shift, showing that expansion of the Universe has accelerated
Highest red-shift supernovae are more distant and so fainter than if expansion had occurred at steady rate

DETECTING DARK ENERGY

The key evidence for dark energy comes from Type Ia supernovae – explosions triggered when a white dwarf star collapses to become a neutron star, which reliably releases the same amount of energy and can be detected over vast cosmic distances. The brightness of these supernovae directly reveals their distance from Earth and, because the speed of light is limited, their age. This allows astronomers to identify supernova host galaxies at a variety of distances and time periods, as well as to measure how the red shift of light from distant galaxies (caused by the expansion of space) has changed over the lifetime of the Universe.

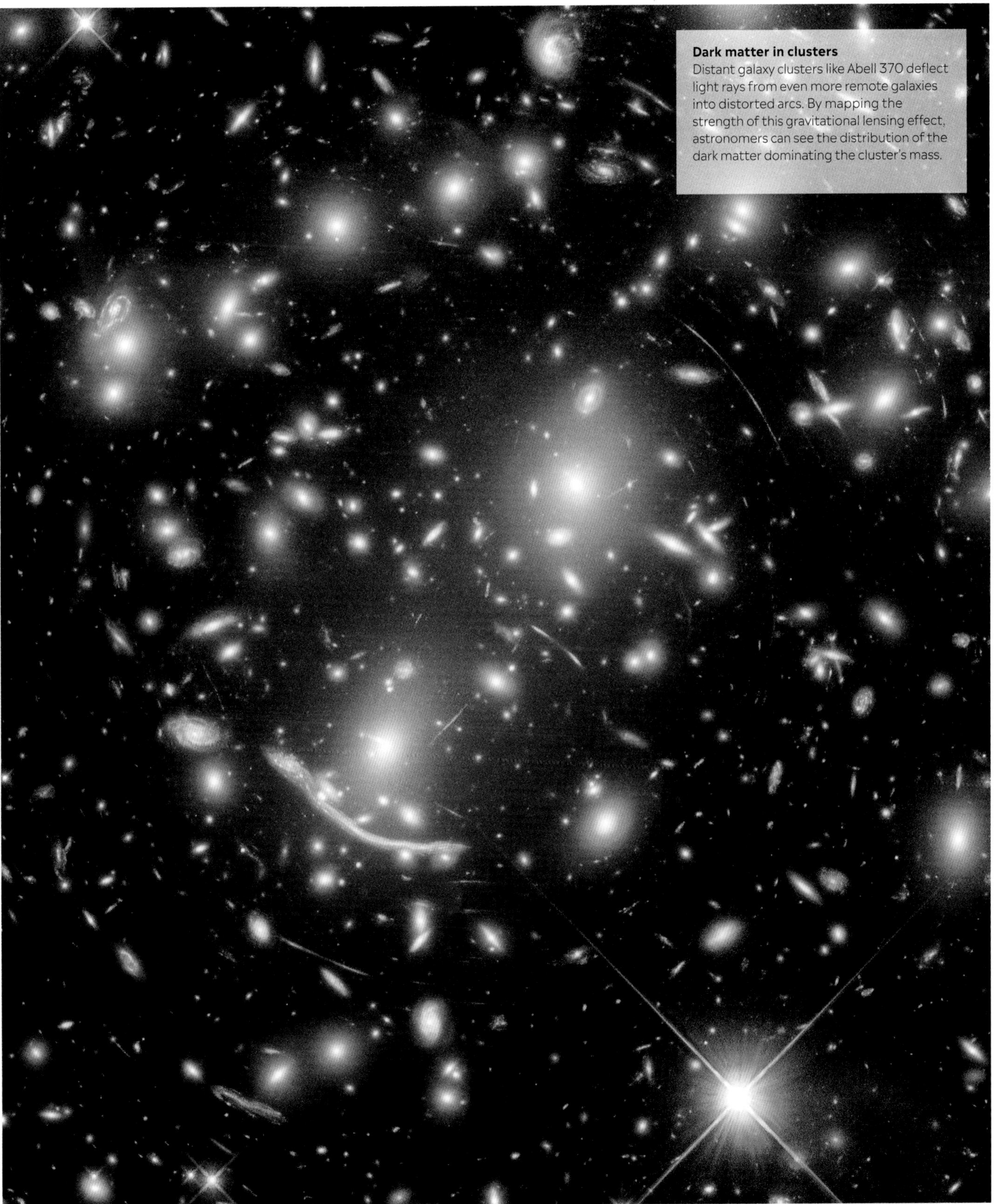

Dark matter in clusters
Distant galaxy clusters like Abell 370 deflect light rays from even more remote galaxies into distorted arcs. By mapping the strength of this gravitational lensing effect, astronomers can see the distribution of the dark matter dominating the cluster's mass.

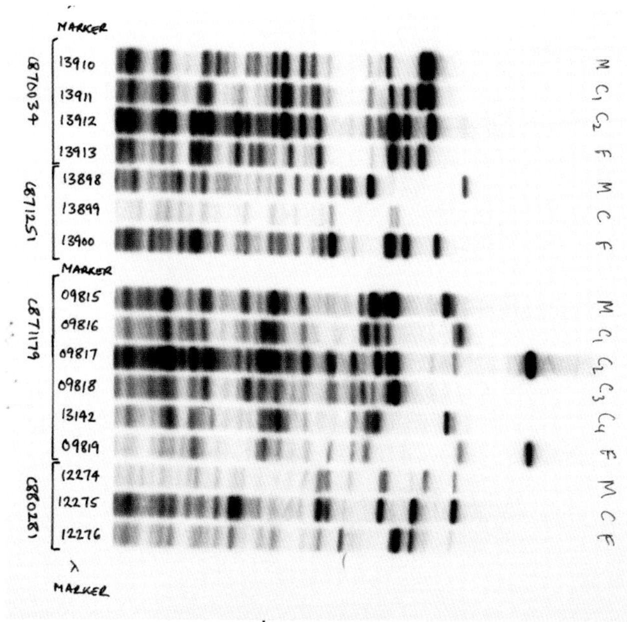

The odds that two people will have the same 13-locus DNA profile is about one in a billion

1984

GENETIC FINGERPRINTING

British geneticist Alec Jeffries was able to detect variations in DNA sequences that were unique to each individual and so produce a telling "fingerprint" by separating the sequences from one another using electric fields to "pull" them through a gel medium. The technique could identify an individual from traces of blood, hair, or saliva, or detect family relatedness between two samples.

◁ **DNA bands** in a fingerprinting test

1984

1984 US zoologist Katy Payne discovers that elephants communicate over long distances using low-frequency infrasound

1984 The first human baby is born from a previously frozen embryo

1984

DNA RELATIONS

US biologists Charles Gald Sibley and Jon Edward Ahlquist presented results of their studies of the DNA of monkeys and apes, which used a method known as DNA hybridization to shows the percentage similarity between species. Their study indicated that the order of divergence from oldest to most recent was gibbons, orangutans, gorillas, chimpanzees, and then humans.

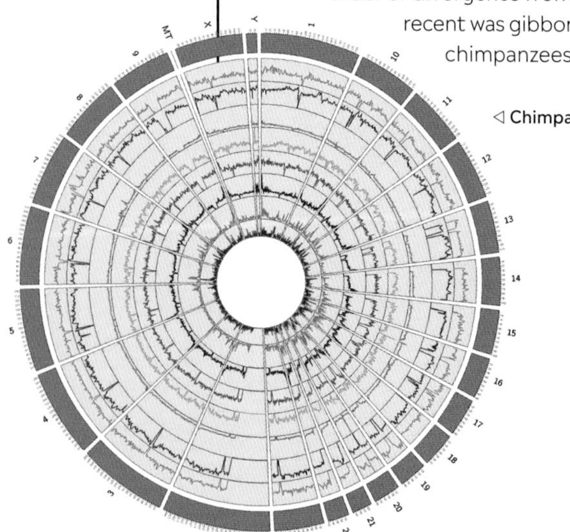

◁ **Chimpanzee** genome map

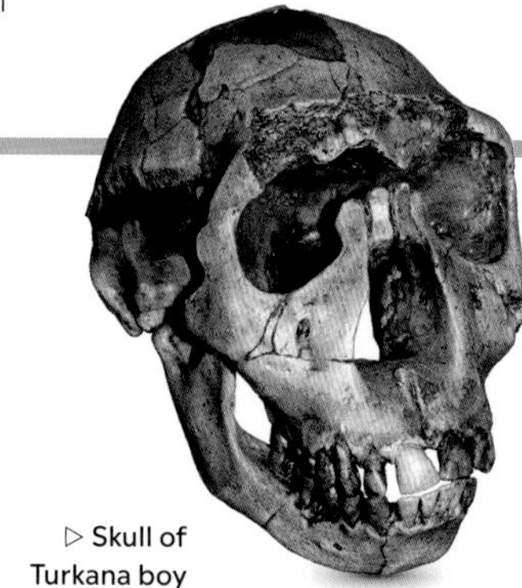

▷ **Skull of Turkana boy**

1984

TURKANA BOY

A team working with Kenyan anthropologist Richard Leakey discovered a fossil skeleton near Lake Turkana in Kenya. This specimen of *Homo erectus* – the first *Homo* species thought to have spread widely in Africa and through Asia – was dated at about 1.5 million years old. The fossil, which became known as "Turkana boy", was remarkable for its integrity, which allowed for new insights into the anatomy and biology of this human ancestor.

NANOTECHNOLOGY

The science and engineering of creating and utilizing matter at an atomic and molecular level is known as nanotechnology. This means working at sizes of up to 100 nanometres (nm). By way of comparison, the width of a human hair is 25,000 nm. At the nanoscale, materials exhibit different and useful properties when in the form of tubes, wires, and particles. Nanotechnology is used to create catalysts, improve batteries, and make fabrics stain-resistant.

Nanotubes and nanoparticles
Carbon exists in multiple forms at the nanoscale, including hollow balls and tubes, as well as sheets (graphene). Silicon wires at this size are solid and are being developed for use in new types of battery.

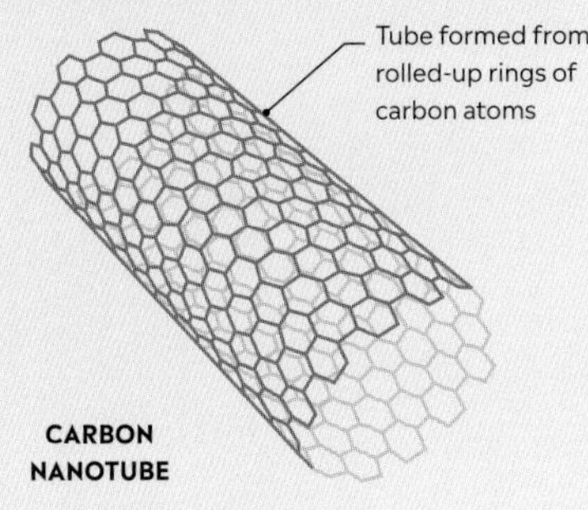
Tube formed from rolled-up rings of carbon atoms
CARBON NANOTUBE

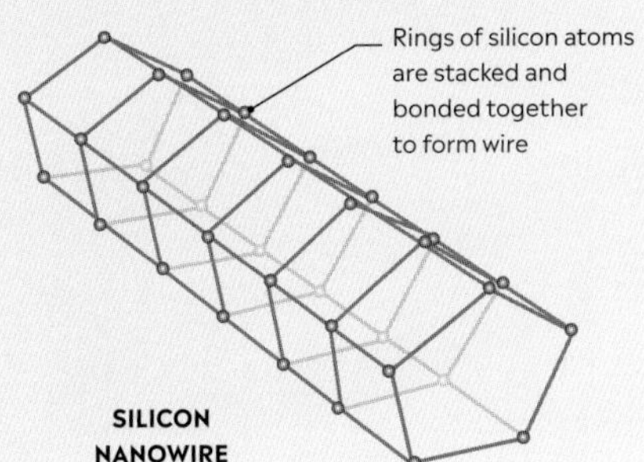
Rings of silicon atoms are stacked and bonded together to form wire
SILICON NANOWIRE

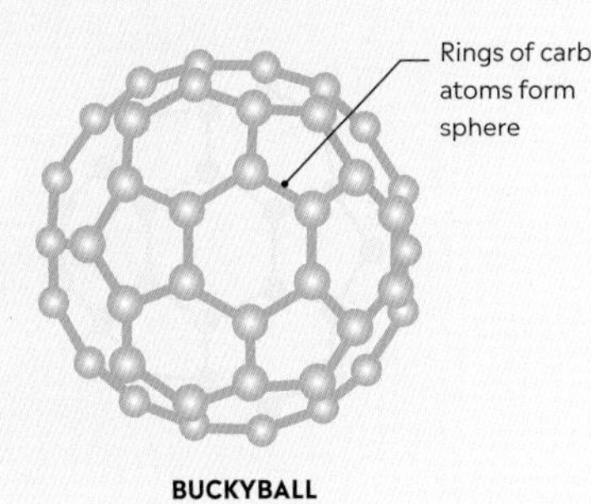
Rings of carbon atoms form sphere
BUCKYBALL

1985

1985 The NSFNET – a network of supercomputers that is a precursor of the internet – is established

1985 Eclipses of Pluto and its moon Charon allow measurement of their dimensions

1985 The Joint Oceanographic Institutions for Deep Earth Sampling (JOIDES) project finds no ocean floor rocks older than 200 million years

△ Buckyball C60 model

1985

BUCKMINSTERFULLERENES

British chemist Harold Kroto and his US colleagues Richard Smalley and Robert Curl discovered a new form of pure carbon made of hollow molecules of 60 atoms. The chemists used a laser to vapourize graphite and detected the allotrope buckminsterfullerene (also known as buckyballs): the molecule is named after the geodesic domes designed by US architect Richard Buckminster Fuller, which they resemble.

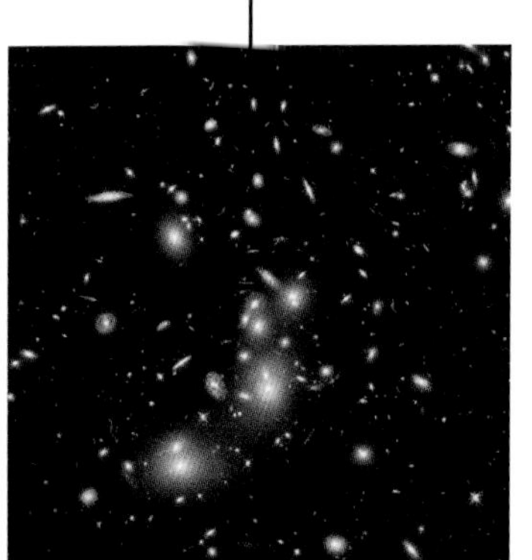
△ The crowded galaxy cluster Abell 2744

1985

ABELL CLUSTERS

The first surveys of large-scale galaxy distribution revealed that galaxies tend to concentrate in matter-rich filaments around seemingly empty voids. South African cosmologist Neil Turok suggested that the distribution of groups of galaxies known as Abell Clusters is consistent with the theory of cosmic strings – vast hypothetical structures created in the Big Bang that could serve as seeds for the concentration of matter to form stars and galaxies.

△ **Chernobyl nuclear power plant** after the disaster

1986

CHERNOBYL MELTDOWN

The worst nuclear accident in history occurred during a safety test at the Chernobyl nuclear power plant near the city of Pripyat, northern Ukraine, then part of the Soviet Union. The accident, which caused a meltdown in Reactor Number 4, and serious fire throughout the plant, was partly due to design flaws – and partly to human error. Tens of thousands of people were evacuated, and radioactive pollution was carried across Europe by prevailing winds.

1986

1986 Voyager 2 flies past Uranus, returning close-up images of the planet and its rings and largest moons

1986 The USSR begins assembly of Mir, the first space station made from multiple linked modules

1986 The US federal government establishes a framework for regulating the use and safety of genetically modified organisms (GMOs) and biotechnology

1986 US scientists harness the ability of the plant pathogen *Agrobacterium tumefaciens* to transfer genes, thereby extending the potential of genetic engineering

1986

SUPERCONDUCTING CERAMICS

Superconductivity – zero resistance to the flow of electric current – was discovered in 1911. Until 1986, it had only ever been observed in metallic materials held at very low temperatures, close to absolute zero. However, in 1986, German physicist Georg Bednorz and Swiss physicist Alex Müller discovered a new class of ceramic materials based on copper oxide, in which superconductivity was achieved at 35 °C (63 °F) above absolute zero, opening new applications for superconductors.

▷ **Ceramic superconductor** photographed under polarized light

1987

SUPERNOVA 1987A

The eruption of a supernova in the Large Magellanic Cloud (a satellite galaxy of the Milky Way) was the brightest such event since the invention of the telescope. Astronomers employed a huge variety of instruments to study the eruption over the following months and learned much about the death of massive stars.

1987

CONDOR CONSERVATION

The California condor (*Gymnogyps californianus*), one of the world's largest flying birds, once bred from British Columbia south to Baja California, Mexico. By the 1980s, populations had crashed to the brink of extinction with only around 20 wild individuals remaining. By 1987, the last of these birds had been captured, and a programme of captive breeding was initiated. The programme was successful, with the first birds being released to the wild in 1992, since when numbers have steadily increased.

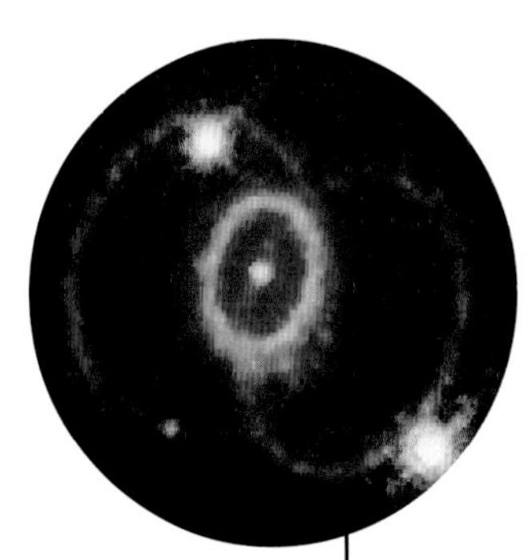

△ **Rings of debris** around Supernova 1987A

△ **California condor** in flight

1987

1987 A bacterium designed to help strawberry plants withstand frost is the first genetically modified organism (GMO) to be released into the wild

1987 An international treaty signed in Montreal, Canada, aims to restrict the release of gases that destroy atmospheric ozone

△ **Challenger** explodes after lift-off

1986

CHALLENGER DISASTER

The space shuttle *Challenger* was destroyed in an explosion 73 seconds after launch, killing all seven astronauts on board. A subsequent inquiry found that seals on one of the shuttle's booster rockets failed due to cold weather, releasing a jet of hot gas with catastrophic effects on the rest of the spacecraft. Shuttle launches were delayed for 32 months to address safety issues.

> "For a successful technology, reality must take precedence over public relations, for nature cannot be fooled."
>
> *RICHARD FEYNMAN, MEMBER OF THE COMMISSION ON THE* CHALLENGER *DISASTER, 1986*

1988

HUMAN GENOME PROJECT

The Human Genome Project (HGP) was begun with James D. Watson (*see p.234*) as Director. Its aim is to map the human genome and determine the sequence of all its 3.2 billion "letters". The HGP provides the possibility of reading the complete genetic blueprint for a human, transforming many aspects of biological science.

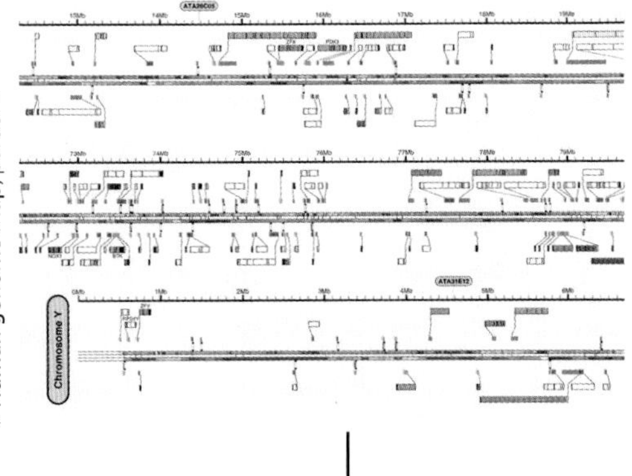

▷ Human genome map, partial Y

1989

GALILEO PROBE

NASA launched its Galileo spacecraft to Jupiter by releasing it from the cargo hold of the space shuttle *Atlantis*. After a complex journey of more than six years, the probe entered orbit around the giant planet in December 1995. It returned images and data about Jupiter and its moons for almost eight years.

▷ **Galileo** is pulled from *Atlantis*'s hold

1988

1988 The World Meteorological Organization and the UN Environment Programme establish an Intergovernmental Panel on Climate Change (IPCC)

1988 British astronomer Simon Lilly discovers the oldest known galaxy, light from which has travelled 12 billion years to reach Earth

RISING TEMPERATURES

The modern way of measuring climate change combines Earth's average surface temperature with the surface temperature of the ocean. This system has been in use since the late 1800s, when the rise of industrialization led to the release of climate-warming greenhouse gases. Estimates differ slightly between organizations, but all show an increase in global surface temperature, with the average for 2006–15 being around 0.87°C (1.57°F) above the average for the second half of the 19th century.

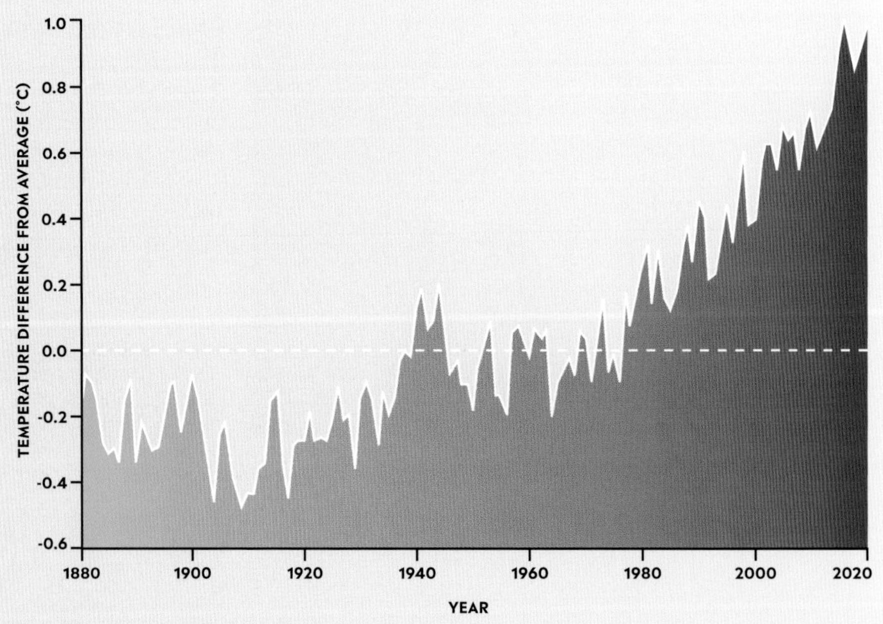

Recording change

Starting in the late 19th century, this chart compares global surface temperatures against the long-term average and illustrates the sharp increase that began in the 1980s.

"The Web does not just connect machines, it connects people."

TIM BERNERS-LEE, SPEECH BEFORE KNIGHT FOUNDATION, US, 2008

1989

THE WORLD WIDE WEB

British computer scientist Tim Berners-Lee developed the World Wide Web while working at the high-energy physics research centre CERN. In 1989, he wrote a paper proposing the use of an existing technology – hypertext – to provide clickable links between documents on the CERN computer network. The system was first put into action at CERN in 1990 and released into the public domain in 1993.

▷ **The first World Wide Web** server

1989 Carbon dioxide concentration in Earth's atmosphere reaches 353 parts per million

1989 US astronomers fire radio signals at Saturn's largest moon, Titan, revealing a reflective surface covered in ice and liquid hydrocarbons

1989 Voyager 2 flies past Neptune, discovering icy geysers on its giant moon Triton

1989

1989

THE BURGESS SHALE

The American palaeontologist Stephen J. Gould published an account of the unique assemblage of fossils first discovered in 1909 in deposits known as the Burgess Shale, in Canada's Rocky Mountains. About 65,000 specimens of more than 120 species were recovered, dating to about 508 million years ago, in the middle Cambrian Period. Many were seemingly unrelated to any known groups.

△ ***Waptia***, a Cambrian arthropod from the Burgess Shale

1989

HOPES FOR COLD FUSION

Hopes of harnessing nuclear fusion on Earth rest on devices that can create and contain the extremely high temperatures needed for the process to occur. However, in 1989, British chemist Martin Fleischmann and US electrochemist Stanley Pons reported that a tabletop electrolysis experiment had generated energy, and had produced neutrons. However, their claims to have achieved "cold fusion" – nuclear reactions at room temperature – could not be replicated.

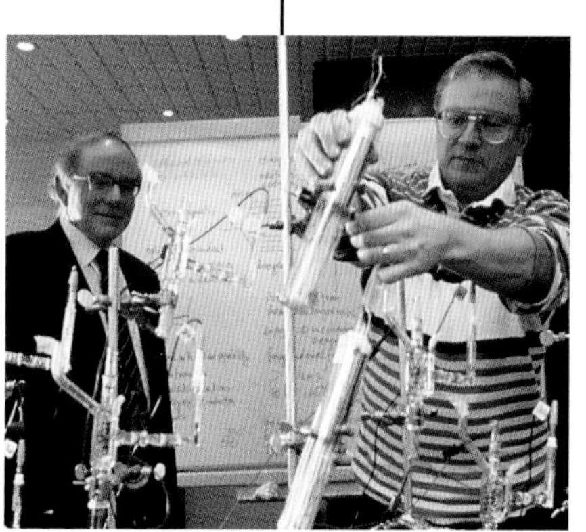

△ **Professors** Fleischmann (l) and Pons (r)

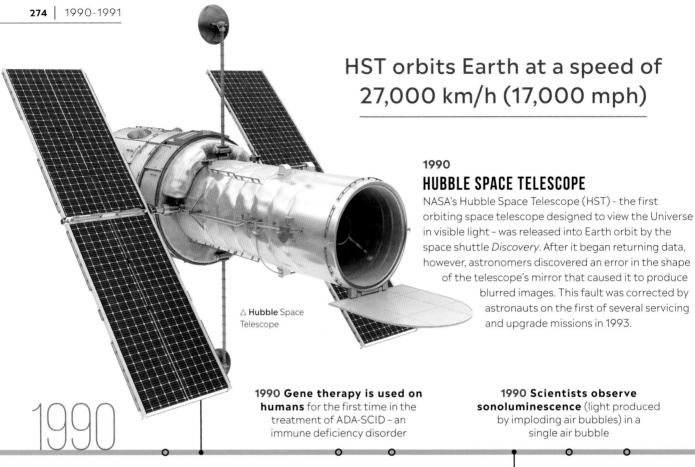

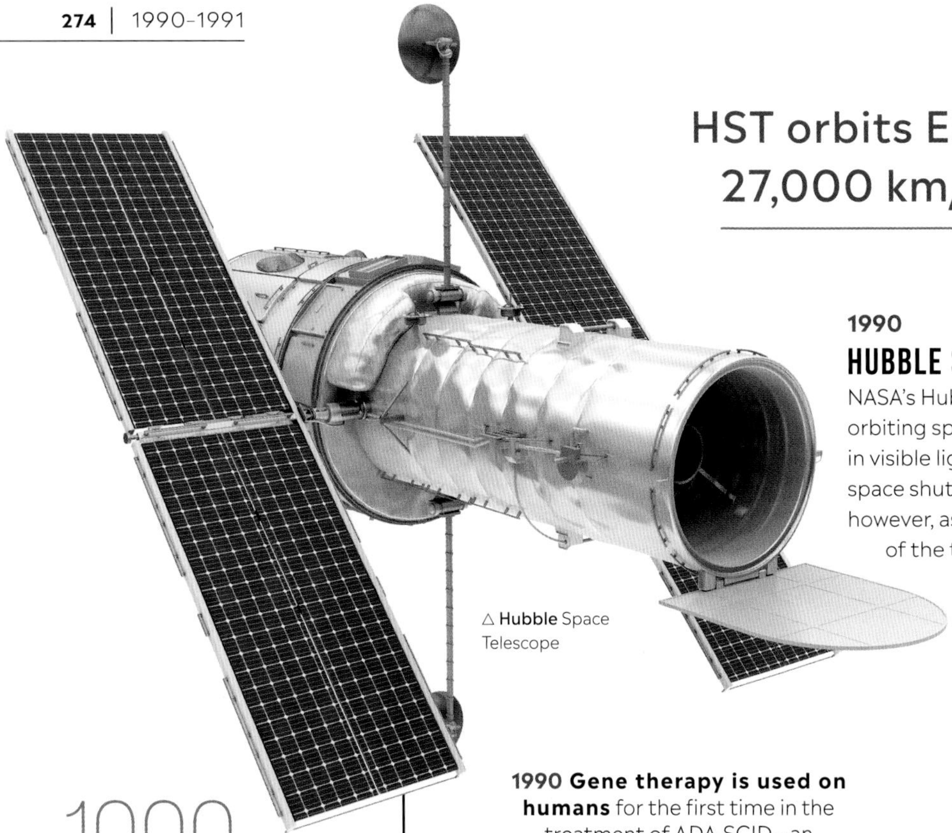

HST orbits Earth at a speed of 27,000 km/h (17,000 mph)

1990

HUBBLE SPACE TELESCOPE

NASA's Hubble Space Telescope (HST) – the first orbiting space telescope designed to view the Universe in visible light – was released into Earth orbit by the space shuttle *Discovery*. After it began returning data, however, astronomers discovered an error in the shape of the telescope's mirror that caused it to produce blurred images. This fault was corrected by astronauts on the first of several servicing and upgrade missions in 1993.

△ **Hubble** Space Telescope

1990

1990 Freshwater geothermal vents discovered in Lake Baikal suggest the crust is spreading to form a new ocean

1990 Gene therapy is used on humans for the first time in the treatment of ADA-SCID – an immune deficiency disorder

1990 US geneticist Mary-Clare King reports that a gene, BRCA1, is responsible for a high inherited risk of developing breast cancer

1990 Scientists observe sonoluminescence (light produced by imploding air bubbles) in a single air bubble

1990 British scientists Robert Winston and Alan Handyside screen embyros for genetic disease before implantation in IVF

b.1946

MARY-CLARE KING

King is best known for her work on gene mutations in breast cancer. She also discovered that gene regulation is primarily responsible for the differences between humans and chimpanzees, which share around 99 per cent of their genes.

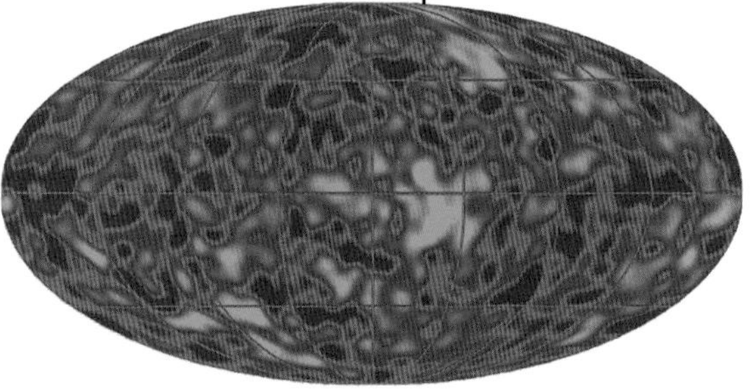

△ **COBE** map of background radiation

1990

COSMIC BACKGROUND EXPLORER

NASA's Cosmic Background Explorer (COBE) satellite began to collect measurements of cosmic microwave background radiation (CMBR) – the afterglow of the Big Bang that created the Universe. From 1990–92, it mapped slight variations in the temperature of the CMBR. These confirmed that clumps of matter of different densities were present in the Universe from the earliest times, rather than developing in the billions of years since, and confirmed the Big Bang theory.

1991

MOLECULAR MECHANICS

The British-US scientist Fraser Stoddart made key breakthroughs in the development of molecule-scale assemblies capable of acting as mechanical switches. These assemblies – one class of which are called rotaxanes – consist of ring-shaped molecules threaded onto molecular "axles". They are able to move in response to external factors such as light or heat. These developments in nanotechnology have varied application, such as in drug-delivery systems and sensor technology.

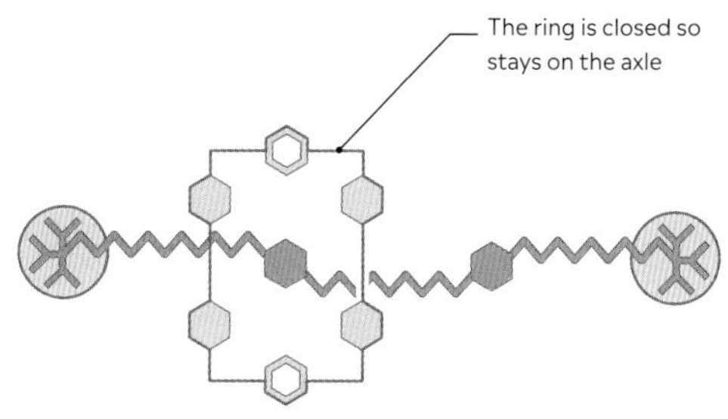
The ring is closed so stays on the axle

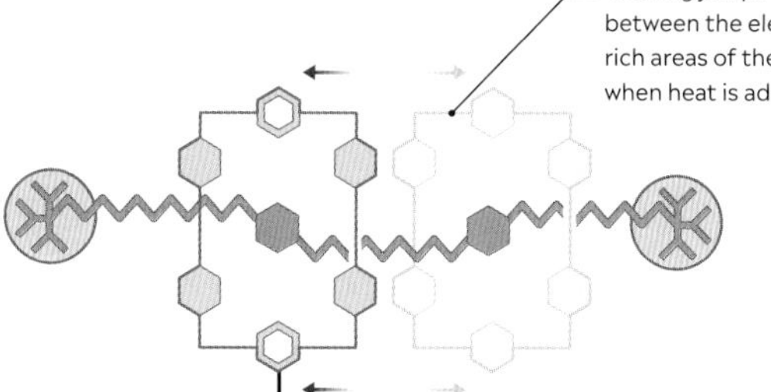
The ring jumps between the electron-rich areas of the axle when heat is added

Ring and axle
A molecular ring slides along a thin axle; dumbbell-shaped ends hold the ring in place. The ring moves between two electron-rich areas on the axle when stimulated by light, acids, solvents, or ions.

1991 Astronomers find evidence that galaxy NGC 6240 may contain one or more enormous black holes in its unusually bright core

1991 Astronomers detect variations in the signal from a pulsar caused by the orbiting objects – evidence for planets beyond the Solar System

1991

1991 The Jupiter-bound Galileo craft makes the first of two asteroid flybys, delivering close-up views of the asteroid Gaspra

1991

NANOTUBES

Carbon nanotubes were discovered by Japanese scientist Sumio Iijima. These tubes are molecules up to 1000 times longer than they are wide, with diameters as small as one nanometer. Carbon nanotubes are stronger and lighter than steel, and can conduct heat and electricity exceptionally well. They have been used in semiconductors and in light-emitting diodes.

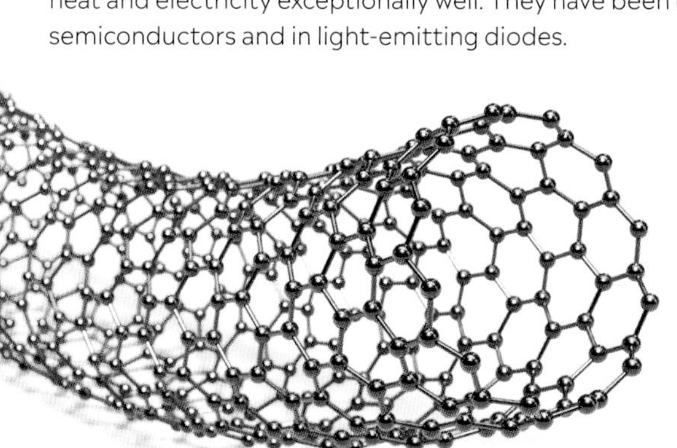

△ **Carbon** nanotube

△ **Mt Pinatubo** erupting

1991

MOUNT PINATUBO ERUPTS

The 1991 eruption of Mount Pinatubo in the Phillippines was the second largest volcanic eruption of the 20th century. Following a major earthquake, magma rose some 20 km (12½ miles) through Earth's crust to explode and blast around 5 cubic km (1 cubic mile) of ash, pumice, and gas some 35 km (22 miles) into the air, with the ash travelling around the world.

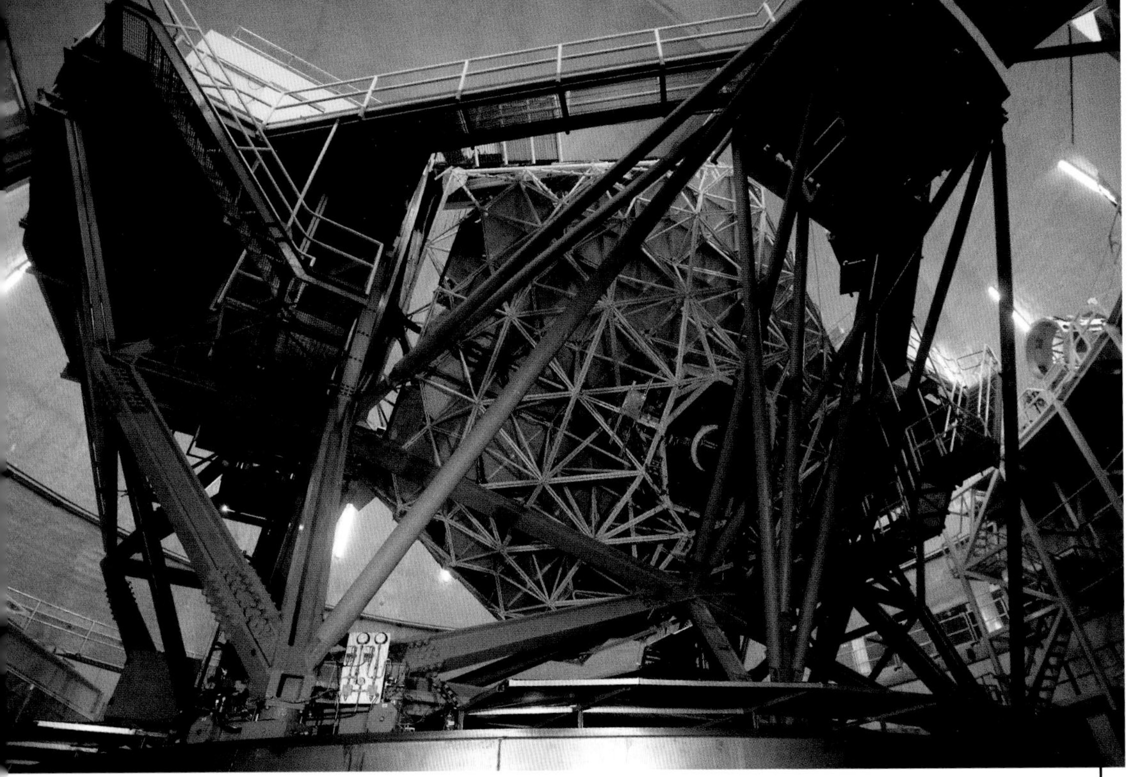

1993
KECK TELESCOPES
The 10-m (33-ft) Keck I telescope on Mauna Kea, Hawaii is completed, becoming the world's largest. Its enormous main mirror is built from hexagonal segments, kept in alignment as a perfect reflecting surface by computer-driven motors. Keck I is joined by its twin Keck II in 1996; working together, the two telescopes are able to detect details equivalent to a single instrument with a diameter of 85 m (279 ft).

◁ **Keck telescope** in its cradle

1992

1992 A United Nations Conference, the UNCED Earth Summit in Rio, Brazil, calls for action on sustainable development and protection of the environment

1992 The US/French TOPEX/Poseidon satellite is launched on a mission to monitor the oceans

1992 US astrophysicist Thomas Gold suggests that microbial life is widespread in Earth's crust down to depths of several km (miles)

▷ *Armillaria bulbosa* fruiting bodies

1992
LARGEST LIVING ORGANISM
Genetic analysis of around 250 samples of the fungus *Armillaria bulbosa* taken over an area of 39 hectares (91 acres) indicated that it was a single organism. Discovered in a forest in Michigan, US, the fungus is estimated to weigh 400 tonnes (440 tons) and to have lived for 2,500 years.

▷ *Mononykus* therapod dinosaur

1993
BIRD ANCESTOR
US palaeontologists described the remains of a new kind of small bird-like dinosaur found in 75-million-year-old Late Cretaceous strata in Mongolia. Named *Mononykus*, the dinosaur stood around 1 m (3 ft) tall, and had short, strong arms ending in a single claw, and a keeled breast bone. It was thought to be a primitive flightless bird, which demonstrated that different kinds of birds evolved earlier than once thought. Later finds suggest that it was a small feathered theropod dinosaur.

1994

COMET SHOEMAKER-LEVY

Fragments from Comet Shoemaker-Levy 9 crashed into the planet Jupiter, creating huge fireballs and forming temporary stains in its cloud layers. Astronomers used the impact to study material dredged from Jupiter's interior. The impact also confirmed the huge gravitational influence wielded by Jupiter – the comet, discovered in 1993, had been torn into fragments up to 2 km (1.2 miles) wide during a previous close encounter.

▷ **A chain of comet fragments** heads towards Jupiter

1994

FLUORESCENT ROUNDWORMS

US biochemist Martin Chalfie and colleagues transferred the gene Gfp for green fluorescent protein into nematode worms. The team first inserted the DNA sequence for the gene into a bacterium and then the roundworm. Using ultraviolet light, illumination of the fluorescent gene aids the study of gene expression in the nematode.

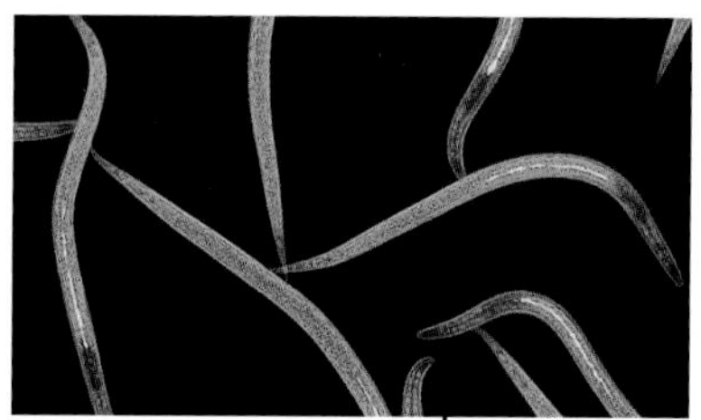

△ **Fluorescing** nematodes

1994

1993 US astronomer Douglas Lin proposes that an extensive halo of "dark matter" extends beyond the visible limits of the Milky Way

Fewer than 100 specimens of the Wollemi Pine exist in the wild

1994

LIVING FOSSIL

Australian wildlife officer David Noble discovered some unusual pine trees in a remote canyon 200 km (120 miles) northwest of Sydney. Identified as evergreen conifers called *Wollemia*, they were thought to be extinct for more than two million years until the few living specimens were found. The specimens are thought to be between 500 and 1,000 years old.

△ ***Wollemia*** pine

1995

GALILEO PROBES JUPITER'S ATMOSPHERE

NASA's Galileo craft reached Jupiter after a six-year journey and released a probe into the giant planet's atmosphere. Protected from the extreme heat by a conical shield, the probe descended into the clouds on a parachute. It returned data for almost an hour, analysing the atmosphere's chemical composition and measuring wind speeds of around 2,900 km/h (1,800 mph).

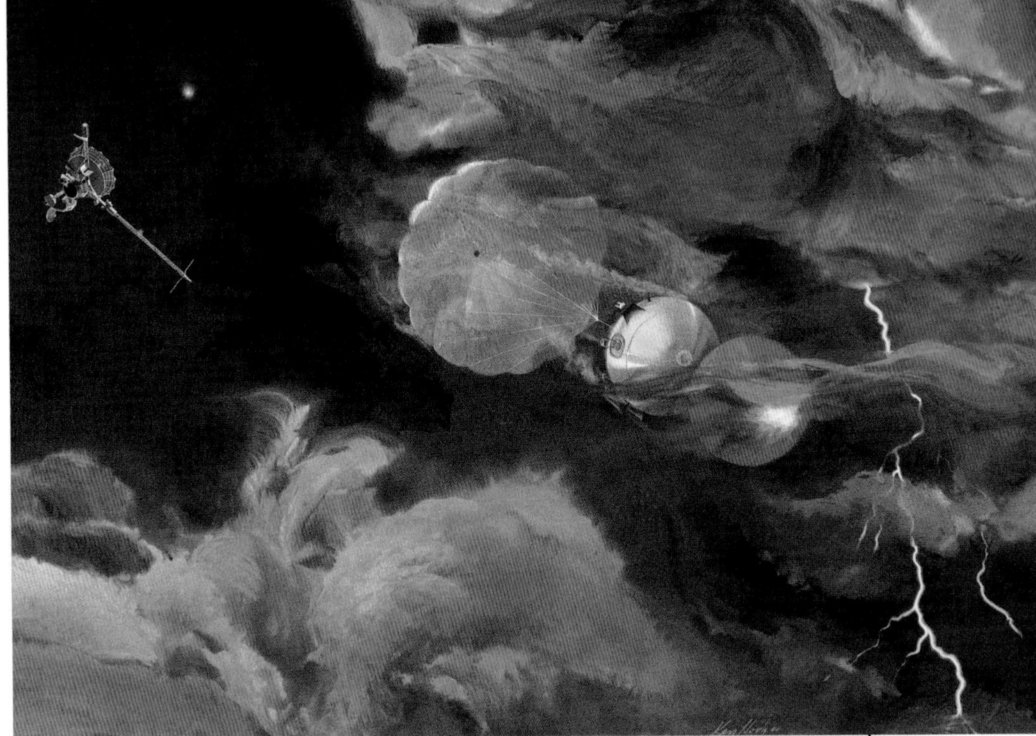

△ **The Galileo** atmospheric probe

1995

1995 One of the largest dinosaurs, a 12.5-m (41-ft) predator named *Giganotosaurus*, is described by Argentinian palaeontologists

1995 Danish biologists discover a small organism, *Symbion pandora*, attached to a lobster's mouth. It belongs to a entirely new phylum of animals called the Cycliophora

1995 Swiss astronomers Michel Mayor and Didier Queloz discover 51 Pegasi b, the first known exoplanet, roughly half the size of Jupiter, orbiting a sun-like star

▽ **Rubidium atoms** (left) condense into a superatom (middle) and evaporate (right)

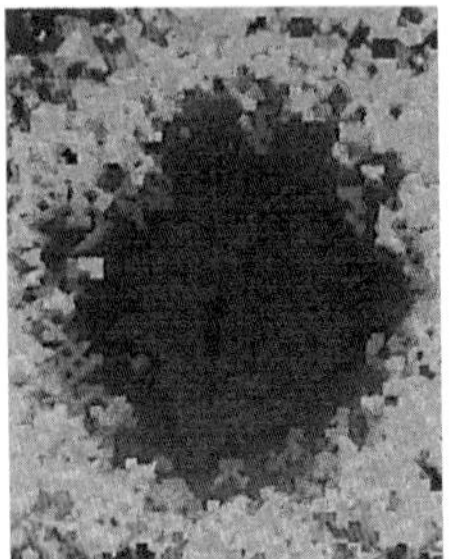

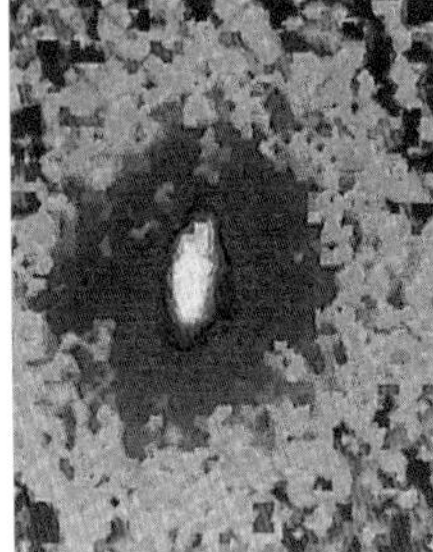

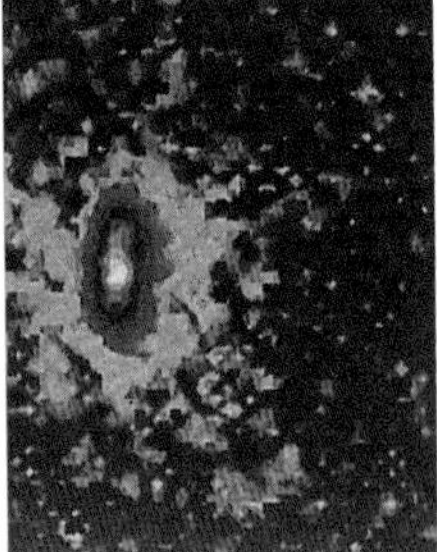

1995

A FIFTH STATE OF MATTER

In 1995, American physicists Eric Cornell and Carl Wieman cooled a small sample of rubidium atoms very close to absolute zero, inducing them to take the form of a "superatom", which behaved more like waves than particles. This new state of matter, known as Bose-Einstein condensate, was predicted in the 1920s and 30s by Indian physicist Satyendra Bose and Albert Einstein.

1995

TOP QUARK OBSERVED

Quarks are elementary particles, the existence of which was first predicted in the 1960s. Six quarks (in three pairs) were predicted in all, and quarks were indeed observed in high-energy particle collisions from 1968 onwards. The last quark to be discovered, the "top" or "truth" quark, is the most massive, and hardest to detect. Physicists at Fermi National Accelerator Laboratory (Fermilab), in Illinois, US, used the Tevatron – a powerful particle accelerator – to produce top quarks in very high-energy collisions of protons and antiprotons.

▷ **Top quark** evidence

1995

SOLAR OBSERVATION

NASA launched the European-built Solar and Heliospheric Observatory (SOHO) mission into an orbit where it could constantly monitor the Sun. Equipped with visible and ultraviolet cameras as well as particle detectors for measuring the solar wind, SOHO returned stunning images of the solar surface, offering new insights into the Sun's structure.

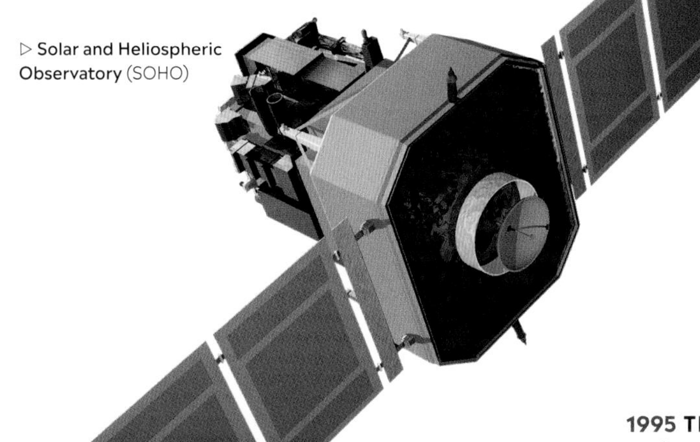

▷ **Solar and Heliospheric Observatory** (SOHO)

b. 1946

JOHN CRAIG VENTER

Venter is among the pioneers of biotechnology in the US. He founded The Institute for Genomic Research and the company Celera Genomics, which led the way in decoding the genomes of many organisms, from bacteria to humans.

1995 The whole genome of an organism - the bacterium *Haemophilus influenzae* - is sequenced at The Institute for Genomic Research, Maryland, US

1995

Data from SOHO helps to predict "space weather" events that affect our planet

EXOPLANETS

NASA estimates that there are at least as many planets in the Milky Way galaxy as there are stars. Astronomers have identified more than 5,000 exoplanets - that is, planets that orbit stars other than the Sun; these include entire planetary systems that are similar in scale to our own Solar System. Several distinct classes of exoplanet have been established, based on properties of the planets themselves, their orbits around their parent stars, and comparison with our Solar System. Some of these - for example, planets orbiting in regions where they cannot have originally formed - reveal the complex ways that planetary systems can evolve over time.

Hot Jupiters
These gas-giant planets orbit very close to their stars, often with extremely hot, expanded atmospheres.

Chthonian planets
These are the solid, exposed cores of hot Jupiters whose outer atmospheres have been blown away into space.

Mega-Earths
These rocky planets are at least ten times Earth's mass. Their conditions vary depending on distance from their star.

Super-Earths
Planets with less than ten times the mass of Earth are unlikely to be gas or ice giants, so they are known as super-Earths.

Water worlds
Some exoplanets have the right conditions for deep-water oceans, but these cannot yet be directly detected.

Exo-Earths
Around 22 per cent of Milky Way stars are thought to have planets of Earth-like mass with conditions that may support life.

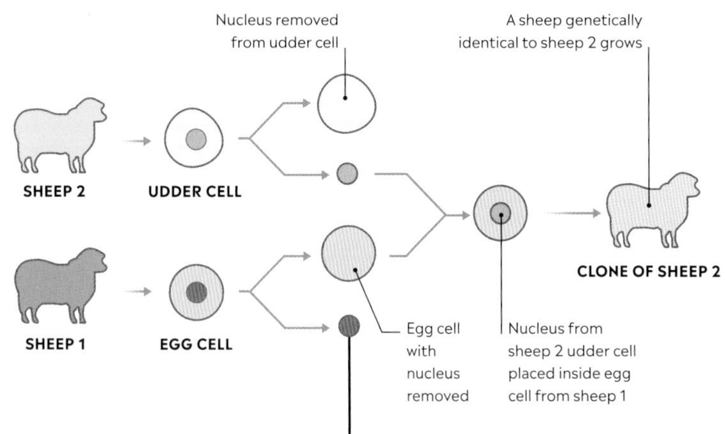
Nucleus removed
from udder cell
A sheep genetically
identical to sheep 2 grows
SHEEP 2
UDDER CELL
CLONE OF SHEEP 2
SHEEP 1
EGG CELL
Egg cell
with
nucleus
removed
Nucleus from
sheep 2 udder cell
placed inside egg
cell from sheep 1

1996

DOLLY THE SHEEP

Scientists at the Roslin Institute, part of the University of Edinburgh in Scotland, cloned a sheep from an adult, rather than an embryo cell, the first time this had been achieved in a mammal. The resulting cell produced was implanted into a surrogate mother to develop. The sheep born through this process, named Dolly, was genetically identical to the original adult cell.

Cloning Dolly

A cell from the udder of one sheep was taken and its nucleus removed. This nucleus was implanted into an enucleated egg cell (one from which the nucleus had been removed) from another sheep.

1996

1996 US scientist Charles Keeling notices that the timing of seasonal activities of plants is shifting due to climate change

1996 US scientist Gustav Arrhenius finds evidence in Greenland rocks for the emergence of life 3.8 billion years ago

1996 Palaeontologists find evidence that the *Eosimias*, a primate that lived some 40 million years ago, may have been the ancestor of apes and humans

1996–2003

DOLLY THE SHEEP

Dolly began life in a test tube as a single cell from a Finn Dorset sheep and an egg cell from a Scottish Blackface Sheep. The embryo was transferred to a surrogate mother and was born on 5 July 1996. Dolly lived at the Roslin Institute and had six lambs.

△ Martian meteorite

1996

LIFE ON MARS?

A team of NASA scientists announced the discovery of possible signs of ancient life in a Martian meteorite – a rock blasted off the Martian surface that later landed in Antarctica. The claims, based on chemical traces and supposed "microfossils", were fiercely debated. Other scientists later proposed ways in which the chemicals and structures observed could have been produced without biological activity.

1997

CARBON SHARING

Canadian biologist Suzanne Simard used isotope tracers to show that the root systems of separate trees – even those of different species – could be interconnected through mycorrhizal fungi. The connections help plants to share and redistribute essential elements such as carbon, nitrogen, and phosphorus.

▷ **Forest in Charlevoix region,** Quebec, Canada

1997 More than 150 countries adopt the Kyoto Protocol to reduce greenhouse gas emissions

1997

1997 NASA's Mars Pathfinder makes the first successful landing on Mars in more than 20 years

1997 IBM's Deep Blue chess computer beats the world champion, Russian grandmaster Garry Kasparov

△ **EHT image** of black hole

1996

GALACTIC CENTRE

By tracing the orbits of fast-moving stars at the heart of the Milky Way, teams led by astronomers Reinhard Genzel and Andrea Ghez demonstrated the influence of a compact, dark object with the mass of at least two million Suns at the very centre of our galaxy - a supermassive black hole. The object was eventually imaged in 2022 by the Event Horizon Telescope (EHT).

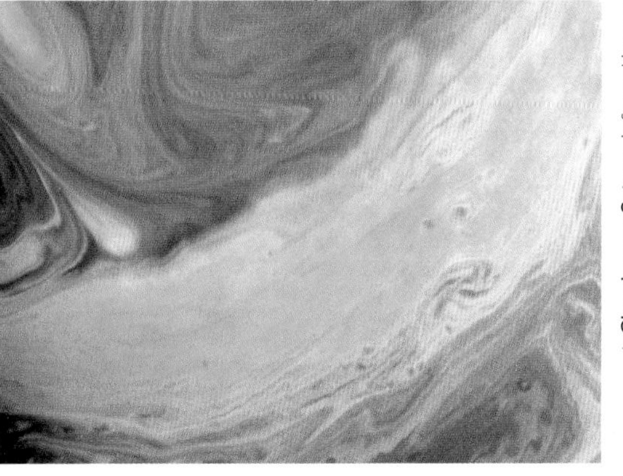

△ **Clouds over Saturn,** infrared image

1997

MISSION TO SATURN

NASA launched its Cassini spacecraft – an ambitious interplanetary mission carrying with it the European-built Huygens lander, which was designed to investigate the giant moon Titan. Cassini's instruments studied Saturn for 13 years , while the Huygens lander returned data from Titan for around 90 minutes.

DARK ENERGY

Despite several lines of evidence now supporting the concept of dark energy, its nature remains a mystery. One popular theory sees it as a "cosmological constant" – a property of space-time itself that only becomes apparent on vast scales. This may explain why the strength of dark energy seems to be increasing over time as the Universe expands. An alternative idea is that dark energy is a fifth fundamental force that opposes gravitational attraction over very large distances.

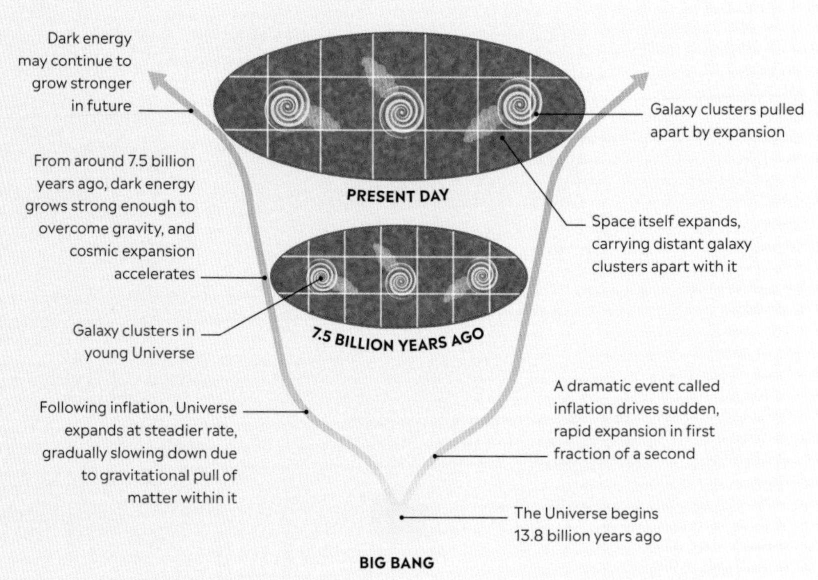
Dark energy may continue to grow stronger in future
From around 7.5 billion years ago, dark energy grows strong enough to overcome gravity, and cosmic expansion accelerates
Galaxy clusters in young Universe
Following inflation, Universe expands at steadier rate, gradually slowing down due to gravitational pull of matter within it
PRESENT DAY
7.5 BILLION YEARS AGO
BIG BANG
Galaxy clusters pulled apart by expansion
Space itself expands, carrying distant galaxy clusters apart with it
A dramatic event called inflation drives sudden, rapid expansion in first fraction of a second
The Universe begins 13.8 billion years ago

Dark energy and expansion of the Universe
Since the initial explosion of the Big Bang, the rate of cosmic expansion has depended on the shifting balance between the gravitational pull of matter and the strength of dark energy.

1998

1998 Astronomers detect a powerful burst of gamma rays from a source 12 billion light years away. Initially, the event is thought to be the most powerful explosion since the Big Bang

1998 Two independent teams of astronomers find evidence that an unknown force called dark energy is accelerating the expansion of the Universe

1998 Images from the Galileo mission show that Jupiter has four distinct rings rather than three

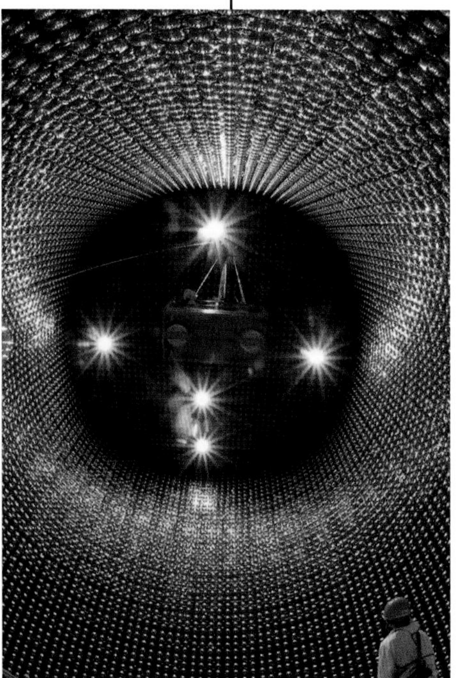

1998

MASS OF NEUTRINOS

Neutrinos are elementary particles first predicted in 1930. There are three types, or "flavours", of neutrino, and observations suggested that neutrinos "oscillate" between these flavours as they travel through space. This would require that they had mass, something that was at odds with the Standard Model (*see pp.176–77*). In 1998, scientists using the Super-Kamiokande neutrino observatory in Japan measured the mass of the neutrino as about one ten-millionth the mass of an electron.

◁ **Neutrino detector, Japan**

▽ **The completed** International Space Station

1999

MOLECULAR MOTORS

The first molecule-scale motor was synthesized by the Dutch chemist Bernard Feringa. His breakthrough was to create a molecule that rotates in one direction only - at a speed of 12 million rpm - in response to an external stimulus, such as light. He developed proof-of-concept machines, such as a "nanocar" made of four molecular "motors" to move it on a surface. Other possible applications include new types of catalysts, self-assembled nanomaterials, and molecular-scale electronics.

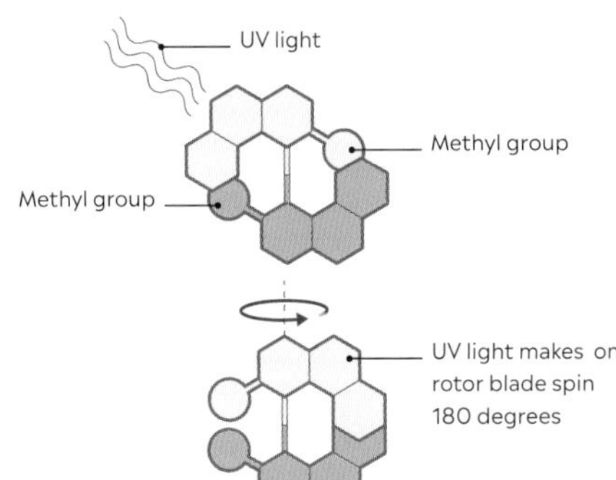

Controlled rotation
Irradiation with UV light causes the molecule, composed of two "paddles", to rotate in one direction only as the attached methyl groups stop it from rotating back.

1999

WETLAND RESTORATION

An assessment of damage to the unique Florida Everglades ecosystem showed that natural flows of water had been altered by drainage for agriculture and human consumption. Declining water quality and the growth of invasive plants had caused losses of wildlife. In response to these findings, a multibillion dollar project was initiated by state and federal agencies with the aims of restoring and conserving the precious wetland.

△ Everglades National Park

1999

1999 The nematode worm *Caenorhabditis elegans* is the first multicellular organism to have its genome sequenced

1999 NASA loses two space probes - the Mars Climate Orbiter and Mars Polar Lander - during their final approaches to Mars

1998

THE INTERNATIONAL SPACE STATION

Russia launched Zarya, the first module of the new International Space Station (ISS) - a collaboration between US, Russian, Japanese, European, and Canadian space agencies. Two weeks later, the first US module, Unity, was launched on the space shuttle Endeavour, and linked to Zarya in the first stage of an assembly programme that continued for 12 years and saw the station grow into a huge orbiting laboratory.

The ISS has been occupied continuously since November 2000

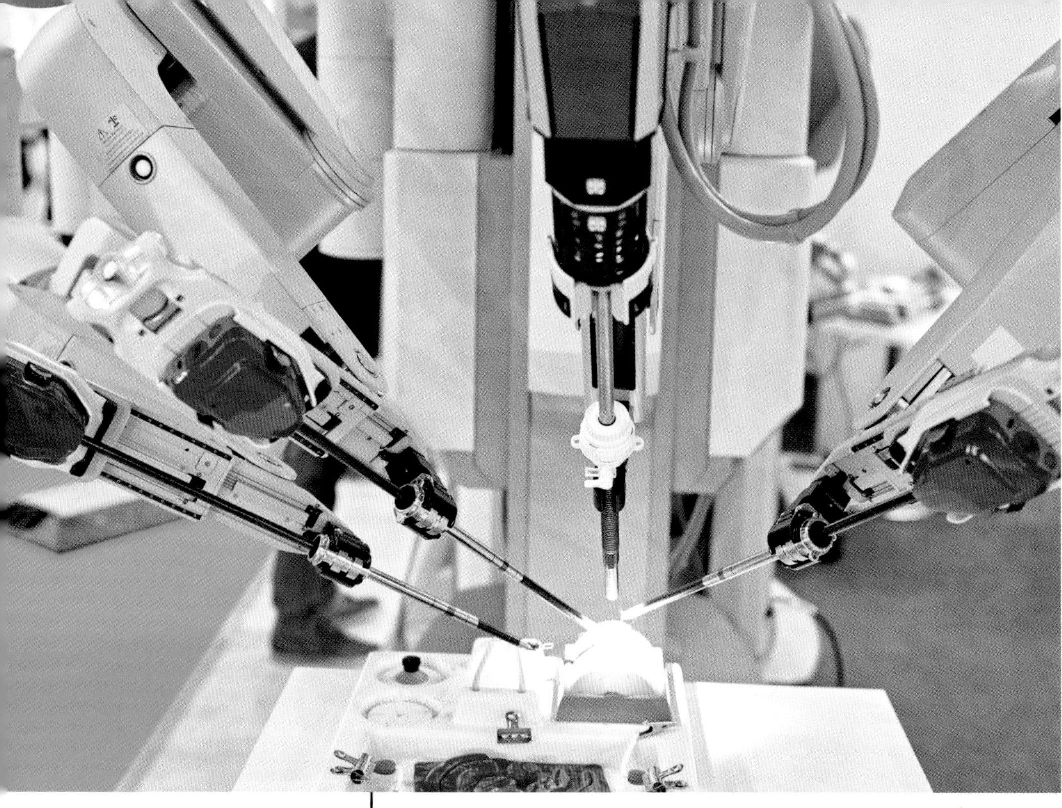

2000
ROBOT SURGERY
The notion that a semi-autonomous robot could assist in surgical operations had become a reality in the 1980s and 90s: several different systems were used, in limited trials, in several countries. The best-known and most widely used surgical robot today, the da Vinci Surgical System, gained Food and Drug Administration approval in the US in 2000.

◁ **A da Vinci** surgical robot

2000

2000 Personal computers are owned by 50 per cent of US households

2000 A genetically modified strain of rice is developed that produces beta carotene, a precursor of vitamin A

2000
GENOME RESULTS
The initial sequencing of 90 per cent of the human genome was completed by the publicly financed international Human Genome Project and the private Celera Genomics Corporation. Scientists studying the draft genome found that humans have only 30,000 genes - about the same as a nematode worm and fewer than initially expected.

▷ **Automated** DNA sequencer

"Without a doubt this is the most important, most wondrous map ever produced by humankind."

US PRESIDENT BILL CLINTON ON THE RESULTS OF THE HUMAN GENOME PROJECT, 2000

▽ Millennium Seed Bank vault

2000
SEED BANK

The Millennium Seed Bank Partnership was established in the UK. With two in five plant species threatened with extinction, the Seed Bank was designed to preserve plant genetic diversity by providing a long-term insurance policy against species loss. It now holds a collection of more than 2.4 billion dried and frozen seeds from over 40,000 species.

2001
STEM CELL RESEARCH

Stem cells are the body's "master cells" from which all other specialized cells are produced and used to regenerate and repair damaged tissues. However, in 2001, US President George Bush banned the federal funding of research on certain human stem cells derived from embryos because of ongoing ethical concerns.

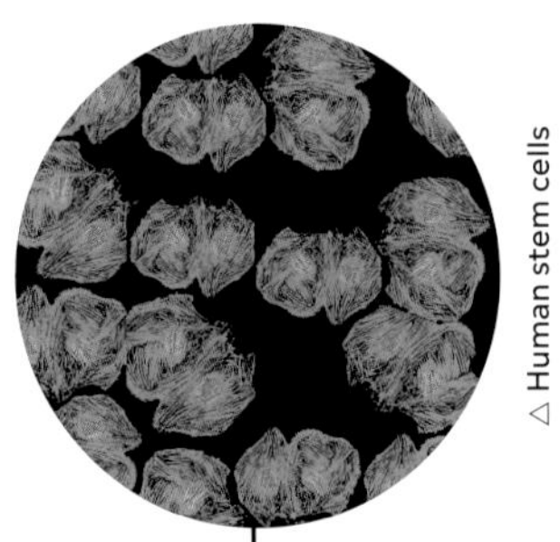
△ Human stem cells

2001 NASA's NEAR-Shoemaker probe touches down on the asteroid Eros following a year-long survey from orbit

2001

2001 The IPCC's third report details growing evidence for the warming of Earth's climates

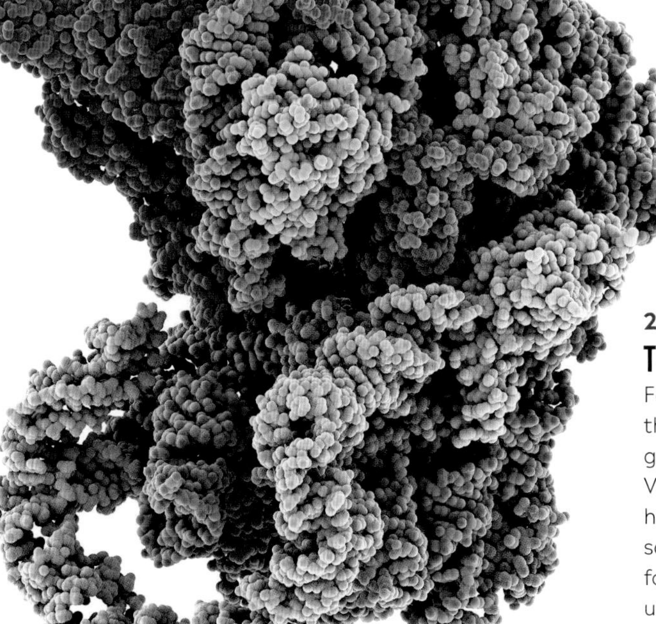
△ Ribosome bound to mRNA

2000
THE RIBOSOME

Found within all cells, ribosomes are structures that perform the vital process of directing protein assembly from the genetic code. Research carried out by Indian-born biologist V. Ramakrishnan and Israeli crystallographer Ada Yonath helped to clarify the fine structure of the ribosome, allowing scientists to work out how it binds to messenger RNA, follows the amino-acid recipe it carries, and binds these units together to produce proteins.

2002

MARS ODYSSEY

NASA announced that its Mars Odyssey orbiter, which had been launched in 2001, had found vast amounts of water ice in the upper layers of Martian soil. The mission's gamma-ray spectrometer detected the characteristic emissions produced when hydrogen (within the water ice) is bombarded by high energy cosmic rays. Odyssey continued to map the ice in detail, showing that it is a key component of the Martian soil, especially at high latitudes.

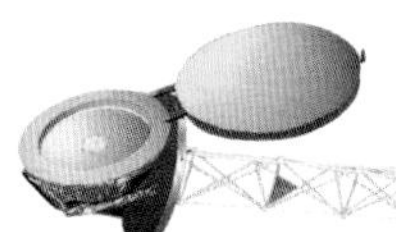

▷ **Mars Odyssey** probe

2002

2002 US geneticists sequence the genome of the protozoan *Plasmodium falciparum*, a parasite that causes the deadliest form of malaria

2002 An icy Kuiper Belt world named Quaoar is the largest new Solar System object discovered since Pluto

2002 Physicists at CERN combine thousands of antiprotons and anti-electrons (positrons) to make a gas of anti-hydrogen

2002 Argentinian-British zoologist Alex Kacelnik demonstrates that the Caledonian crow can bend wire into a hook – a remarkable feat of toolmaking

2002

ICE SHELF COLLAPSE

Unusually warm temperatures in the Weddell Sea caused dramatic surface melting and crevasse development in the Antarctic Peninsula's ice shelf. Within a month, almost the entire Larsen B Ice Shelf – an area of 3,250 sq km (1,250 sq miles) – splintered, broke away, and disintegrated.

◁ **Larsen B ice shelf** satellite image

The Larsen B Ice shelf had been stable for 10,000 years before 2002

2003
MAPPING THE UNIVERSE

NASA released the first detailed map from its Wilkinson Microwave Anisotropy Probe (WMAP). The orbiting spacecraft carried instruments to map cosmic microwave background radiation from the edge of the visible Universe and so reveal the early seeds of large-scale cosmic structure. From these studies, scientists were able to estimate the date of the Big Bang at around 13.8 billion years ago.

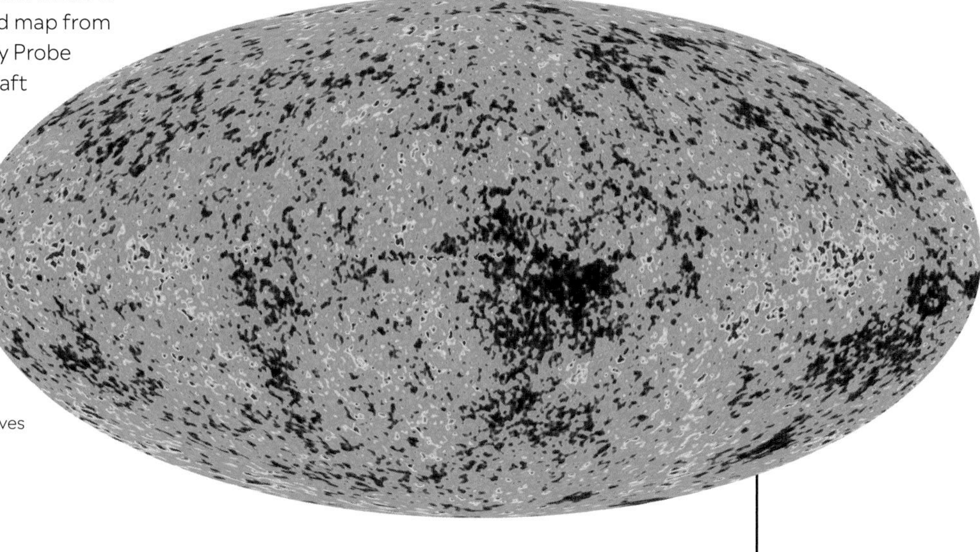

▷ **WMAP's all-sky map** of cosmic microwaves

2003

2003 The space shuttle *Columbia* breaks up while re-entering Earth's atmosphere

2003 The Human Genome Project is largely completed, having sequenced some 92 per cent of the human genome

2003 An epidemic of Severe Acute Respiratory Syndrome (SARS) breaks out in China and spreads to 26 countries

▽ *Mantophasma zephyra*

2002
NEW INSECT ORDER

A new order of insects was recognized. Mantophasmids – also known as heelwalkers because they hold the extreme ends of their legs off the ground when walking – are wingless carnivorous insects found in South Africa and Namibia. They have subsequently been classified together with mole crickets in the order Notoptera.

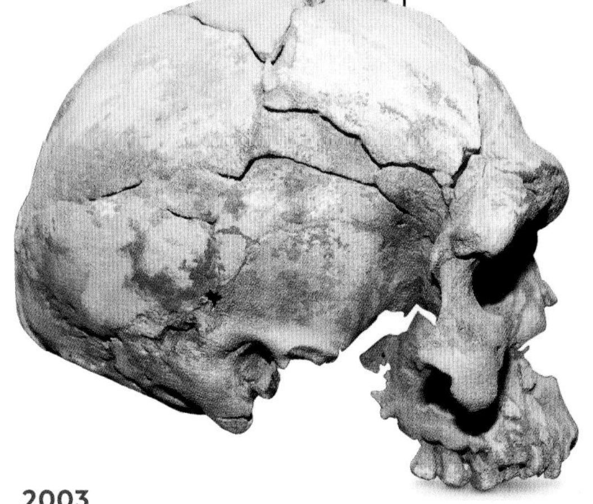

△ **Skull** from Ethiopia

2003
HUMAN EVOLUTION IN AFRICA

An international team working in Ethiopia unearthed the 160,000-year-old remains of two adults and one child. They were some 60,000 years older than the oldest previously known specimen of *Homo sapiens*, and so supported the theory that modern humans evolved as a single species in Africa, rather than as the result of interbreeding with other human precursors, especially the European Neanderthals.

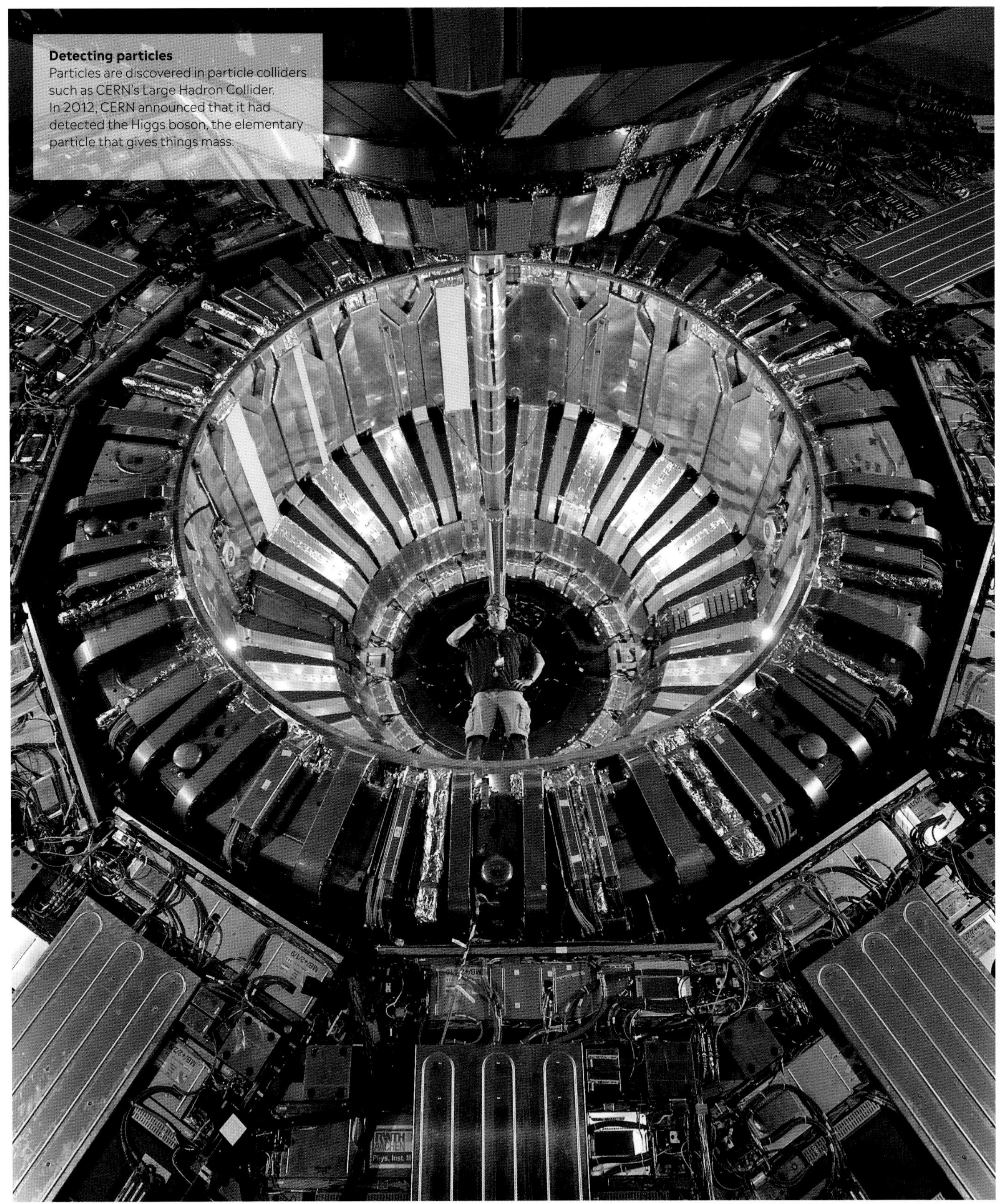

Detecting particles
Particles are discovered in particle colliders such as CERN's Large Hadron Collider. In 2012, CERN announced that it had detected the Higgs boson, the elementary particle that gives things mass.

THE STANDARD MODEL OF PARTICLE PHYSICS

The most complete theory of particle physics so far, the standard model was developed through the latter half of the 20th century by both experimental and theoretical physicists. It reflects the current understanding that everything in the Universe is made up of a limited number of building blocks of matter (elementary particles) and governed by four fundamental forces.

The standard model classifies all known elementary particles by their properties - including electric charge and mass, as well as less tangible properties such as spin - and describes three of four fundamental forces.

In the model, matter particles (fermions) are separated into quarks and leptons. There are six quarks and six leptons. They can be further paired into three "generations" with increasing mass.

Each fermion has a corresponding antiparticle with identical mass but opposite charge - for example, an electron's antiparticle is a positron.

Fermions are influenced by force-carrying particles known as bosons: electromagnetism is carried by photons; the strong force (which holds particles together) is carried by gluons; and the weak force (which causes nuclear reactions) is carried by W and Z bosons.

As successful as the standard model has been - for example, it predicted the properties of W and Z bosons - it is not a "theory of everything". Among other shortcomings, it fails to incorporate gravity, as described successfully by the general theory of relativity (*see p.197*), or to include any particle that could comprise the dark matter that outweighs ordinary matter in the Universe (*see p.266*).

Hunting for quarks
The existence of the heaviest quark, called the top quark, was predicted by the standard model. It was discovered in a particle collider at Fermilab (in Illinois, USA) in 1995.

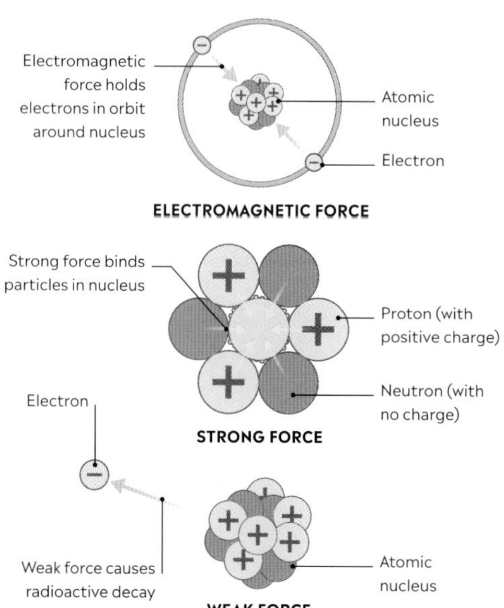
Electromagnetic force holds electrons in orbit around nucleus
Atomic nucleus
Electron
ELECTROMAGNETIC FORCE
Strong force binds particles in nucleus
Proton (with positive charge)
Neutron (with no charge)
STRONG FORCE
Electron
Weak force causes radioactive decay
Atomic nucleus
WEAK FORCE

FUNDAMENTAL FORCES

The four fundamental forces are gravitational, electromagnetic, strong, and weak. Three of the four have known force-carrying particles, such as electrons for the electromagnetic force, that give rise to interactions between other particles.

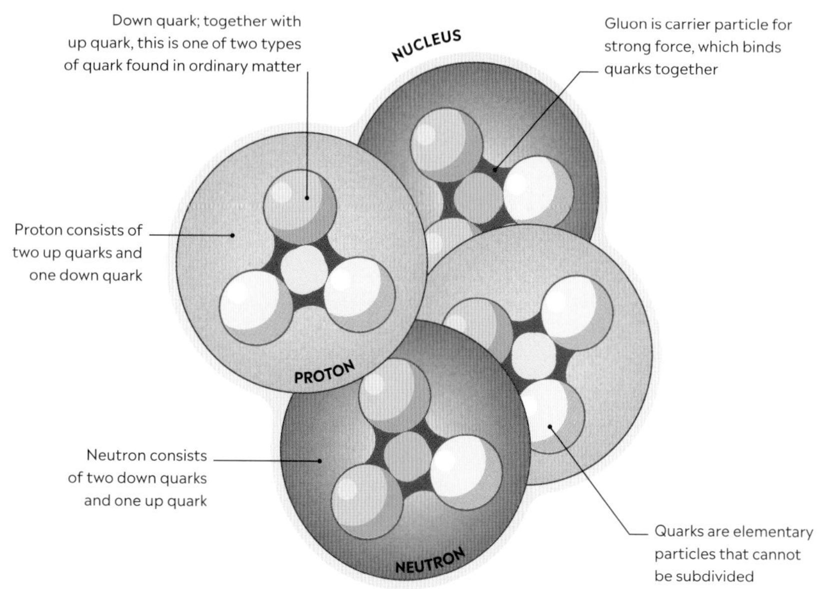
Down quark; together with up quark, this is one of two types of quark found in ordinary matter
NUCLEUS
Gluon is carrier particle for strong force, which binds quarks together
Proton consists of two up quarks and one down quark
PROTON
Neutron consists of two down quarks and one up quark
NEUTRON
Quarks are elementary particles that cannot be subdivided

ELEMENTARY PARTICLES

All matter in the Universe is made up of elementary particles. These are particles that cannot be divided any further. Matter particles, known as fermions, are divided into two families: quarks and leptons. Leptons can exist alone, but quarks can only exist when bound to other quarks by the strong force to form composite particles.

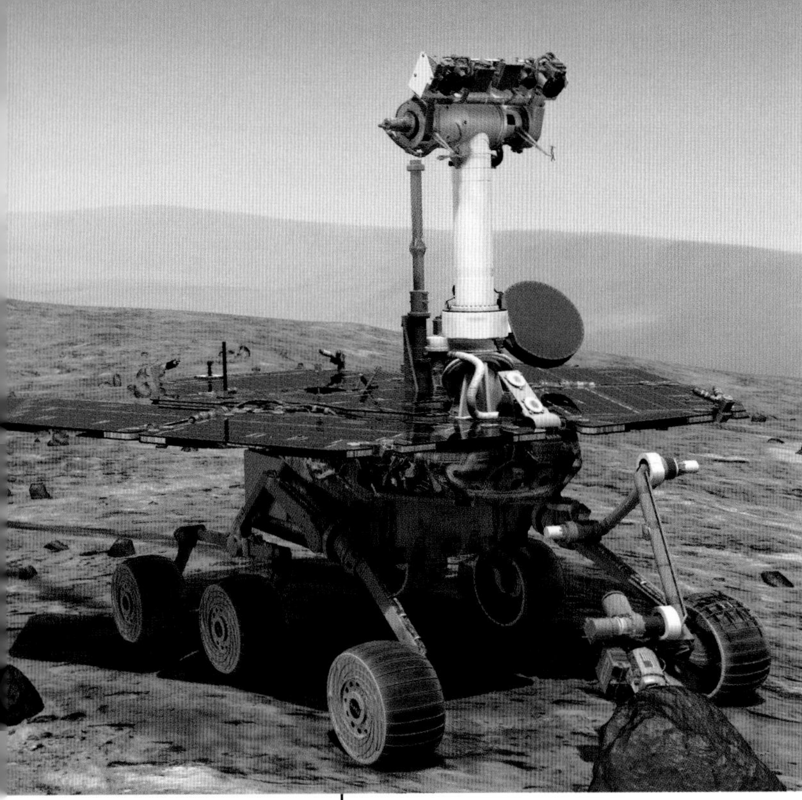

2004

MARS EXPLORATION ROVERS

NASA's twin Mars Exploration Rovers, named Spirit and Opportunity, landed on the Martian surface. Each weighing 185 kg (408 lb) and standing 1.5 m (4.9 ft) tall, the solar-powered vehicles carried a variety of instruments for imaging the surface and analysing rocks. Spirit explored the crater Gusev until 2010, while Opportunity studied features on Mars's Meridiani plain until 2018.

◁ Mars Exploration Rover

2004

2004 US researchers find that oceans absorb and store almost half of human-generated atmospheric carbon dioxide

2004 The Cassini mission enters orbit around Saturn, deploying its Huygens lander on to Titan in 2005

2004

DEVASTATING TSUNAMI

One of the deadliest tsunamis in recorded history killed some 230,000 people in a dozen countries around the Indian Ocean. It was caused by a powerful earthquake and seabed faulting off northern Sumatra that suddenly displaced a enormous volume of seawater. This created waves that travelled at over 700 km/h (435 mph) in opposite directions, one crossing the Indian Ocean towards Africa and the other moving towards Indonesia and Thailand.

△ **Buildings destroyed** in Thailand by the 2004 tsunami

2004

GRAPHENE

Dutch-British physicist Andre Geim and Russian-British scientist Konstantin Novoselov discovered a simple method by which to isolate graphene, a new form of carbon. Graphene is a one-atom-thick layer of carbon atoms arranged in a hexagonal (honeycomb) lattice, making it the first two-dimensional crystal structure discovered. It is stronger than steel, conducts electricity and heat, and is transparent.

△ Professor André Geim

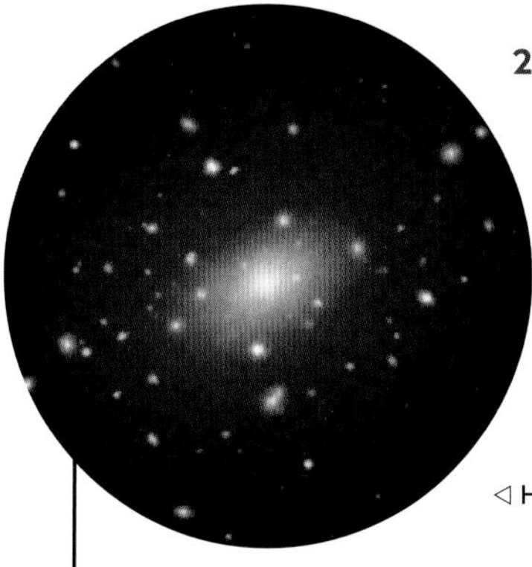

2004
DARK ENERGY CONFIRMED

Astronomers using the Chandra X-ray Observatory studied the hot gas of galactic clusters billions of light years from Earth. Analysis of their data suggested that the expansion of the universe had gradually been slowing down until about 6 billion years ago, when it started to accelerate. This change was thought to be due to the repulsive effect of dark energy.

◁ **Hot gas** in galaxy cluster Abell 2029

2005–2020
CRISPR

Clustered Regularly Interspaced Short Palindromic Repeats (CRISPR) are DNA sequences found in many bacteria. When combined with the enzyme Cas9, they can be used to target, cut, and edit genes with great precision. Early work in 2005 was developed by French biochemist Emmanuelle Charpentier and US biochemist Jennifer Doudna into a technique that has been used in a range of biological and medical applications.

△ Jennifer Doudna (l) and Emmanuelle Charpentier (r)

2005

2005 The Deep Impact space probe fires an impactor at Comet Tempel 1 and studies its effects

2005 US biologist Peter Andolfatto shows that so-called "junk DNA", which makes up over 95% of human DNA, is of evolutionary significance

2005 Eris and Makemake, two large, icy dwarf planets, are discovered beyond Neptune: Eris is roughly equal in size to Pluto

2005 A map of the human genome known as the HapMap (haplotype map), is made, allowing researchers to locate genes and variations that may affect health and disease

GENE EDITING

Technological advances have made it possible to edit a genome by removing, replacing, or adding gene sequences. The latest technique relies on CRISPR – DNA sequences that bacteria use to fight viruses – and a bacterial enzyme called Cas9 that cuts through the genomic DNA sequence at a precise location. The applications of this technology stretch from targeted gene modification in plants – for example, creating crops that are resistant to pathogens – to removing or replacing genes that cause diseases, such as cancer, in humans.

How CRISPR-Cas9 works

CRISPR guides the bacterial enzyme Cas9 to specific locations in the DNA sequence, where it cuts through the DNA strand. The CRISPR-Cas9 technology has been given the nickname "genetic scissors".

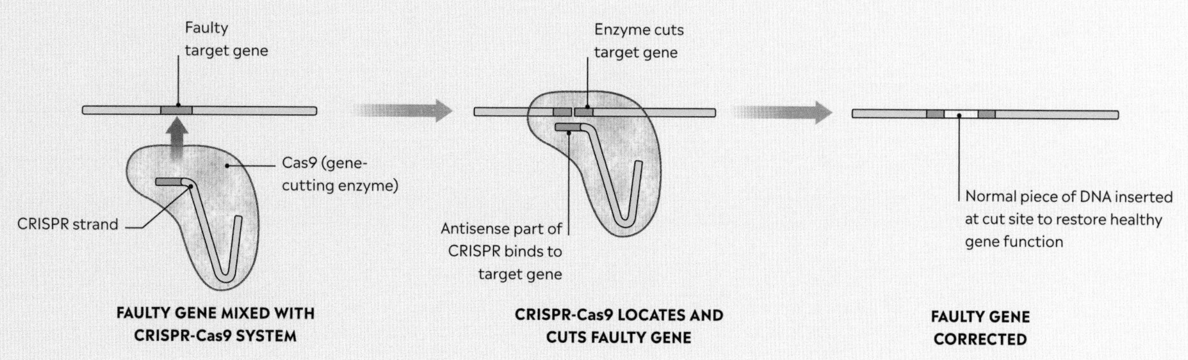
Faulty target gene
Cas9 (gene-cutting enzyme)
CRISPR strand
FAULTY GENE MIXED WITH CRISPR-Cas9 SYSTEM
Enzyme cuts target gene
Antisense part of CRISPR binds to target gene
CRISPR-Cas9 LOCATES AND CUTS FAULTY GENE
Normal piece of DNA inserted at cut site to restore healthy gene function
FAULTY GENE CORRECTED

2006

COLLECTING STARDUST

A capsule of dust from comet Wild 2 parachuted back to Earth after release from NASA's Stardust spacecraft. Launched from Earth in 1999, Stardust flew past the comet in 2004, collecting dust from its coma on a low-density material called aerogel. The spacecraft continued its mission to rendezvous with another comet, Tempel 1, and study the aftermath of the Deep Impact mission.

▽ Comet Wild 2

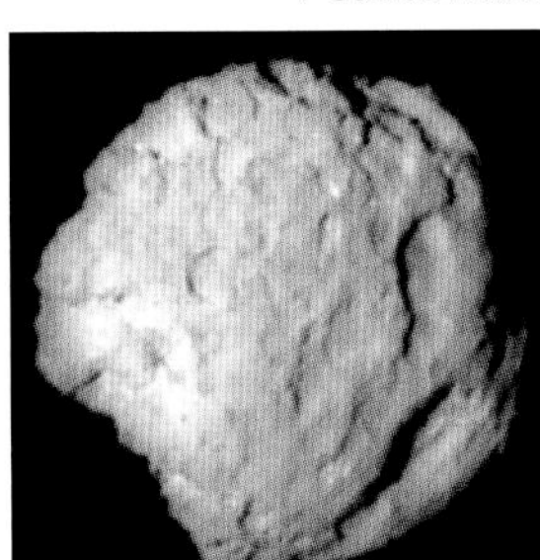

2007

RETREAT OF GLACIERS

Scientists revealed that summit ice cover on Mount Kilimanjaro in Tanzania had shrunk by 1 per cent a year from 1912 to 1953 and around 2.5 per cent a year from 1989 to 2007. This ice loss was similar to that of other low latitude glaciers. At this rate of loss, Kilimanjaro's ice fields and glaciers would disappear within the next few decades.

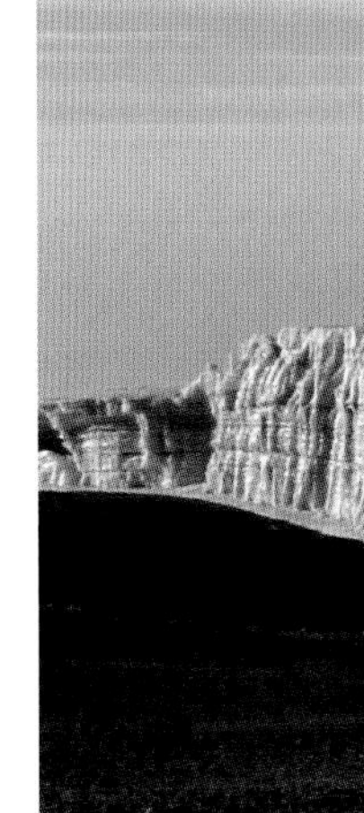

▷ **Mt Kilimanjaro** glacier

2006

2006 With large objects discovered beyond Neptune. Pluto is relegated to a new class of objects called "dwarf planets"

2006 Scientists warn that extinctions are occurring at 100 to 1,000 times the normal background rate

2007 Genome-wide association studies accelerate research on diseases by comparing DNA samples from thousands of affected people with normal DNA

2006 The Wide Angle Search for Planets (WASP) discovers exoplanets by looking for dips in brightness as they transit their stars

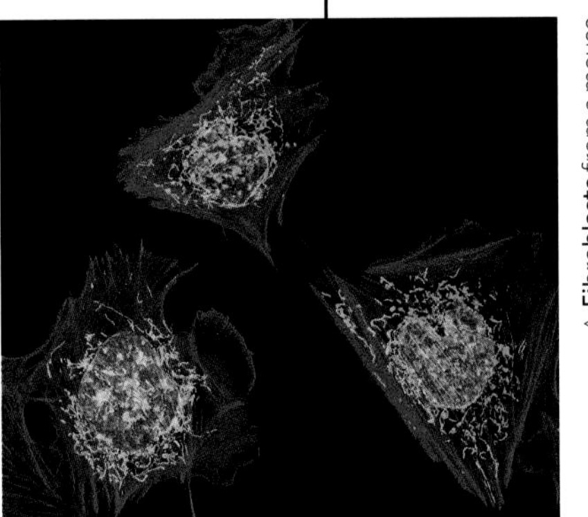

△ **Fibroblasts** from a mouse

2006

MAKING STEM CELLS

Stem cells are undifferentiated cells that can develop into a wide range of tissues. Japanese embryologist Shinya Yamanaka made a breakthrough in showing how stem cells can be generated from adult cells. Known as induced pluripotent stem cells (IPS or IPSCs) they can be produced in quantity and may give rise to any other type of cell.

▽ *Tiktaalik roseae*

2006

MISSING LINK

The discovery in Arctic Canada of the partial skeleton of a 375-million-year-old Devonian age fossil called *Tiktaalik* showed an evolutionary link between fish and four-limbed (tetrapod) animals that first moved on land. Although still water-dwelling, with a crocodile-like skull and bony scales, *Tiktaalik* also had a pair of muscular front fins. These resembled limbs capable of pushing the animal forwards on land.

In 2007, the IPCC reported that sea levels had risen by 3.1 mm (⅕ in) per year since 1993

2007

2007 US geneticist Craig Venter creates the first synthetic organism by designing, synthesizing, and assembling the bacterium *Mycoplasma mycoides*, in a process involving the replacement of the genome of one bacterial species with that of another

△ Zebrafish

2007
HEART REGENERATION

Mammals, including humans, have very little capacity to repair damaged heart tissue. Fish and amphibians, however, are able to do so. US biologists Robert Major and Kenneth Poss investigated the processes involved in heart regeneration in zebrafish – research that may lead to the goal of stimulating heart regeneration in our own species.

△ Bleached coral

2007
CLIMATE CHANGE REPORT

The Intergovernmental Panel on Climate Change (IPCC) presented definitive evidence for climate warming and predicted increases in rates of icecap melting, rising sea levels, and coral bleaching. It reported that global temperatures had risen by 0.77 °C (1.4 °F) over the last 100 years and that 11 of the 12 preceding years (1995–2006) had been among the warmest on record.

2008

LARGE HADRON COLLIDER

The world's most powerful particle accelerator, the Large Hadron Collider at CERN, began operating on 10 September 2008. Its main component is a ring with a circumference of 27 km (17 miles) that straddles the border between France and Switzerland. Inside, supercooled magnets guide and accelerate beams of particles and antiparticles to speeds close to that of light, and detectors observe what happens when these beams collide.

▷ Large Hadron Collider

CERN's Large Hadron Collider is the largest machine ever built by humans

2008

2008 The MESSENGER spacecraft flies within 200 km (125 miles) of Mercury during the first of three flybys prior to entering orbit

2008 China's Shenzhou 7 mission sees its first three-person crew and first spacewalk

2008 IBM's Roadrunner computer breaks the petaflop barrier, carrying out 1,000 trillion operations (one petaflop) per second

2008 European astronomers detect a sugar called glycoaldehyde in a star-forming region of the Milky Way - evidence that the building blocks of life are widespread

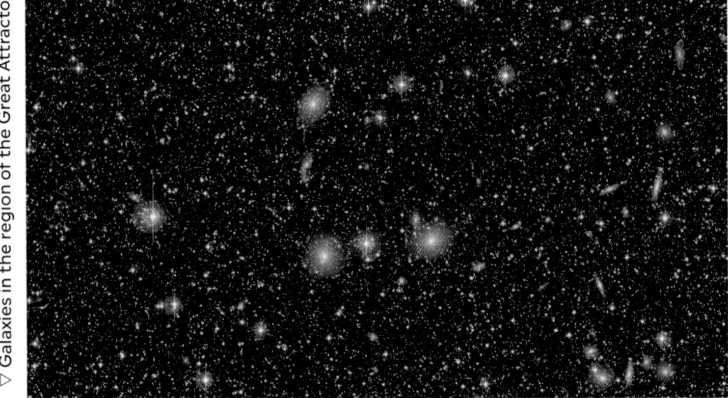

▷ Galaxies in the region of the Great Attractor

2008

THE GREAT ATTRACTOR

Astronomers led by Latvian-born NASA scientist Alexander Kashlinsky suggested that clusters of galaxies are attracted towards a certain patch of the southern sky. They proposed that this "dark flow" may be due to conditions initiated in the Big Bang . Others argue that the movement is caused by the "Great Attractor" - a concentration of galaxies in that region that exert a huge gravitational pull.

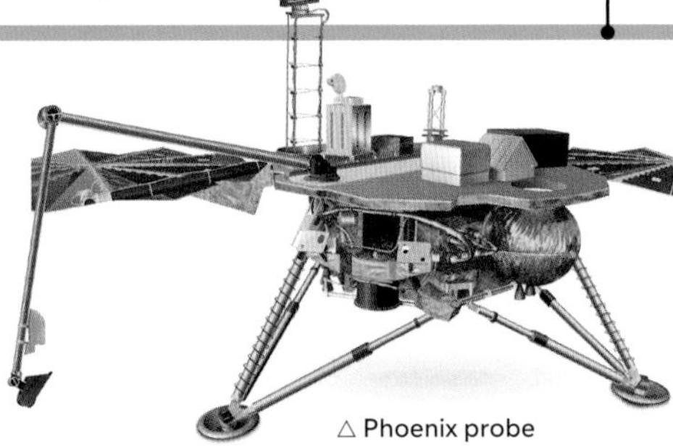

△ Phoenix probe

2008

PHOENIX ON MARS

A NASA lander called Phoenix touched down in the northern polar region of Mars. The craft used its robot arm to dig a trench and expose the Martian permafrost (a mix of ice and soil) for study, confirming the presence of water ice. Further analyses identified salts in the Martian soil, including some that could react with sunlight to make the surface layers of soil hostile to life.

2008
HISTORY IN ICE

As fallen snow compacts into dense layers of ice, it traps tiny bubbles of atmospheric gas, which are then buried beneath the surface. Analysis of ice cores from the Antarctic showed that today's levels of the greenhouse gases carbon dioxide and methane are respectively 28 per cent and 134 per cent higher than at any time in the last 800,000 years.

△ **Ice core** collected by scientists

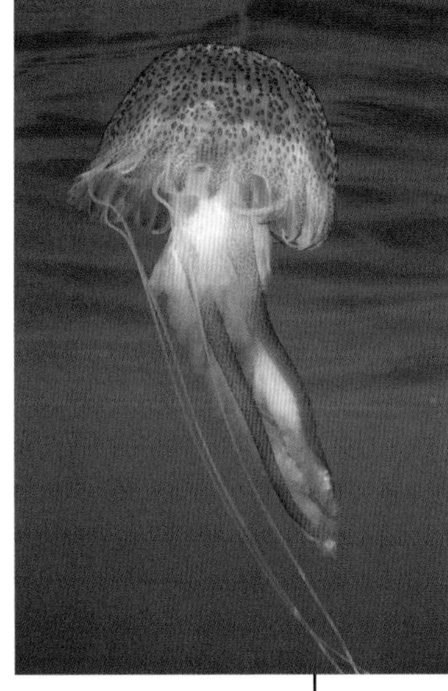

◁ **Jellyfish** displacing water

2009
MARINE MIXING

Oceanic life depends on the movement of water to transport vital nutrients from one place to another. For many years, it was thought that wind and tidal motion were responsible for the distribution of nutrients. However, research in the 2000s showed that perturbations of the water caused by the movement of fish and other marine animals made a significant contribution to nutrient movement.

2009 *Ardipithecus ramidus*, or "Ardi", a 4.4 milllion-year-old hominid that could walk upright and climb trees, is discovered in Ethiopia

2009

2009 The Kepler space telescope is launched and begins its search for exoplanets

2009
FEATHERED DINOSAUR

The discovery of a remarkably well-preserved feathered dinosaur in China showed that bird-like dinosaurs were present on Earth over 150 million years ago in Late Jurassic times. The crow-sized *Anchiornis huxleyi* had wing-like feathers on its arms and legs, and also on its feet and tail. The fossil slightly predates *Archaeopteryx*, which was previously thought to be the oldest bird-like dinosaur.

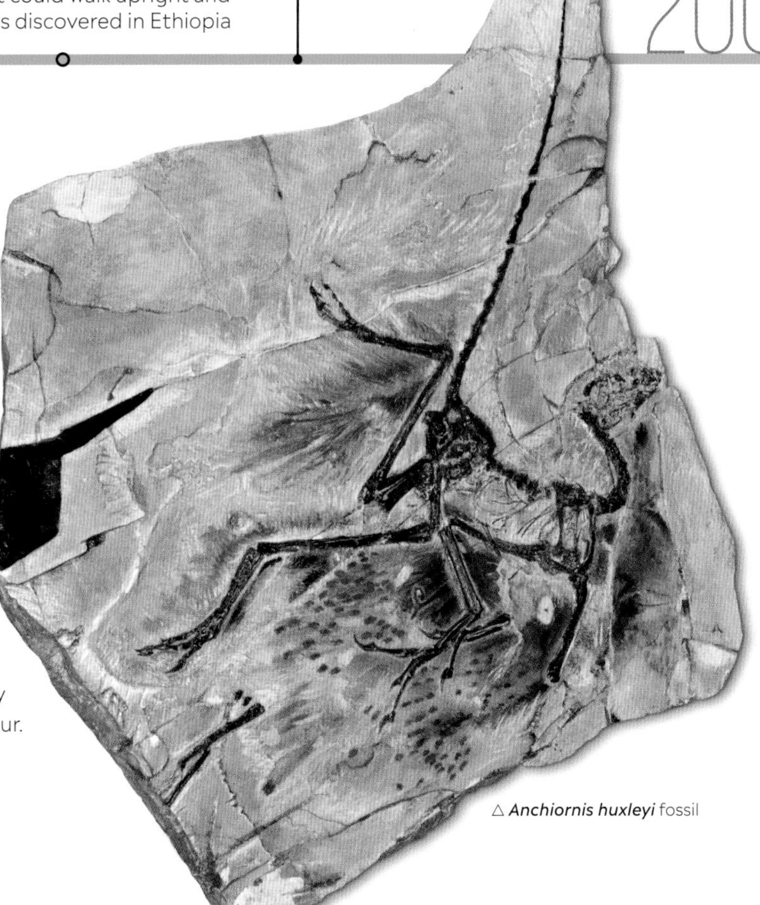

△ *Anchiornis huxleyi* fossil

2010
NEANDERTHAL GENES

Swedish geneticist Svante Pääbo and colleagues recovered DNA from 38,300–44,400-year-old Neanderthal bones found in a Croatian cave. They sequenced 60 per cent of the Neanderthal genome and found evidence that modern humans living outside Africa interbred with Neanderthals between 45,000 and 80,000 years ago.

△ **Svante Pääbo**, geneticist

▷ *Sinosauropteryx*

2010
DINOSAURS IN COLOUR

Palaeontologists discovered colour-bearing cell structures, called melanosomes, in exceptionally well-preserved fossils of feathered dinosaurs and some of the earliest birds from the Early Cretaceous strata of China. They were able to reconstruct the original colours and patterning of some of these long-extinct animals. The small theropod *Sinosauropteryx*, for example, had a long tail covered in reddish-brown stripes of small filament-like feathers.

2010

2010 A mixture of glacial floodwaters and magma erupts steam and ash from the Icelandic volcano of Eyjafjallajökull

2010 NASA launches the Solar Dynamics Observatory to study and help predict space weather

2010 Researchers at the University of California, Santa Barbara, build a device capable of exhibiting a pure quantum state that is big enough to be seen by the naked eye

2010
ENVIRONMENTAL DISASTER

An explosion on the BP oil drilling rig Deepwater Horizon killed 11 workers and triggered the largest offshore oil spill in the history of marine oil exploration. Over 4 million barrels of oil polluted the Gulf of Mexico and surrounding coastal areas. The cost of compensation and clean-up exceeded US$ 60 billion.

▷ **Deepwater Horizon** in flames

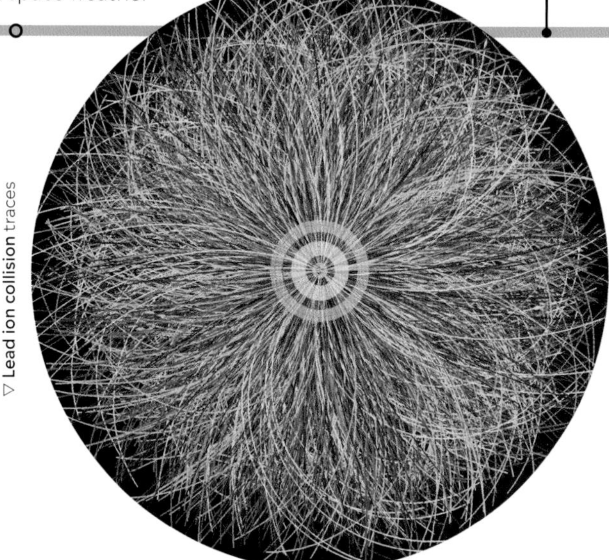

▷ **Lead ion collision** traces

2010
RECREATING THE BIG BANG

CERN's Large Hadron Collider (*see p.294*) is typically used to collide protons and antiprotons. However, for a few weeks in 2010, physicists used it to accelerate and collide lead ions instead. These ions have much higher mass than protons, so the energies involved in their collisions were much greater. The experiments gave physicists the opportunity to create and study "quark-gluon plasmas" – similar conditions to those in the first few moments after the Big Bang.

△ **Fermilab Tevatron** particle accelerator

2010

MATTER AND ANTIMATTER

One of the major challenges for particle physics is to explain why our Universe is dominated by matter, rather than antimatter. In 2010, researchers using the Tevatron particle accelerator at Fermilab, Illinois, US, found an important clue related to this asymmetry. They observed that particles called B mesons decay more often into muons than into antimuons.

"... a phenomenal discovery in the course of human history."

US ASTRONOMER GEOFF MARCY, ON THE DISCOVERY OF KEPLER-22B, 2011

2011

2011 A severe earthquake and tsunami kills over 19,000 people and disables Japan's Fukushima nuclear power plant

2011 NASA announces the discovery of Kepler-22b, the first known planet in the habitable zone of a Sun-like star

2011 Earth's human population hits seven billion just 12 years after reaching six billion

2011 China launches a prototype space station, Tiangong-1: it is visited by two crewed missions in 2012 and 2013

2011

THE DENISOVANS

Analysis of fossil DNA recovered from a tiny finger bone found in Denisova Cave, Siberia revealed the existence of a previously unknown genome belonging to an extinct group of humans. They became known as the Denisovans because not enough of their anatomy was known to name them scientifically. The Denisovans lived between 500,000 and 30,000 years ago across central Asia. According to genetic analysis they encountered and interbred with both the Neanderthals and the ancestors of modern Papua New Guineans before becoming extinct.

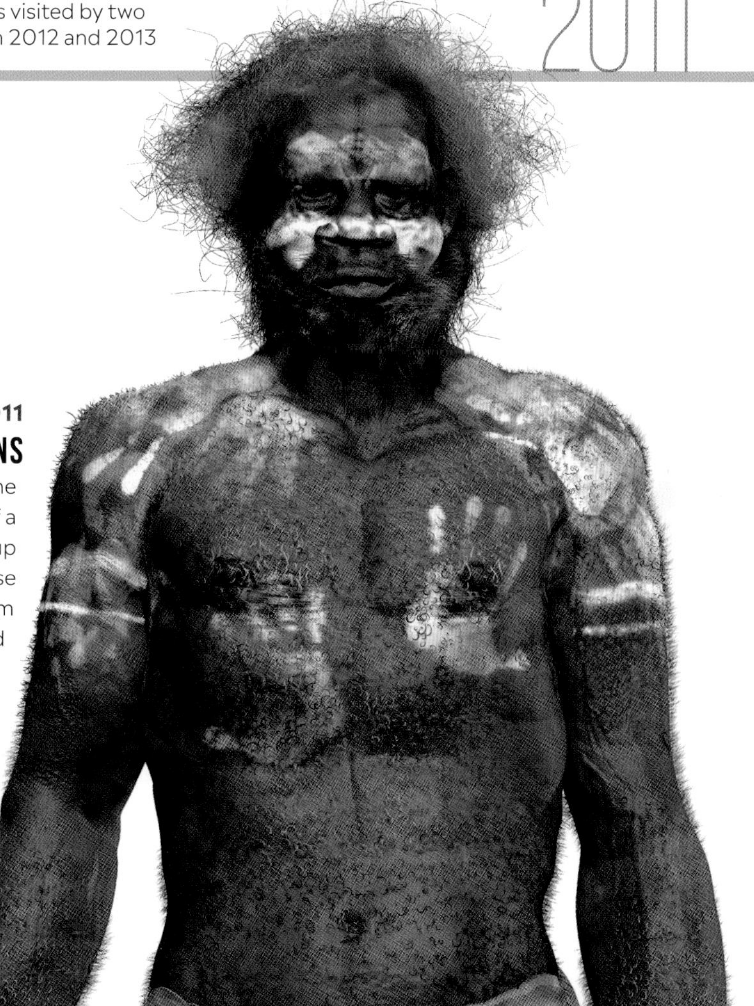

▷ **Male Denisovan,** highly speculative artist's recreation

2012

HIGGS BOSON

On 4 July 2012, physicists at CERN announced they had detected the Higgs boson, a particle associated with the Higgs field, which gives fundamental particles their mass. This was confirmation of an earlier theory proposed by scientists, including British physicist Peter Higgs. So important was the Higgs boson to the model of particles and interactions that some dubbed it "the God particle".

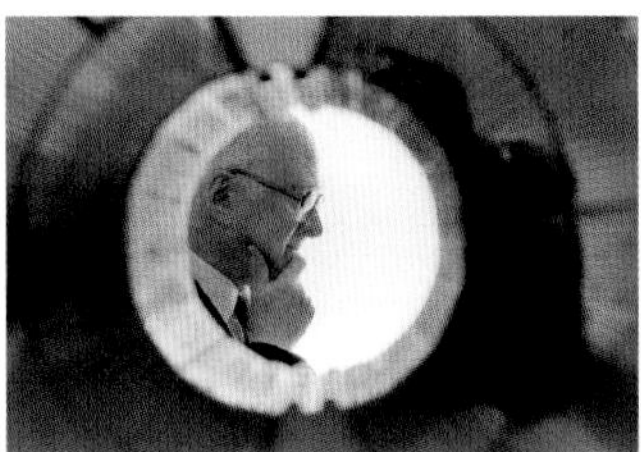

△ Peter Higgs

"It's very nice to be right sometimes ... it has certainly been a long wait."

PETER HIGGS AT A PRESS CONFERENCE, 2012

2012

WATER ON MERCURY

Data from the MESSENGER spacecraft, which entered orbit around the innermost planet in 2011, confirmed the presence of water ice at Mercury's north pole. Despite searing daytime temperatures, the planet's orientation allows craters near the planet's poles to stay permanently in shadow. Here, ice from comets that strike the surface has accumulated over billions of years, along with organic (carbon-based) chemicals.

▷ Mercury's poles

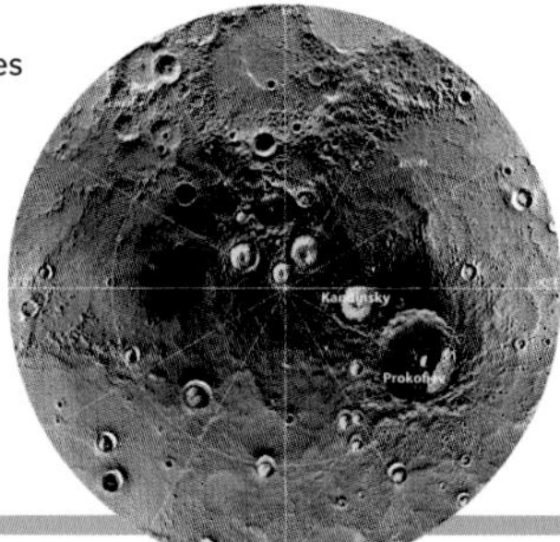

2012

2012 Arctic sea ice is reduced to 3.4 million square km (1.3 million square miles), its lowest extent since satellite observations began in 1979

2012 The Voyager 1 spacecraft, launched in 1977, crosses the heliopause and enters interstellar space

2012

CURIOSITY ON MARS

NASA's Mars Science Laboratory mission deployed the Curiosity rover on the surface of Mars. This car-sized vehicle carried instruments such as lasers, sample drills, and onboard laboratories to analyse Martian rocks. Results showed that the landing site (in Gale crater) was underwater in the distant past. Curiosity also detected organic molecules in the soil, suggesting past conditions that could have been suitable for life.

◁ The Curiosity rover

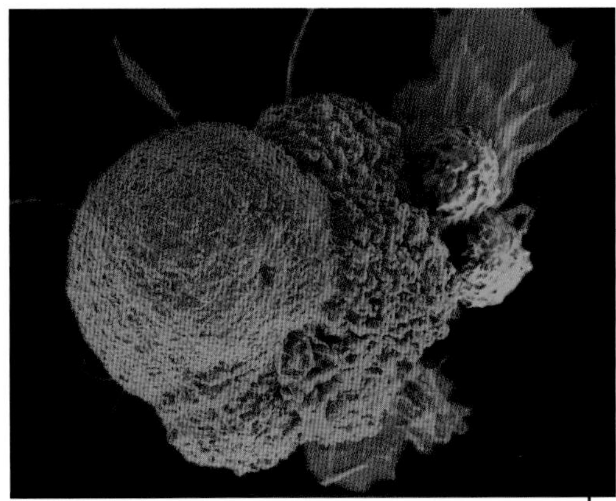
△ **A cancer cell (red)** is attacked by two T cells (blue)

2013

GENE THERAPY

A new treatment for certain forms of leukaemia based on the genetic modification of T cells (a type of immune cell), was developed by US scientists. T cells extracted from patients' blood were given a new gene for a protein that directed them to attack leukaemia cells. The T cells were then injected back into the patients, two-thirds of whom experienced remission.

2013

STRONGEST STORM RECORDED

Typhoon Haiyan, the most powerful tropical storm on record, hit the Phillippines on 8 November killing over 6,300 people. Formed in the Pacific Ocean, the storm moved northwest to strike the Southern Phillippines with wind speeds rising to 315 km/h (196 mph). Coastal storm surges were worsened by climate-induced high sea levels.

△ **Typhoon** Haiyan

2012 A SpaceX Dragon vehicle makes the first commercial cargo delivery to the ISS

2013 Global levels of atmospheric carbon dioxide pass 400 ppm for the first time in 400,000 years

2013

2012 Russian scientists drill 3.8 km (2.2 miles) through Antarctic ice to reach the ancient freshwater Lake Vostok

2013 Using the IceCube detector at the South Pole, scientists detect neutrinos so energetic that they must have originated beyond the solar system

2013

BRAIN INITIATIVE

A group of neuroscientists proposed the creation of the Brain Activity Map (BAM), seeking to understand better how the brain produces perception, action, memories, thoughts, and consciousness. The ambitious project sought to explain the circuitry involved in the most complex of human organs, and called for a technological revolution comparable to that required for the Human Genome Project.

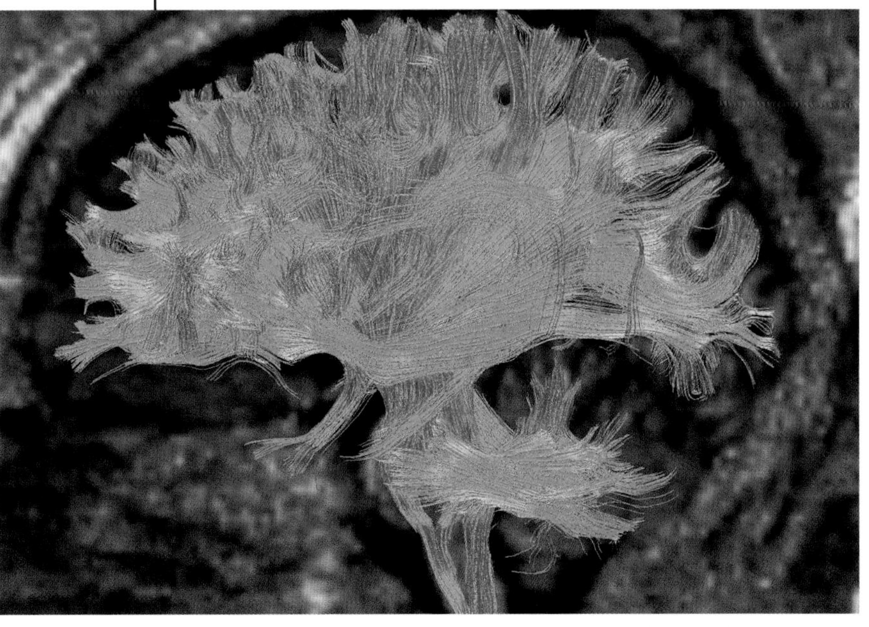
△ **Diffusion MRI** of the human brain

2014
CHASING A COMET

After a ten-year journey, the European Space Agency's Rosetta craft entered orbit around Comet 67P Churyumov-Gerasimenko. Its aim was to study the comet as it approached the Sun from beyond the orbit of Mars, warming up and becoming active. Rosetta's Philae lander touched down on the surface but lost contact because it landed in a deep crevice. However, two years later, at the end of its mission, Rosetta itself made a successful descent and touchdown.

▷ **Philae's landing site** on Comet 67P

2014

2014 Studies confirm that melting of the West Antarctic Ice Sheet is unstoppable and will add 4 m (13 ft) to global sea levels

2014 The Orbiting Carbon Observatory is the first Earth satellite able to map regional carbon dioxide emission and absorption

2015 Teixobactin becomes the first antibiotic to be discovered in 30 years

2014 An engraving found on a 500,000-year-old shell in Java indicates that *Homo erectus* may have been capable of symbolic thought

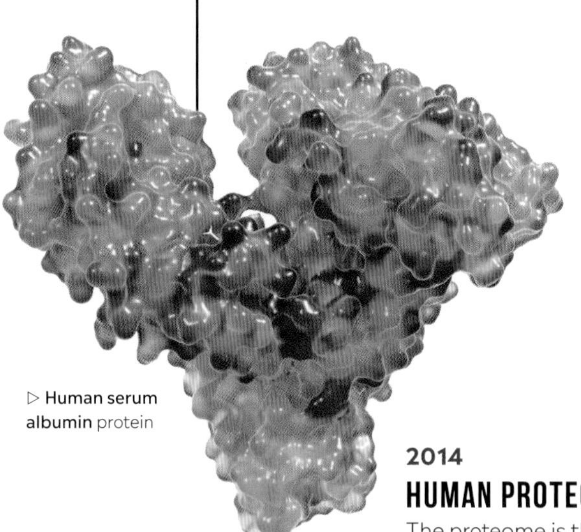

▷ **Human serum albumin** protein

2014
FROZEN VIRUS

French biologists discovered a virus that had lain dormant for 30,000 years in the permafrost of Siberia. After thawing out, the virus, which they named *Pithovirus sibericum*, became infectious again. The researchers warned that as climate warming defrosts more soil, further dormant viruses could be released.

△ **Pleistocene permafrost** sample

2014
HUMAN PROTEOME MAP

The proteome is the complete set of all proteins present in the human body. An international team of researchers used techniques including mass spectroscopy to construct the Human Proteome Map (HPM) – the protein equivalent of the Human Genome Project (HGP). They found proteins encoded by 17,294 genes — about 84 per cent of all the genes predicted to encode proteins.

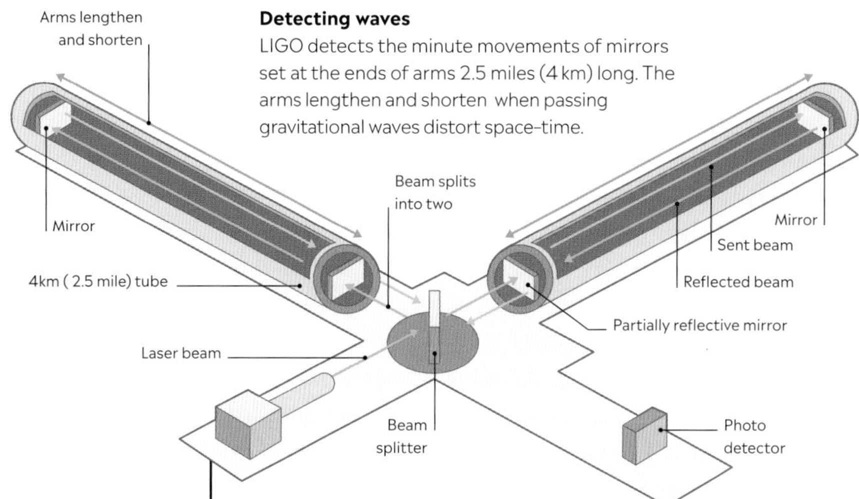

Detecting waves
LIGO detects the minute movements of mirrors set at the ends of arms 2.5 miles (4 km) long. The arms lengthen and shorten when passing gravitational waves distort space-time.

2015

GRAVITATIONAL WAVES

Gravitational waves are ripples in space-time caused by the acceleration of masses – just as electromagnetic waves are caused by the acceleration of electric charges. In 2015, the Laser Interferometer Gravitational-Wave Observatory (LIGO) detected gravitational waves for the first time, in this case from a distant merging of black holes. US physicist Rainer Weiss conceived of the detector in the 1960s, and he shared the 2017 Nobel Prize in Physics with US physicists Kip Thorne and Barry Barish.

"Gravitational waves will bring us exquisitely accurate maps of black holes – maps of their space-time."

KIP THORNE, LIGO SCIENTIST

2015

2015 The discovery of teeth in a 2.8 million-year-old jawbone from Ethiopia fills an evolutionary gap between humans and australopithecine ancestors

2015 Adding rubidium to buckyballs converts the insulator into a conductor and creates a new state of matter called a Jahn-Teller metal

2015 The Dawn and New Horizons probes return close-up views of Ceres and Pluto

2015

NEW SPECIES OF HOMO

Hundreds of bones were found in a South African cave. They were later determined to belong to a new human species – *Homo naledi* which lived between 236,000 and 335,000 years ago. The species had some modern features in their hands and feet but retained more ancient features in their brain and skull size.

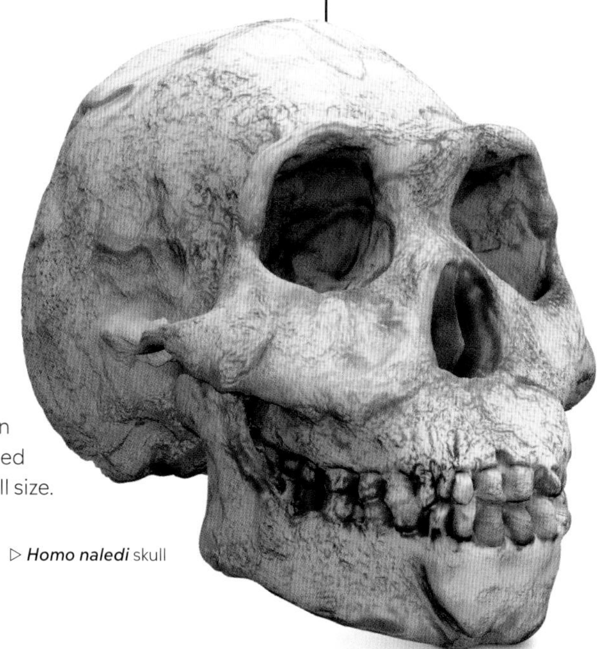
▷ ***Homo naledi*** skull

2017

NEUTRON STARS COLLIDE

The US-based LIGO Laboratory (*see p.301*) and the VIRGO interferometer in Italy observed the gravitational waves produced when two superdense neutron stars, 130 million light years away, spiralled together and merged. This extraordinary, cataclysmic event was also observed by a range of telescopes at different wavelengths. Observations confirmed that neutron star mergers are the source of short gamma-ray bursts from deep space.

▷ **LIGO facility,** Livingstone, Louisiana, US

2016

2016 The world's first three-parent baby is born in Mexico from the father's sperm, the mother's cell nucleus, and an enucleated egg cell

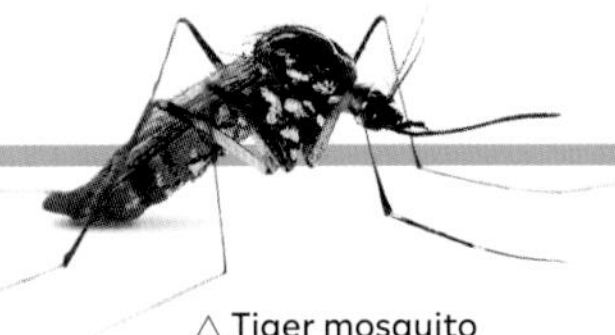

△ Tiger mosquito

2016 An exoplanet, Proxima Centauri b, is discovered orbiting our nearest neighbouring star

2016 Human trials begin using CRISPR-Cas9 technology (*see p.291*) to create genetically altered immune cells to attack cancer

2016

ZIKA VIRUS

In February 2016, the World Health Organization (WHO) declared an outbreak of the Zika virus in Brazil a public health emergency. The virus, named after Zika Forest in Uganda where it was first isolated in 1947, spread rapidly and was associated with birth defects in newborn babies. It is spread by mosquitoes, mainly the yellow fever mosquito *Aedes aegypti*, and the tiger mosquito *A. albopictus*.

▽ **Oumuamua** in the Solar System, artist's impression

2017

VISITOR FROM OUTER SPACE

Canadian astronomer Robert Weryk discovered a comet-like object that proved to be the first confirmed interstellar visitor to the Solar System. Named Oumuamua (the Hawaiian word for "scout"), the object is up to 1 km (0.6 miles) long but much narrower, has a reddish surface, and tumbles end over end. Its hyperbolic orbit, which helps it avoid capture by the Sun's gravity, cannot be explained by past encounters with any of the planets.

2017
QUANTUM COMPUTING

Quantum computers are able to carry out calculations far faster than conventional computers, using qubits (quantum bits), which can have many values simultaneously, rather than bits (binary digits) which can only be 1 or 0. The IBM Q system reached a milestone in 2017, handling 50 qubits at the same time.

△ **Cryostat,** which cools the IBM Q close to absolute zero

2018
CLIMATE CRISIS

In a year that saw wildfires rage in the western US, record heatwaves in southern Europe, and huge storms in the eastern Pacific, an Intergovernmental Panel on Climate Change (IPCC) report warned that limiting global warming to 1.5 °C by 2100 would require far-reaching and unprecedented changes to society. It also warned that failure to limit warming would have devastating consequences.

△ **Wildfire** in the US

2018 The General Conference on Weights and Measures approves a new definition of the kilogram, in terms of fundamental constants. This replaces the 1889 definition, which was the mass of a particular metal cylinder

2018

2017 Biologists name a new species of orangutan, *Pongo tapanuliensis*, on the basis of DNA studies; only 800 remain in the wild

△ ***Dicksonia*** fossil

2018
MOLECULAR PALEONTOLOGY

Chemical analysis of 550-million-year-old *Dickinsonia* fossils from Australia revealed traces of cholesterol-like organic molecules, which are found only in animals. The presence of the distinctive biomarkers supported the classification of *Dickinsonia* not only as an animal but also as the oldest known animal in the fossil record.

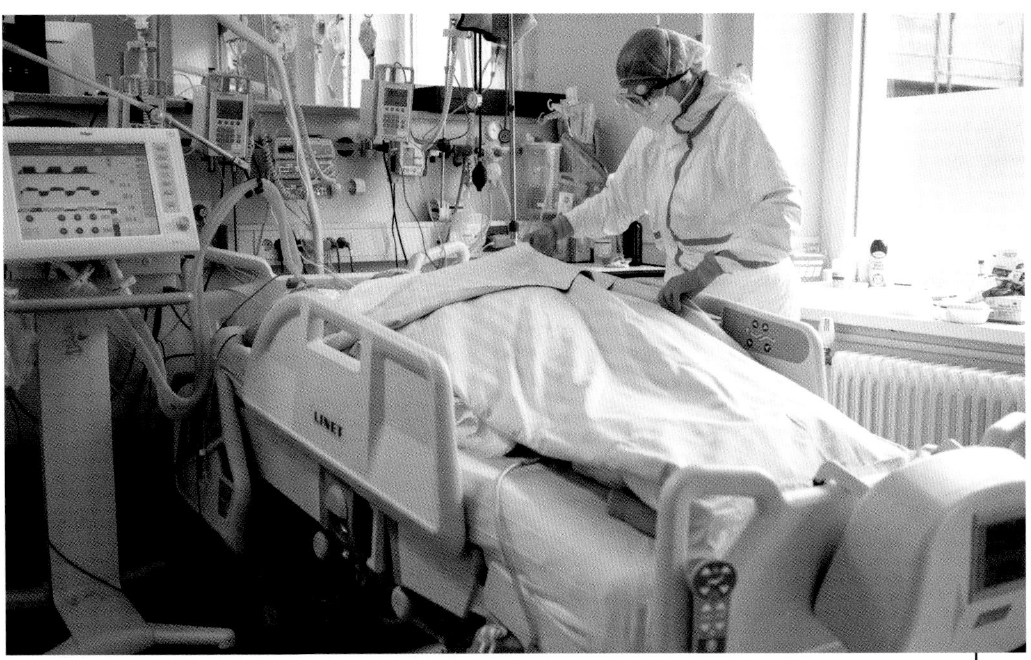

2020

COVID-19

In December 2019, the Health Commission in Wuhan Province, China reported a cluster of pneumonia-like cases. Their cause was found to be a novel coronavirus, 2019-nCoV. In March 2020, the World Health Organization declared that Covid-19 should be regarded as a pandemic. Effective vaccines were developed in Europe and the US, in some cases in less than a year. However, by the end of 2022 Covid-19 had claimed the lives of 6.6 million people worldwide.

◁ **Intensive care** for a Covid-19 patient

2019

2019 Google achieve "quantum supremacy", claiming their quantum computer carries out a calculation in 200 seconds that would take a normal computer many years

2019 NASA's InSight Mars lander detects the first seismic tremor or "Marsquake" on another planet

2019 A global radio telescope collaboration images the silhouette of a supermassive black hole in the distant galaxy Messier 87

2019

MOST DISTANT FLYBY

The New Horizons probe, which launched from Earth in 2006 and visited Pluto in 2015, flew past a small object in the Kuiper Bell known as Arrokoth. This was the most distant encounter to date between a spacecraft and a space object. Images revealed Arrokoth to be a dumbbell-shaped "contact binary" 36 km (22 miles) across, formed by a pair of icy objects spiralling together and gently colliding. It is covered in complex carbon-based chemicals formed by long exposure to solar and cosmic radiation.

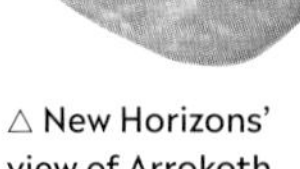

△ New Horizons' view of Arrokoth

2020

ARTIFICIAL INTELLIGENCE

Google's DeepMind algorithm outperformed human radiologists in spotting breast cancer from X-ray images, and an artificial intelligence (AI) system trained by doctors at the University of Pittsburgh bettered humans in identifying cancer from images of biopsy tissue. AI continued to find new applications in other areas, too. These included improving search engine results, facial recognition, playing games, and driving autonomous vehicles.

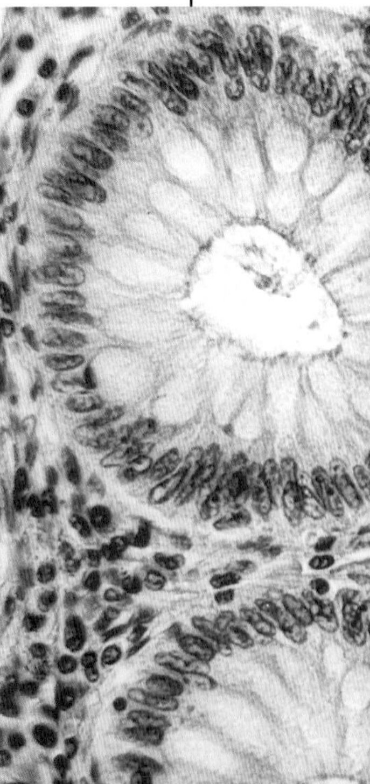

2020
PLASTICS POLLUTION

An Australian study that used a robot to sample sediments found that the ocean floor had become a sink for microplastics. This debris is produced by the breakdown of plastics dumped in ocean waters and amounts to an estimated global total of 14 million tonnes (15.5 million tons). Sources of such microplastics include the microbeads in scrubs, shower gels, and toothpastes.

◁ **Microplastic particles** in a drop of seawater

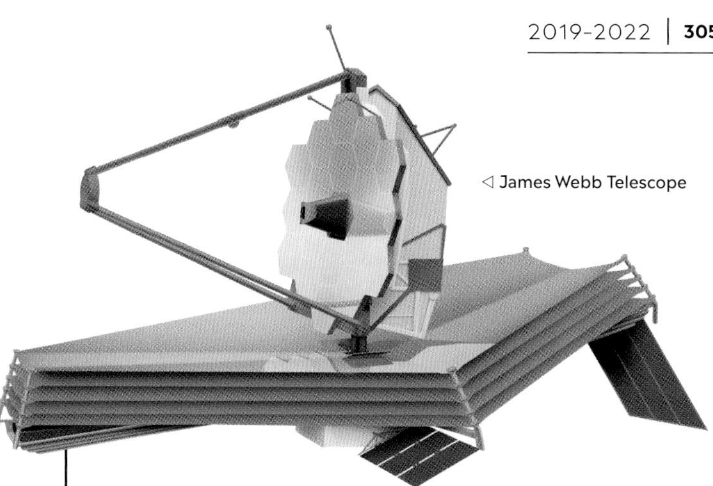

◁ James Webb Telescope

2021
JAMES WEBB TELESCOPE LAUNCHED

NASA launched the James Webb Space Telescope (JWST) – a giant infrared telescope built to detect heat radiation from faint sources such as the earliest stars and galaxies in the Universe, and to identify exoplanets orbiting stars in our own galaxy.

2021 Three different missions arrive at Mars, having launched during a close Earth-Mars alignment in 2020

2022

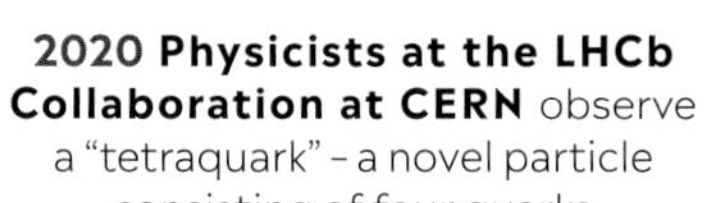

2020 Physicists at the LHCb Collaboration at CERN observe a “tetraquark” – a novel particle consisting of four quarks

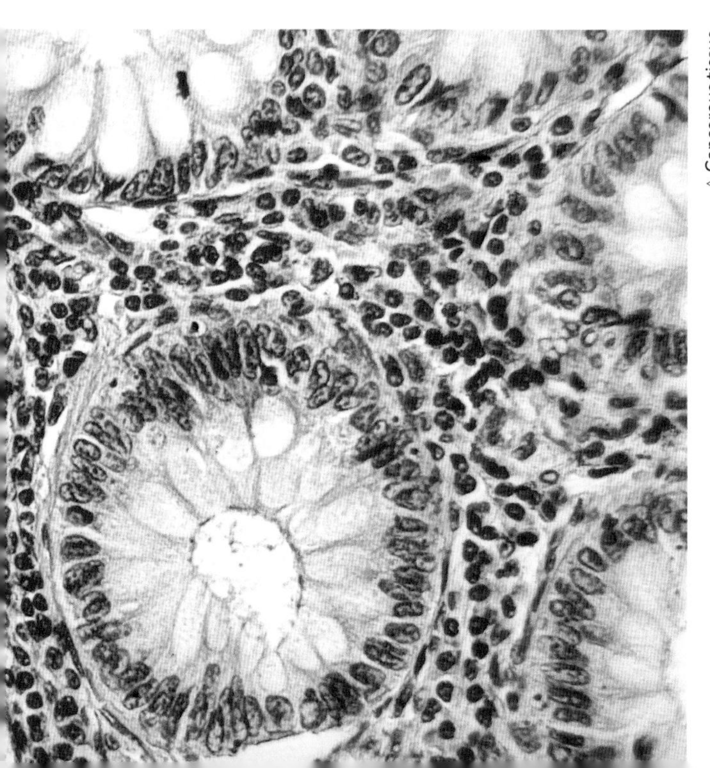

△ Cancerous tissue

△ Great roundleaf bat (*Hipposideros armiger*)

2022
HUMAN ACTIVITY AND EVOLUTION

Scientists from Deakin University, Australia, and Brock University, Canada, investigated some 30 different animal species to see how they had evolved adaptations to human-induced environmental changes. They discovered a number of adaptations to rising temperatures in the enlargement of body parts concerned with heat loss and thermal regulation, finding for example, that Australian parrots increased bill size, rabbits increased ear size, mice tail length, and roundleaf bats wing size.

INDEX

Page numbers in **bold** refer to main entries.

C

D

I

J

ACKNOWLEDGMENTS

The publisher would like to thank Janet Mohun for her comments on the text and Victoria Pyke and Diana Vowles for proof-reading.

The publisher would like to thank the following for their kind permission to reproduce their photographs:

(Key: a-above; b-below/bottom; c-centre; f-far; l-left; r-right; t-top)

1 Dorling Kindersley: Gary Ombler / Whipple Museum of History of Science, Cambridge (c). **2 Science Photo Library:** SCIENCE SOURCE (c). **4-5 Alamy Stock Photo:** Bartlomiej K. Wroblewski (t). **6 Alamy Stock Photo:** CBW (cr); GRANGER - Historical Picture Archive (cl). **Bridgeman Images:** Museum of Science and Industry, Chicago / Photo © 2014 J.B. Spector (r); The Stapleton Collection (c). **The Metropolitan Museum of Art:** Rogers Fund, 1930 (l). **7 Alamy Stock Photo:** BSIP SA (c); Science History Images (l); Science Photo Library (cl); LWM / NASA / LANDSAT (cr). **Science Photo Library:** SHEFFIELD UNIVERSITY, DRS P. WARD & T. BUTTON (r). **8-9 Alamy Stock Photo:** ADC PICTURES (c). **10 Alamy Stock Photo:** MET / BOT (tr); Mlouisphotography (bl); The Natural History Museum (tc). **11 Alamy Stock Photo:** Arterra Picture Library (tl); UPI (c). **12 Alamy Stock Photo:** Claudio Rampinini (tr); The Natural History Museum (tl). **13 Alamy Stock Photo:** mer Kele (bl); Wirestock, Inc. (bc). **Dorling Kindersley:** Dreamstime.com: Darryl Brooks / Dbvirago (tr). **14-15 Alamy Stock Photo:** funkyfood London - Paul Williams (tc). **14 Alamy Stock Photo:** Kutsal Lenger (bl). **The Metropolitan Museum of Art:** Gift of Valdemar Hammer Jr., in memory of his father, 1936 (br). **15 Alamy Stock Photo:** Nigel Spooner (cl); PA Images (tr); www.BibleLandPictures.com (bl). **16 Alamy Stock Photo:** Evgeni Ivanov (bl); Science History Images (tr); World History Archive (br). **17 akg-images:** Andr Held (br). **Alamy Stock Photo:** Artokoloro (tr); Mike Goldwater (bl). **18 akg-images:** Interfoto (br). **Alamy Stock Photo:** GRANGER - Historical Picture Archive (bl). **The Metropolitan Museum of Art:** Rogers Fund, 1930 (tl). **19 Alamy Stock Photo:** Adam Jn Fige (bl); Eraza Collection (tr); Science History Images (tl). **20 Alamy Stock Photo:** World History Archive (tl); World History Archive (tc). **Bridgeman Images:** Pictures from History (bl). **21 Alamy Stock Photo:** Classic Image (br); WBC ART (tl); GRANGER - Historical Picture Archive (cl). **22 Alamy Stock Photo:** A. Astes (tl); Artokoloro (tr); Antiqueimages (bl); World History Archive (br). **23 Alamy Stock Photo:** Album (bl). **Bridgeman Images:** Archives Charmet (cl). **24 Alamy Stock Photo:** Science History Images (br); The Granger Collection (tl). **The Metropolitan Museum of Art:** Rogers Fund, 1914 (c). **25 Alamy Stock Photo:** PRISMA ARCHIVO (br); Science History Images (tl). **26 Alamy Stock Photo:** Artokoloro (tl); imageBROKER (cr). **27 Alamy Stock Photo:** Chroma Collection (tr); Heritage Image Partnership Ltd (br). **Getty Images:** Universal History Archive (cl). **28 Alamy Stock Photo:** Chronicle (tr); Photo 12 (bl); Stock Montage, Inc. (br). **28-29 Alamy Stock Photo:** Stocktrek Images, Inc. (tc). **29 Alamy Stock Photo:** IanDagnall Computing (bl). **30 Alamy Stock Photo:** GRANGER - Historical Picture Archive (bl). **Bridgeman Images:** Universal History Archive / UIG (tl). **Getty Images:** LOUISA GOULIAMAKI / Stringer (br). **31 akg-images:** Science Source (c). **Alamy Stock Photo:** Ancient Art and Architecture (tl); tom pfeiffer (br). **32 Alamy Stock Photo:** CPA Media Pte Ltd (cr); Panther Media GmbH (tl); Historic Collection (tr). **Dorling Kindersley:** Clive Streeter / The Science Museum, London (bl). **33 Alamy Stock Photo:** GRANGER - Historical Picture Archive (br); Science History Images (tr). **Dorling Kindersley:** John Lepine / Science Museum, London / John Lepine / Science Museum, London / Dorling Kindersley (tc, tc). **34 Alamy Stock Photo:** The History Collection (bl). **Bridgeman Images:** Stefano Bianchetti (tl). **Getty Images:** DE AGOSTINI PICTURE LIBRARY / Contributor (tr). **35 Alamy Stock Photo:** Album (t). **Getty Images / iStock:** boris_1983 (br). **36 Alamy Stock Photo:** INTERFOTO (bc). **Getty Images:** NurPhoto / Contributor (tl). **Getty Images / iStock:** yuriz (br). **37 Alamy Stock Photo:** Alan Dyer / VWPics (bl); The History Collection (br). **Bridgeman Images:** British Library Board. All Rights Reserved (t). **38 Alamy Stock Photo:** Album (tl); World History Archive (tc). **39 Alamy Stock Photo:** Abu Castor (tl); FLHC K (bl). **Getty Images:** Science & Society Picture Library / Contributor (tr). **40 Alamy Stock Photo:** Aclosund Historic (tl); The Natural History Museum (tr); Volgi archive (bc). **40-41 Alamy Stock Photo:** Panther Media GmbH (b). **41 Alamy Stock Photo:** Heritage Image Partnership Ltd (tr); The History Collection (br). **Getty Images:** Pictures from History / Contributor (tc). **42 akg-images:** © NYPL / Science Source / SCIENCE SOURCE (tc). **Alamy Stock Photo:** Science History Images (bl). **Bridgeman Images:** Bridgeman Images (tl). **43 Alamy Stock Photo:** CPA Media Pte Ltd (br); World History Archive (bc). **44 Alamy Stock Photo:** Classic Image (bl); zhang jiahan (tc); Science History Images (br). **Bridgeman Images:** Giancarlo Costa (tl). **45 Alamy Stock Photo:** Album (tl). **Getty Images:** Sino Images (tr). **46 Alamy Stock Photo:** Album (bl); CPA Media Pte Ltd (br). **Bridgeman Images:** Archives Charmet (tr). **47 Alamy Stock Photo:** Magite Historic (bl); Pictorial Press Ltd (tl). **48 Bridgeman Images:** NPL - DeA Picture Library / M. Seemuller (tl). **Getty Images:** Science & Society Picture Library / Contributor (bl). **49 Alamy Stock Photo:** Azoor Photo Collection (tr); Magite Historic (tl); Chronicle (bl). **50 akg-images:** Fototeca Gilardi (cr). **Alamy Stock Photo:** GRANGER - Historical Picture Archive (bc); Tim Brown (tl). **51 Alamy Stock Photo:** Album (bl); ART Collection (tl); Realy Easy Star / Toni Spagone (tr). **52 Alamy Stock Photo:** Artokoloro (br); The History Collection (tl). **Bridgeman Images:** Bridgeman Images (bc). **53 akg-images:** Science Source (cr). **Alamy Stock Photo:** Azoor Collection (c); Science History Images (br). **54 Alamy Stock Photo:** GRANGER - Historical Picture Archive (tc); World History Archive (bl); The Granger Collection (bc). **Bridgeman Images:** © Christie's Images (tl). **55 Alamy Stock Photo:** CPA Media Pte Ltd (tc); The Natural History Museum (tl). **56 Alamy Stock Photo:** Artokoloro (cl); The Granger Collection (tl); The Picture Art Collection (cr). **56-57 Alamy Stock Photo:** Art Collection 3 (b). **57 Alamy Stock Photo:** GRANGER - Historical Picture Archive (bc); World History Archive (br). **58-59 Alamy Stock Photo:** Science History Images (t). **58 Alamy Stock Photo:** PhotoStock-Israel (br); The Granger Collection (tl). **Science Photo Library:** CNRI (bl). **59 Alamy Stock Photo:** Album (bc); Russell Mountford (br). **60 Alamy Stock Photo:** GL Archive (tl); The Print Collector (tr); The History Collection (bl). **Bridgeman Images:** © NPL - DeA Picture Library / M. Seemuller (br). **61 Alamy Stock Photo:** GRANGER - Historical Picture Archive (br); The Print Collector (bl). **62 Alamy Stock Photo:** Chronicle (cr); Jonathan Orourke (bl). **Bridgeman Images:** The Stapleton Collection (tl). **63 Alamy Stock Photo:** GL Archive (br); Science History Images (tr). **Bridgeman Images:** University of St. Andrews Library (bl). **64 akg-images:** Rabatti & Domingie (tr). **Alamy Stock Photo:** Diego Barucco (tc); Tibbut Archive (br). **Bridgeman Images:** Universal History Archive / UIG (bl). **65 Alamy Stock Photo:** gameover (tl). **Science & Society Picture Library:** Science Museum (tr). **66 Alamy Stock Photo:** AF Fotografie (br); GRANGER - Historical Picture Archiv (cr); IanDagnall Computing (bl). **Wellcome Collection:** 4.0 International (CC BY 4.0) (tc). **67 Alamy Stock Photo:** Andreas Huslbetz (cl); Heritage Image Partnership Ltd (br). **68 Wellcome Collection:** (c). **69 Science Photo Library:** Zephyr (tr). **70 Alamy Stock Photo:** Heritage Image Partnership Ltd (bl); The Picture Art Collection (tl); Science History Images (tr); Science History Images (br). **71 Alamy Stock Photo:** Well / BOT (tl). **Bridgeman Images:** NPL - DeA Picture Library / G. Cigolini (br). **72-73 Alamy Stock Photo:** Chronicle (tl). **72 Alamy Stock Photo:** Album (bl). **73 akg-images:** Collection Joinville (bc). **Alamy Stock Photo:** Atlaspix (cl); Hamza Khan (tr). **74-75 Alamy Stock Photo:** Science History Images (t). **74 Alamy Stock Photo:** History & Art Collection (cr); Science History Images (tl); IanDagnall Computing (bl). **Getty Images:** Science & Society Picture Library / Contributor (c). **75 Alamy Stock Photo:** B.A.E. Inc. (tr). **76 NASA:** Solar Dynamics Observatory (c). **77 Science Photo Library:** CHARLES D. WINTERS (tl). **78 Alamy Stock Photo:** GRANGER - Historical Picture Archive (tr); Science History Images (tl); Pictorial Press Ltd (br). **79 Alamy Stock Photo:** PRISMA ARCHIVO (br); Science History Images (tl); Science History Images (bc). **80 Alamy Stock Photo:** Alfio Scisetti (tr); Eraza Collection (br). **Bridgeman Images:** Iberfoto (tl). **81 Alamy Stock Photo:** ACTIVE MUSEUM / ACTIVE ART (bl); Heritage Image Partnership Ltd (tl); The Granger Collection (cr); The Natural History Museum (br). **82 Science Photo Library:** NASA / JPL (tl). **83 Dreamstime.com:** Dmitrydesigner (c). **84 Alamy Stock Photo:** Science History Images (cl); World History Archive (bc). **85 Alamy Stock Photo:** Florilegius (tc); Phanie (bc). **Getty Images:** Science & Society Picture Library / Contributor (tl). **86 Alamy Stock Photo:** Ron Giling (bl); The Granger Collection (tr). **87 Alamy Stock Photo:** Dinodia Photos (tl); Science History Images (tr); World History Archive (bl); Science History Images (br). **88 Alamy Stock Photo:** Bjrn Wylezich (tl); The Picture Art Collection (tr); GRANGER - Historical Picture Archive (br). **89 Alamy Stock Photo:** The Natural History Museum (bl); The Picture Art Collection (bc). **90 Alamy Stock Photo:** GL Archive (br). **Bridgeman Images:** Muse Cond, Chantilly (tr). **Getty Images:** Science & Society Picture Library / Contributor (c). **91 Alamy Stock Photo:** Eraza Collection (br); Wim Wiskerke (tl); Svintage Archive (tr). **92 Alamy Stock Photo:** Artokoloro (bl); CBW (tl). **Bridgeman Images:** Philadelphia Museum of Art / Gift of Mr. and Mrs. Wharton Sinkler (br). **93 Alamy Stock Photo:** GRANGER - Historical Picture Archive (tc). **Bridgeman Images:** Lorio / Iberfoto (tl). **94 Alamy Stock Photo:** Science History Images (tl); Science History Images (tr). **94-95 Alamy Stock Photo:** INTERFOTO (b). **95 Alamy Stock Photo:** Album (bc); The History Collection (tl). **Getty Images:** Science & Society Picture Library / Contributor (tr). **96 Alamy Stock Photo:** Chronicle (tl); Science History Images (tr). **Dorling Kindersley:** Dave King / The Science Museum (bl). **97 Alamy Stock Photo:** The Picture Art Collection (br). **Bridgeman Images:** Granger (tr). **Getty Images:** Science & Society Picture Library / Contributor (bl). **98 Alamy Stock Photo:** agefotostock (br); GL Archive (bl). **99 Alamy Stock Photo:** Science History Images (bl); The Granger Collection (br). **100 Alamy Stock Photo:** Artokoloro (bl). **Getty Images:** Science & Society Picture Library (br). **100-101 Bridgeman Images:** Giancarlo Costa (t). **101 Alamy Stock Photo:** Central Historic Books (br); PWB Images (bl). **102 Alamy Stock Photo:** incamerastock (tr); Science History Images (bl); Science History Images (br). **Getty Images:** Print Collector (tl). **103 Alamy Stock Photo:** NASA Photo (tl). **Getty Images:** Science & Society Picture Library (tr). **104 Alamy Stock Photo:** Jason Smith (tr); The Picture Art Collection (tl). **105 akg-images:** (tl). **Alamy Stock Photo:** Florilegius (tr). **Science Photo Library:** ROYAL INSTITUTION OF GREAT BRITAIN (bl). **106 Alamy Stock Photo:** Science History Images (bl); Soren Klostergaard Pedersen (bc). **107 Alamy Stock Photo:** ACTIVE MUSEUM / ACTIVE ART (b); Dirk Daniel Mann (tl); North Wind Picture

Archives (tr). **108 Alamy Stock Photo:** Science History Images (tr); The History Collection (tl). **Bridgeman Images:** Natural History Museum, London (br). **109 Alamy Stock Photo:** CBW (tr); GRANGER - Historical Picture Archive (bc); Pictorial Press Ltd (bl). **Dorling Kindersley:** Clive Streeter / The Science Museum, London (tl). **110 Alamy Stock Photo:** Science History Images (tr). **Dorling Kindersley:** Gary Ombler / Whipple Museum of History of Science, Cambridge (tl). **111 Alamy Stock Photo:** Christophe Coat (c); Heritage Image Partnership Ltd (bl); Hamza Khan (br). **112 Alamy Stock Photo:** Matteo Chinellato (bc); SuperStock (t). **Getty Images:** Science & Society Picture Library (bl). **113 Alamy Stock Photo:** World History Archive (bl). **114 Alamy Stock Photo:** Chronicle (bl); Science History Images (tl); The Print Collector (tr). **Getty Images:** Bildagentur-online (br). **115 Alamy Stock Photo:** Heritage Image Partnership Ltd (br); The Natural History Museum (bl). **Bridgeman Images:** PVDE (tr). **116 Alamy Stock Photo:** Iuliia Nemchinova (c). **118 Alamy Stock Photo:** Artokoloro (tr); RGB Ventures / SuperStock (tl); The Natural History Museum (br). **119 Alamy Stock Photo:** Album (br). **Getty Images:** Joseph Niepce / Stringer (tr). **120 Alamy Stock Photo:** Science History Images (tl); Science Photo Library (tr); World History Archive (br). **121 Alamy Stock Photo:** Chronicle (br); Science History Images (tl). **Dorling Kindersley:** Harry Taylor / Sedgwick Museum of Geology, Cambridge (tr). **122 Alamy Stock Photo:** GL Archive (bl); The Natural History Museum (bl); INTERFOTO (tr). **123 Alamy Stock Photo:** Everett Collection Historical (bl). **Bridgeman Images:** Bridgeman Images (tr). **Science Photo Library:** SCIENCE STOCK PHOTOGRAPHY (br). **124 Alamy Stock Photo:** Classic Collection 3 (tl); Science History Images (tr); World History Archive (cl); Scenics & Science (br). **125 Alamy Stock Photo:** AC NewsPhoto (br); GRANGER - Historical Picture Archive (tl). **Getty Images:** Science & Society Picture Library (bl). **126 Alamy Stock Photo:** Pictorial Press Ltd (bl); The Granger Collection (tr); Stocktrek Images, Inc. (br). **Getty Images:** Science & Society Picture Library (tl). **127 Alamy Stock Photo:** rico ploeg (bl). **Getty Images:** Science & Society Picture Library (t). **128 Alamy Stock Photo:** Science History Images (tl). **Getty Images:** DE AGOSTINI PICTURE LIBRARY (bl). **Science Photo Library:** DR JEREMY BURGESS (tr). **128-129 Alamy Stock Photo:** GRANGER - Historical Picture Archive (bc). **129 Alamy Stock Photo:** Science History Images (br). **130 Alamy Stock Photo:** INTERFOTO (tr). **Getty Images:** Science & Society Picture Library (tl). **131 Alamy Stock Photo:** Alan Dyer / VWPics (c); The Reading Room (tl). **Getty Images:** Science & Society Picture Library (br). **132-133 Alamy Stock Photo:** Sueddeutsche Zeitung Photo (c). **132 Science Photo Library:** SHEILA TERRY (tl). **134 Alamy Stock Photo:** NASA Pictures (br); Science History Images (t). **135 Alamy Stock Photo:** Ben Queenborough (tc); Henri Koskinen (tr). **Bridgeman Images:** (br). **Dorling Kindersley:** Dreamstime.com: Jan Martin Will (tl). **136 Alamy Stock Photo:** Allstar Picture Library Limited (bl); Chronicle (tl); Science History Images (br); Nathaniel Noir (cr). **137 Alamy Stock Photo:** Central Historic Books (tl); Realy Easy Star / Toni Spagone (tr). **Bridgeman Images:** Museum of Science and Industry, Chicago / Photo © 2014 J.B. Spector (br). **138 Alamy Stock Photo:** Pictorial Press Ltd (bl); Universal Images Group North America LLC / DeAgostini (br). **139 Alamy Stock Photo:** Pictorial Press Ltd (br); The Print Collector (tl). **Science Photo Library:** NATURAL HISTORY MUSEUM, LONDON (tr). **140 Science Photo Library:** ARGONNE NATIONAL LABORATORY (c). **141 NASA:** JPL / Caltech (tr). **142 Alamy Stock Photo:** GL Archive (bl); Science History Images (tr). **143 Alamy Stock Photo:** Phil Degginger (br); Science Photo Library (tl). **Science Photo Library:** CARLOS CLARIVAN (bl); M.I. WALKER (tr). **144 Alamy Stock Photo:** Ivan Vdovin (bl). **145 Alamy Stock Photo:** Classic Image (cl); The Granger Collection (tl); PRISMA ARCHIVO (bl). **Bridgeman Images:** Look and Learn (br). **146 Science Photo Library:** TED KINSMAN (tl). **147 NASA:** ESA / CSA / STScI / NIRCam (c). **148 Alamy Stock Photo:** Science Photo Library (cl). **Getty Images / iStock:** theasis (tl). **Science Photo Library:** SCIENCE SOURCE (bl); SHEILA TERRY (br). **149 Alamy Stock Photo:** World History Archive (tl). **150 Alamy Stock Photo:** Central Historic Books (br); Chronicle (tl). **Science Photo Library:** JOHN READER (tc). **151 Alamy Stock Photo:** Monika Wisniewska (tr). **Science Photo Library:** JOSE CALVO (br). **152 Getty Images:** Science & Society Picture Library (c). **154 Alamy Stock Photo:** History and Art Collection (br); Pictorial Press Ltd (tl); Old Books Images (tr); NASA Image Collection (bl). **155 Alamy Stock Photo:** Alexandros Lavdas (br); GRANGER - Historical Picture Archive (tl); World History Archive (cr). **156 Alamy Stock Photo:** Library Book Collection (tr). **157 Alamy Stock Photo:** GRANGER - Historical Picture Archive (br). **Getty Images / iStock:** Patrick Jennings (bl). **Getty Images:** Science & Society Picture Library (tr). **158 Alamy Stock Photo:** Library Book Collection (br); North Wind Picture Archives (tl). **159 Alamy Stock Photo:** Johann Schumacher (tl); Sari O'Neal (cl); Science History Images (tr). **Getty Images:** Science & Society Picture Library (br). **160 Alamy Stock Photo:** incamerastock (tl); Science History Images (br). **Getty Images:** Science & Society Picture Library (bl). **161 Alamy Stock Photo:** Book Worm (br); Cultura Creative Ltd (tr); IanDagnall Computing (bl). **Dorling Kindersley:** Ruth Jenkinson / Holts Gems (cl). **162 Alamy Stock Photo:** leonello calvetti (tr); Science History Images (br). **163 Alamy Stock Photo:** Hi-Story (bl). **Getty Images:** Bettmann (br). **164 Alamy Stock Photo:** Helen Cowles (tl); pittawut junmee (tr). **165 Alamy Stock Photo:** GRANGER - Historical Picture Archive (br); World History Archive (tl); Science History Images (cr). **166 Alamy Stock Photo:** Paul Fearn (tr); RBM Vintage Images (bl); Science History Images (br). **167 Alamy Stock Photo:** colaimages (tl); Stocktrek Images, Inc. (br). **Getty Images:** Universal History Archive (tr). **168 Alamy Stock Photo:** Science History Images (bl). **169 Alamy Stock Photo:** Leighton Collins (tl); Oliver Smart (tr); The Granger Collection (bl); Maidun Collection (br). **170 Alamy Stock Photo:** Book Worm (bl); Science Photo Library (tr). **Getty Images:** Science & Society Picture Library (br). **171 Getty Images:** Lazy_Bear (br); Universal History Archive (tl); Universal History Archive (bl). **172 Alamy Stock Photo:** Arterra Picture Library (tr); ephotocorp (c). **173 Alamy Stock Photo:** North Wind Picture Archives (tr); Science History Images (br). **174 Alamy Stock Photo:** Bjrn Wylezich (tl). **175 Science Photo Library:** NASA (c). **176 Getty Images:** Science & Society Picture Library (br). **177 Alamy Stock Photo:** FLHC 220C (cl); Grant Heilman Photography (tl); The Print Collector (tr); Scott Camazine (br). **178 Alamy Stock Photo:** Science History Images (br). **Getty Images:** Print Collector / Contributor (tr). **179 Alamy Stock Photo:** Chronicle (br); The Book Worm (tl). **180 Science Photo Library:** CHRISTIAN LUNIG (c). **181 Science Photo Library:** ARSCIMED (tr). **182 akg-images:** n / a (tl). **Alamy Stock Photo:** Everett Collection Inc (tr); The Print Collector (br). **183 Alamy Stock Photo:** Hilary Morgan (t); Pictorial Press Ltd (bl). **Science Photo Library:** ADRIAN T SUMNER (br). **184 Alamy Stock Photo:** Giulio Ercolani (bl); Science Photo Library (tl). **Dorling Kindersley:** Clive Streeter / The Science Museum, London (tr). **185 Alamy Stock Photo:** IanDagnall Computing (tr). **Getty Images:** Science & Society Picture Library / Contributor (bl). **186 Bridgeman Images:** Israel Museum, Jerusalem / Gift of the Jacob E. Safra Philanthropic Foundation (c). **187 Getty Images:** Science & Society Picture Library (tl). **188-189 Alamy Stock Photo:** The History Collection (t). **189 Alamy Stock Photo:** World History Archive (bl). **Science Photo Library:** MARTIN SHIELDS (br). **190 Alamy Stock Photo:** GRANGER - Historical Picture Archive (tr). **Dorling Kindersley:** Clive Streeter / The Science Museum, London (tc). **Getty Images / iStock:** flyparade (bl). **191 Alamy Stock Photo:** GRANGER - Historical Picture Archive (tr); The History Collection (br). **192 Science Photo Library:** PHILIPPE PLAILLY (c). **193 Getty Images:** Science & Society Picture Library (tr). **194 Alamy Stock Photo:** Album (br); Science History Images (tl). **195 Alamy Stock Photo:** Life on white (bl); Science History Images (bc). **196 Alamy Stock Photo:** molekuul.be (bc). **Dorling Kindersley:** Frank Greenaway / Natural History Museum, London (br); Gary Ombler, Oxford University Museum of Natural History (bl). **197 Getty Images:** Science & Society Picture Library / Contributor (tr). **Science Photo Library:** R.BIJLENGA / DEPT. OF MICROBIOLOGY, BIOZENTRUM (tl). **198 Getty Images / iStock:** CasarsaGuru (tl). **199 Science Photo Library:** EQUINOX GRAPHICS (c). **200 Alamy Stock Photo:** blickwinkel (bl); Science History Images (tl); Gainew Gallery (tr). **201 Alamy Stock Photo:** agefotostock (br). **Science Photo Library:** MICHEL DELARUE, ISM (bl). **202 Alamy Stock Photo:** Chronicle (bc). **Getty Images:** Science & Society Picture Library / Contributor (tl). **203 Alamy Stock Photo:** Sueddeutsche Zeitung Photo (tr). **204 Alamy Stock Photo:** Chronicle (tr); ruelleruelle (tl). **205 Alamy Stock Photo:** IanDagnall Computing (tl); Photo 12 (tr); INTERFOTO (bl). **206 Science Photo Library:** PHILIPPE PSAILA (tl). **206-207 Getty Images:** Gamma-Rapho / Patrick Aventurier (bc). **208 Alamy Stock Photo:** Alpha Historica (tr); GL Archive (tl). **Getty Images:** Bettmann / Contributor (br). **209 Alamy Stock Photo:** NASA Pictures (tr). **Science Photo Library:** SOUTHERN ILLINOIS UNIVERSITY (tl). **210 Alamy Stock Photo:** J Marshall - Tribaleye Images (t). **Getty Images:** Krista Few (b). **211 Alamy Stock Photo:** Archive PL (cl); Science History Images (tr). **Science Photo Library:** NASA (tl). **212 Alamy Stock Photo:** imageBROKER (tr); Science History Images (b). **Dorling Kindersley:** Clive Streeter / The Science Museum, London (c). **213 Alamy Stock Photo:** Pictorial Press Ltd (bl). **214 Alamy Stock Photo:** Everett Collection Inc (tl); NASA Photo (tr); RBM Vintage Images (bl). **215 Alamy Stock Photo:** Andy Thompson (br). **Science Photo Library:** Science Photo Library (bl). **216 Alamy Stock Photo:** Thomas Lehtinen (tl). **217 Alamy Stock Photo:** Classic Picture Library (c). **218 Alamy Stock Photo:** Keystone Press (tl); Universal Images Group North America LLC (bl). **Bridgeman Images:** British Library Board (tr). **219 Alamy Stock Photo:** Science History Images (br); World History Archive. **Dorling Kindersley:** 123RF.com: Corey A Ford (bl). **220 Alamy Stock Photo:** Imago History Collection (tr). **Getty Images:** Universal History Archive (tl). **221 Alamy Stock Photo:** GRANGER - Historical Picture Archive (t). **222 Alamy Stock Photo:** Chronicle (br); Tango Images (tl); Science History Images (tr); RGB Ventures / SuperStock (bl). **223 Alamy Stock Photo:** JSM Historical (bl); UPI (br). **Getty Images:** Bletchley Park Trust (tr). **224-225 Getty Images / iStock:** Filipp Borshch (bc). **224 Science Photo Library:** PATRICK LANDMANN (tl). **226 Alamy Stock Photo:** Rudmer Zwerver (cl); Science History Images (tl); SuperStock (br). **227 Alamy Stock Photo:** Science History Images (bl). **Getty Images:** Bettmann (tl). **228 Alamy Stock Photo:** PBH Images (br); www.BibleLandPictures.com (bl). **Science Photo Library:** C. POWELL, P. FOWLER & D. PERKINS (t). **229 Alamy Stock Photo:** Science History Images (tr); World History Archive (tl). **230 Alamy Stock Photo:** Science History Images (br). **Getty Images:** Daily Herald Archive (bl); Science Museum (tl). **231 Alamy Stock Photo:** Science History Images (tl). **Science Photo Library:** NATIONAL LIBRARY OF MEDICINE (bl). **232 Alamy Stock Photo:** BSIP SA (bl). **Getty Images / iStock:** alex-mit (tr). **233 Alamy Stock Photo:** Nature Photographers Ltd (cl); Science History Images (br). **234 Alamy Stock Photo:** NGDC / NOAA / Phil Degginger (br). **Getty Images:** Bettmann (tr). **235 Alamy Stock Photo:** Science History Images (tr). **Science Photo Library:** HANK MORGAN (tl). **236 Science Photo Library:** (tl). **237 Science & Society Picture Library:** Science Museum (c). **238 Alamy Stock Photo:** Science Photo Library (tl). **Science Photo Library:** PROF. ERWIN MUELLER (br). **239 Alamy Stock Photo:** Keystone Press (bl); Tango Images (tl); Science History Images (tr); World History Archive (br). **240 Alamy Stock Photo:** Science History Images (bl); Vitaliy Gaydukov (tl); Science History Images (br). **Bridgeman Images:** Sovfoto / UIG (tr). **241 Alamy Stock Photo:** John Frost Newspapers (tr); NASA Photo (tl); REUTERS (br). **242 Alamy Stock Photo:** Historic Images (tl); Reading Room 2020 (tr). **243 Alamy Stock Photo:** Heritage Image Partnership Ltd (tr); Stocktrek Images, Inc. (bl). **244 Getty Images:** Richard Heathcote / Getty Images Sport (c). **245 Science Photo Library:** BIOLUTION GMBH (tr). **246 Alamy Stock Photo:** NASA Image Collection (tl); Universal Art Archive (bc); PhotoSpirit (br). **247 Science Photo Library:** EUROPEAN SOUTHERN OBSERVATORY (tl). **248 Alamy Stock Photo:** blickwinkel (cl); Space prime (tl). **Getty Images:** Sovfoto (br). **249 Alamy Stock Photo:** Gado Images (tl). **Science Photo Library:** CARLOS CLARIVAN (tr); NATIONAL PHYSICAL LABORATORY © CROWN COPYRIGHT (br). **250 Alamy Stock Photo:** imageBROKER (br). **Science Photo Library:** DETLEV VAN RAVENSWAAY (tr). **251 Alamy Stock Photo:** Philip Game (br); Pictorial Press Ltd (bl). **Getty Images:** Daily Herald Archive (tr). **252 Alamy Stock Photo:** LWM / NASA / LANDSAT (br). **Getty Images:** Bettmann (bl). **253 Alamy Stock Photo:** Science History Images (br). **Dorling Kindersley:** Dreamstime.com: Jm73 (tr). **254 Alamy Stock Photo:** Science History Images (tl); WENN Rights Ltd (bl); Stocktrek Images, Inc. (cl); stock imagery (br). **255 Alamy Stock Photo:** PB / YB (tl); Stocktrek Images, Inc. (tr). **256 Alamy Stock Photo:** Science Photo Library (tl). **Getty Images:** Bettmann (cr). **257 Alamy Stock Photo:** NG Images (tl); NG Images (bl); SBS Eclectic Images (cr). **258 Alamy Stock Photo:** CBW (br); Medicshots (tl); dotted zebra (cr). **Science Photo Library:** DETLEV VAN RAVENSWAAY (bl). **259 Alamy Stock Photo:** Auk Archive (bl); Science History Images (br). **260 Alamy Stock Photo:** Maidun Collection (tl); ZUMA Press, Inc. (bl); NASA Image Collection (br). **Science Photo Library:** MILLARD H. SHARP / SCIENCE SOURCE (cr). **261 Alamy Stock Photo:** Science Photo Library (tr). **262 Alamy Stock Photo:** Sueddeutsche Zeitung Photo (tr). **Getty Images:** VW Pics (b). **Science Photo Library:** NASA (tl). **263 Alamy Stock Photo:** agefotostock (tl); Science Photo Library (bl). **Science Photo Library:** EYE OF SCIENCE (br). **264 Science Photo Library:** DETLEV VAN RAVENSWAAY (tl); R.B.HUSAR / NASA (bl); EM UNIT, VLA (br). **265 Alamy Stock Photo:** Image Source (tr); NASA Image Collection (tl). **Science Photo Library:** JIM WEST (br). **266 ESA:** AOES Medialab (tl). **267 NASA:** ESA / Hubble (c). **268 Alamy Stock Photo:** Sabena Jane Blackbird (cr). **Science Photo Library:** DAVID PARKER (tl); NCI / ADVANCED BIOMEDICAL COMPUTING CENTER / SCIENCE SOURCE (bl). **269 Alamy Stock Photo:** CBW (br). **Science Photo Library:** THOM LEACH (bl). **270 Getty Images:** Wojtek Laski (tl). **Science Photo Library:** SHEFFIELD UNIVERSITY, DRS P. WARD & T. BUTTON (br). **271 Alamy Stock Photo:** CBW (bl); Minden Pictures (c). **Getty Images:** Historical (tr). **272 Alamy Stock Photo:** GRANGER - Historical Picture Archive (tr); Martin Shields (tl). **273 Alamy Stock Photo:** Stocktrek Images, Inc. (c). **Science Photo Library:** CERN (tr); PHILIPPE PLAILLY (br). **274 Alamy Stock Photo:** Historic Collection (bl); RGB Ventures / SuperStock (br). **Getty Images / iStock:** Nerthuz (tl). **275 Alamy Stock Photo:** Hemis (br); Thibault Renard (bl). **276 Alamy Stock Photo:** agefotostock (bl); Corey Ford (br). **Getty Images / iStock:** elgol (tl). **277 Alamy Stock Photo:** Kathy deWitt (br); Science History Images (tc). **Science Photo Library:** ROBERT MARKUS (tr). **278 Alamy Stock Photo:** Science History Images (br). **Getty Images:** Historical (tr). **Science Photo Library:** SCIENCE SOURCE (bl). **279 Alamy Stock Photo:** REUTERS (tr). **Science Photo Library:** ALEX LUTKUS / NASA (tl). **280 Alamy Stock Photo:** UrbanImages (bl). **Science Photo Library:** NASA / JSC (br). **281 Alamy Stock Photo:** American Photo Archive (bl); Hemis (tr); NG Images (br). **282 Alamy Stock Photo:** Newscom (bl). **282-283 Alamy Stock Photo:** Geopix (b). **283 Alamy Stock Photo:** Nature Picture Library (tr). **284 Alamy Stock Photo:** Sergey Ryzhov (tl). **Science Photo Library:** DAVID PARKER (br). **285 Alamy Stock Photo:** Edgloris Marys (tr); James King-Holmes (tl). **Science Photo Library:** LAGUNA DESIGN (bl). **286 Getty Images:** Gallo Images (bl); NASA / Handout (tc). **287 Alamy Stock Photo:** agefotostock (br); World History Archive (tr); Minden Pictures (bl). **288 Science Photo Library:** FONS RADEMAKERS / CERN (c). **289 Science Photo Library:** FERMILAB (tr). **290 Alamy Stock Photo:** James King-Holmes (br); Stocktrek Images, Inc. (tl); Trinity Mirror / Mirrorpix (c). **291 Alamy Stock Photo:** dpa picture alliance (tr). **NASA:** Optical: NOAO / Kitt Peak / J.Uson, D.Dale; X-ray: NASA / CXC / IoA / S.Allen et al. (tl). **292-293 Alamy Stock Photo:** Andrew Kimber (tr). **292 Alamy Stock Photo:** BSIP SA (bl); NASA Image Collection (tl); dotted zebra (br). **293 Alamy Stock Photo:** blickwinkel (bl); cbimages (br). **294 Alamy Stock Photo:** Universal Images Group North America LLC (br); ZUMA Press, Inc. (tr). **Science Photo Library:** EUROPEAN SOUTHERN OBSERVATORY (bl). **295 Alamy Stock Photo:** Martin Shields (br); Panther Media GmbH (tr). **Science Photo Library:** BRITISH ANTARCTIC SURVEY (tl). **296 Alamy Stock Photo:** Everett Collection Historical (bl). **Science Photo Library:** CERN (br); MASATO HATTORI (tl); VOLKER STEGER (tr). **297 Alamy Stock Photo:** GRANGER - Historical Picture Archive (tl). **Science Photo Library:** JOHN BAVARO FINE ART (br). **298 Alamy Stock Photo:** GARY DOAK (tl); Science History Images (tr); NG Images (bl). **299 Alamy Stock Photo:** Cultura Creative RF (br); Science History Images (tl); World History Archive (tr). **300 Alamy Stock Photo:** Abaca Press (tr). **Science Photo Library:** GABRIELLE VOINOT / EURELIOS / LOOK AT SCIENCES (br); MEDICAL GRAPHICS / MICHAEL HOFFMANN (bl). **301 Science Photo Library:** JOHN BAVARO FINE ART (br). **302 Alamy Stock Photo:** Christian Offenberg (tr); StellaNature (cl). **302-303 Alamy Stock Photo:** dotted zebra (b). **303 Alamy Stock Photo:** Xinhua (tr). **Science Photo Library:** DR. GILBERT S. GRANT (br); IBM RESEARCH (tl). **304 Alamy Stock Photo:** Science History Images (bc). **Getty Images:** SOPA Images (tl). **304-305 Alamy Stock Photo:** ukasz Szczepanski (b). **305 Alamy Stock Photo:** Nature Picture Library / Alamy (br). **Science Photo Library:** QA INTERNATIONAL / SCIENCE SOURCE (tr); SINCLAIR STAMMERS (tl)